T0276918

The Messier Objects
Second Edition

The bright galaxies, star clusters, and nebulae catalogued in the late 1700s by the famous comet hunter Charles Messier are still the most widely observed celestial wonders in the sky. The second edition of Stephen James O'Meara's acclaimed observing guide to the Messier objects features improved star charts to help you find the objects, a much more robust telling of the history behind their discovery – including a glimpse into Messier's fascinating life – and updated astrophysical facts to put it all into context. These additions, along with new photos taken with the most advanced amateur telescopes, bring O'Meara's first edition more than a decade into the twenty-first century. Expand your universe and test your viewing skills with this truly modern Messier guide. It is a must for all budding night watchers.

Author of several highly acclaimed books, including others in the celebrated *Deep-Sky Companions* series, Stephen James O'Meara is well known among the astronomical community for his engaging and informative writing style and for his remarkable skills as a visual observer. O'Meara spent much of his early career on the editorial staff of *Sky & Telescope* before joining *Astronomy* magazine as its Secret Sky columnist and a contributing editor. An award-winning visual observer, he was the first person to sight Halley's comet upon its return in 1985 and the first to determine visually the rotation period of Uranus. One of his most distinguished feats was the visual detection of the mysterious spokes in Saturn's B-ring before spacecraft imaged them. Among his achievements, O'Meara has received the prestigious Lone Stargazer Award, the Omega Centauri Award, and the Caroline Herschel Award. Asteroid 3637 was named O'Meara in his honor by the International Astronomical Union. In his spare time, he travels the world to document volcanic eruptions. He is a contract videographer for National Geographic Digital Motion and a contract photographer for the National Geographic Image Collection.

For Deborah Carter, who helped me to see the stars through new eyes. Thank you.

DEEP-SKY COMPANIONS

The Messier Objects Second Edition

STEPHEN JAMES O'MEARA

Shaftesbury Road, Cambridge CB2 8EA, United Kingdom

One Liberty Plaza, 20th Floor, New York, NY 10006, USA

477 Williamstown Road, Port Melbourne, VIC 3207, Australia

314–321, 3rd Floor, Plot 3, Splendor Forum, Jasola District Centre, New Delhi – 110025, India

103 Penang Road, #05–06/07, Visioncrest Commercial, Singapore 238467

Cambridge University Press is part of Cambridge University Press & Assessment, a department of the University of Cambridge.

We share the University's mission to contribute to society through the pursuit of education, learning and research at the highest international levels of excellence.

www.cambridge.org
Information on this title: www.cambridge.org/9781107018372

First published 1998
Second edition published 2014
3rd printing 2021

A catalogue record for this publication is available from the British Library

Library of Congress Cataloging-in-Publication data
O'Meara, Stephen James, 1956– author.
The Messier objects / Stephen James O'Meara. – Second edition.
pages cm. – (Deep-sky companions)
Revision of: Th e Messier objects. 1998.
Includes bibliographical references and index.
ISBN 978-1-107-01837-2 (hardback : alk. paper)
1. Astronomy – Charts, diagrams, etc. 2. Astronomy – Observers' manuals. 3. Galaxies – Charts, diagrams, etc. 4. Stars – Clusters – Charts, diagrams, etc. 5. Nebulae – Charts, diagrams, etc. 6. Messier, Charles. Catalogue des nébuleuses et des amas d'étoiles. I. Title. II. Series: Deep-sky companions.
QB65.O44 2013
523.1–dc23 2013019585

ISBN 978-1-107-01837-2 Hardback

Contents

Preface to the second edition

Cambridge University Press first published *Deep-Sky Companions: The Messier Objects* 15 or so years ago. It was the first in what is now a series of five *Deep-Sky Companions* volumes. The other books in the series are subtitled *The Caldwell Objects, Hidden Treasures, The Secret Deep*, and *Southern Gems.* Not only was *The Messier Objects* the first book in this series, but it was also the first deep-sky book I had ever written. As such it stands apart from the other volumes for several reasons.

First, many observers already had in their possession one or more books on the Messier objects, most notably *The Messier Album* by the late John H. Mallas and Evered Kreimer (Sky Publishing, Cambridge, MA, 1978), segments of which first appeared in *Sky & Telescope* magazine in the late 1960s, and *Messier's Nebulae and Star Clusters*, 2nd edition, by the late Kenneth Glyn Jones (Cambridge University Press, 1991). Thus, I realized, my book would need a fresh approach.

For instance, to me, the most outstanding aspects of Glyn Jones's book are the rich histories he presents on Messier and his contemporaries, as well as his summaries of historical observations of the "M" objects. Not wanting to duplicate this effort, I decided to minimize those aspects in my own book. I saw the Mallas and Kreimer book as having three strengths: (1) Kreimer's beautiful photographs of the Messier objects taken through his 12 1/2-inch Cave reflector from Prescott, Arizona; (2) Mallas's pencil drawings of each M object as seen through a 4-inch f/15 Unitron refractor from his backyard in Covina, California; and (3) Harvard historian Owen Gingerich's scholarly biography of Messier and his contemporary Pierre Méchain.

To avoid duplicating these efforts, I asked comet hunter David Levy to write a short history on Messier, who was himself predominantly a comet hunter. I also decided to include photographs by different amateur astronomers that would help to inspire observers. And, although I, too, had a 4-inch telescope, mine was an f/5 refractor, which offered a wider field of view than Mallas's f/15 refractor, so our impressions of these objects would be different. It was this latter impression that I wanted most to share with observers in the first edition of *The Messier Objects.* So I put my heart and soul into that aspect of the book, perhaps even more so than in the other volumes, only because, through a small telescope, the Messier objects offer the most detailed views of those visible from mid-northern latitudes.

Second, while both the Glyn Jones and the Mallas and Kreimer books included finder charts and directions to each Messier object, I tried to improve on them. Now, looking back, I see the charts in the first edition of *The Messier Objects* as "rough" compared with those in my other *Deep-Sky Companions* volumes (which are all hand plotted). The charts in the first edition were computer generated by an old software program, and the stars didn't bin very well. Nevertheless, they still were a slight improvement over those in the Glyn Jones and Mallas and Kreimer works.

And third, when I wrote the first edition of *The Messier Objects*, I decided to include *some* interesting historical and astrophysical facts about these objects, but I didn't treat each with equal concern. I rectified that in the other volumes in the series. Also, while I did include some astrophysical highlights in the first edition, it does not compare, overall, with what I have done for the other *Deep-Sky Companions* guides.

This last point is further exacerbated by the tremendous leaps in telescope technology and imaging (both from orbit and on the ground), which have opened vast new vistas throughout the electromagnetic spectrum to our eyes and offered astronomers new insights into the mechanisms and forces governing the appearance of these glorious objects. Furthermore, when I wrote the first edition, amateur charge-coupled device (CCD) imaging was still in its infancy. No more! Take for instance the beautiful CCD images accompanying each object in Chapter 4, taken by my friend Mario Motta in Gloucester, Massachusetts.

So, much has happened in the 15 years leading up to the publication of this second edition, including my exchanging the 4-inch f/5 Tele Vue Genesis refractor for a Tele Vue 5-inch f/5 refractor, then a 3-inch f/5 Tele Vue refractor, some observations from which have been added in Chapter 4. Still, I have kept the basic structure of the book the same, but I have included a new history (a glimpse into the fascinating life of Charles Messier and the making of his now famous catalogue); new, detailed star charts (hand plotted to show the most efficient way to find objects by star-hopping); and a much more robust telling of historical and astrophysical facts. All these additions bring the twentieth-century first edition of the book more than a decade into the twenty-first century. I hope you enjoy this fresh look at the Messier objects – the most popular deep-sky targets for more than two centuries.

Stephen James O'Meara
Volcano, Hawaii

Preface to the first edition

Two and a half centuries ago a French comet hunter named Charles Messier began compiling a catalogue of nebulous sky objects. He explained his motivation in the French almanac *Connaissance des Temps* for 1801:

> What caused me to undertake the catalogue was the nebula I discovered above the southern horn of Taurus on 12 Sept. 1758, while observing the comet of that year.... This nebula had such a resemblance to a comet, in its form and brightness, that I endeavored to find others, so that astronomers would not confuse these same nebulae with comets just beginning to shine.

Messier died long before twentieth-century astronomers realized the profound nature of these hauntingly diffuse glows. The 110 "Messier objects," it turns out, are an eclectic collection of celestial treasures: 39 galaxies, 57 star clusters, 9 nebulae, a supernova remnant, a swath of Milky Way, a tiny grouping of stars, a double star, and even a duplication. The list includes the most massive and luminous galaxy known, the ghostly remains of a cataclysmic stellar explosion, and an immense cosmic cloudscape that cradles newborn stars in dense cocoons of hydrogen gas.

Every Messier object is within reach of a small telescope, and many are visible with binoculars and the naked eye, especially under clear, dark skies far away from city lights. Amateur astronomers of all ages enjoy tackling the Messier catalogue members, because they represent a good sampling of what's "out there," and because finding them helps to hone observing skills. In a sense, the Messier objects are the testing grounds for budding skywatchers.

The Messier objects entered my life in 1966, when I was 10 years old. They were mentioned in the *Sky Observer's Guide*, a Golden Guide by R. Newton Mayall and Margaret Mayall, whose words also taught me the basics of astronomy. I quickly located the brightest Messier objects from my back porch in Cambridge, Massachusetts. Then one winter evening a friend loaned me his 2-inch refractor. I recall pointing its white enameled tube over a frozen city landscape, beyond the smoking chimneys, and seeing swirls of nebulosity surrounding the Trapezium, a group of young, bright stars in the mighty Orion Nebula (Messier 42, or M42), then the knitted stars of the Pleiades (M45) – Lord Tennyson's "glittering swarm of fireflies tangled in a silver braid." These views were visual poetry, and like Tennyson and countless others before him, I became captivated by the allure of the stars.

Several years later, when I acquired a 4 1/2-inch reflector, another friend gave me a box of *Sky & Telescope* magazines – a monthly publication devoted to the hobby of amateur astronomy. An article immediately caught my eye; it was one in a series by John Mallas and Evered Kreimer spotlighting the Messier objects. Each article featured drawings by Mallas and photographs by Kreimer, as well as visual descriptions and brief histories of the objects. On the first clear night I eagerly set up the scope, and with magazines and red flashlight in hand, I followed along with Mallas and Kreimer as they toured the "M" objects. No longer feeling alone in my pursuits, I began dedicating one clear night to each Messier object – studying it, writing down impressions, and making drawings.

That Mallas and Kreimer series was later compiled into a book called *The Messier Album*, which served the astronomical community for many years. But time has marched on. The optical quality of many commercially available telescopes and eyepieces is superior to the telescope Mallas used a quarter-century ago, providing sharper images and revealing fainter details. We also have more accurate astronomical information today about many of the objects – their sizes, distances, magnitudes, and more – than we did when *The Messier Album* was written. Object positions and maps have been updated to equinox 2000.0 coordinates from the equinox 1975.0 coordinates used by Mallas and Kreimer. Clearly, the time was right for a fresh new look at these classical astronomical specimens.

As observers today, we are not only better equipped but also wiser. Looking back, it is hard to believe that Mallas used a 4-inch f/15 refractor to make his observations. Although a popular telescope type in its day, it was best suited for studying the Moon and planets. Today's deep-sky observers prefer rich-field telescopes over long-focus refractors. It is also amazing that Mallas made his observations from a Los Angeles suburb tainted by light pollution and smog! These problems are even more pervasive today and represent an insidious threat to our continued enjoyment of the heavens. More and more, residents of cities and suburbs must pack their gear and drive some distance from their homes to less populated areas to enjoy a night under the stars. And several times a year thousands of amateur astronomers journey to national conventions and star parties held at first-class, dark-sky sites. In fact, one such journey inspired this book.

For a week in May 1994, I attended the Texas Star Party, a deep-sky observing event held on the Prude Ranch near Fort Davis, Texas, at an altitude of just over 5,000 feet. One sparkling-clear night, Al Nagler, founder of Tele Vue Optics, showed me a wide-field view of the Milky Way – specifically, the region known as M24 – through his 4-inch Genesis refractor. The field was bristling with starlight and threads of dark nebulosity. I didn't want to take my eye from the telescope. Later that week, I borrowed another friend's Genesis and spent three hours studying the Whirlpool Galaxy (M51) in Canes Venatici; I spent many more hours on subsequent nights. The graceful spiral arms, the numerous punctuating knots – all the subtle detail was awesome to behold. I began to wonder what the other Messier objects would look like through a telescope of uncompromising quality, with superior eyepieces, and viewed from the darkest sites on earth. Thus, I decided to revisit, one by one, the deep-sky gems that Charles Messier catalogued more than two centuries ago and that started me on what was to become my latest adventure in an already long and exciting career in astronomical observation, teaching, and writing. The result is this book, which I hope will inspire and inform you as much as Mallas and Kreimer's seminal articles and book did for me and the countless others who have used them.

The purpose of this book is to provide new *and* experienced observers with a fresh perspective on the Messier objects. Chapter 1 is a brief account of the life of Charles Messier and how his catalogue of deep-sky curiosities came to be. Chapter 2 introduces beginners to the basics of skywatching and to some important terms and concepts. It is designed specifically to help newcomers orient themselves to the sky and start locating the brightest Messier objects. In Chapter 3, "The making of this book," I review the methods and the equipment with which I conducted the observations described herein and provide additional information about the book's content.

Of course, the heart of this book is Chapter 4, which looks in detail at each Messier object. Since this book is a "companion," I've used a conversational tone; I speak to you as if I'm with you in the field. Along with the descriptive text, I have provided for each object a list of essential data, including its coordinates, size, brightness, and distance. The equinox 2000.0 finder charts have been carefully drafted to work together with Wil Tirion's all-new, wide-field constellation map at the back of the book; together they will enable you to quickly zero in on your targets. A new and comprehensive translation of Messier's original published catalogue was commissioned and is included here. It supersedes earlier translations, which often were abridged and prone to occasional errors and misinterpretations. I have also included, in Appendix A, the endnotes to Messier's original catalogue, in which he lists a number of objects reported by other observers that he tried, but failed, to find himself.

The detailed drawings I made of the Messier objects are another distinctive feature of this book. Each drawing was based on several hours of observing each object over several extremely transparent nights. I think you will find these illustrations revealing and useful in helping you to see subtle details in the objects that you may not have noticed before (and which may not be apparent in photographs). I have also updated and revised many of the objects' magnitude estimates, offered thoughts about some of the "missing" Messier objects, and distributed observing challenges throughout the chapter.

In Chapter 5, "Some thoughts on Charles Messier," I offer some summarizing thoughts or "analysis" on Messier and his catalogue, which I felt compelled to do having spent so much time thinking about the man and what he saw - and didn't see. I also felt compelled to describe, even in a book devoted to the Messier catalogue, 20 of my favorite *non*-Messier objects, in Chapter 6. Consider them honorable mentions to the catalogue that are conspicuous by their absence from it and certainly deserving of a look while you're out hunting Messier objects.

The appendices contain additional information that you will find useful, including a brief discussion about "Messier marathons" and a guide to navigating the Coma–Virgo Cluster.

Perhaps the most unique aspect of this book, and what I most want to convey, is the *approach* I take to observing. It's an approach based on creative perception and on using the imagination to see patterns and shapes in the subjects seen through the eyepiece. It involves using not just your eye but also your *mind's eye* to associate those patterns and shapes with things that are familiar to you, to create pictures and even stories. Rather than barrage you with just facts (of which you get plenty), I thought you'd also enjoy seeing these objects in new ways - especially the clusters, whose multitude of inherent shapes lend themselves to being seen as celestial Rorschach tests. By using the imagination you can add another dimension to your observing - a highly personal and entertaining one (after all, this is a hobby). Anyone who has read *Hard Times* by Charles Dickens will understand my protest of a diet consisting totally of fact. (By teaching youngsters fact not fancy, conformity not curiosity, Dickens's bleak character Thomas Gradgrind tried to stifle inquisitive minds.) If you have never gazed at the ethereal quality of a Messier object through a telescope, I encourage you to look upon them as you would a painting or a piece of art - and let that art add meaning to your experience.

It is my hope that this book will not only introduce you to the objects themselves - or reintroduce you, as the case may be - but that it will also challenge you to raise your observing skills to a higher level and to push your visual limits. I hope it compels you to search for new and mysterious aspects about these objects, to see them in rich and creative ways, and to grow as an observer.

I know the magic of the Messier objects because I have been under their spell for three decades. Today I see the same magic in my wife's eyes, whenever she raises her binoculars to the sky and happens upon her "comets." May the spell never be broken.

Stephen James O'Meara
Volcano, Hawaii

Acknowledgments to the first edition

"If I have seen so far, it is because I have stood on the shoulders of giants." Like Sir Isaac Newton and others before and after him who used this axiom, I would like to recognize the giants who have helped me in my observational and literary journeys.

Highest tribute goes to the late Walter Scott Houston, who shared his observing experiences and techniques in his Deep-Sky Wonders column in *Sky & Telescope* magazine for so many years. I will never forget the times we spent by the campfire at the annual Stellafane Convention on Breezy Hill in Vermont, or in chaise longues at the Winter Star Party in the Florida Keys, just gazing at the stars and musing on the limits of vision.

I am honored to recognize George Phillips Bond (1825–1865), second director of Harvard College Observatory, whose dedication to unlocking the visual mysteries of the Orion Nebula with the Harvard Observatory's 15-inch refractor led ultimately to his premature death. Reading his diaries two decades ago kept me enchanted on many a cloudy night and taught me how to be a patient and persistent observer.

A deep bow to "envelope pushers" Barbara Wilson and Larry Mitchell, who roped and tied this wild planetary observer at the Texas Star Party and force-fed me deep-deep-deep-sky objects until I became a convert – thank you (I think)! Assisting them was a phalanx of galaxy hunters, including David Eicher, Tippy and Patty D'Auria, and Jack Newton. Brian Skiff introduced me to many visual challenges, including seeing faint globular clusters with the naked eye. Peter Collins first introduced me to the more challenging Messier objects. David Levy was always around to say, "No, Steve, that's a galaxy not a comet." And at several star parties, Tom and Jeannie Clark were incredibly generous with their Tectron telescopes, encouraging me to sweep the Coma Cluster with these enormous Dobsonians until I nearly fainted with delight.

Al Nagler and the editors of *Sky & Telescope* will never fully know how grateful I am to them for helping me to complete this journey. Thanks for your special encouragement and support.

Steve Peters gets the blue ribbon for nurturing the book idea, and kudos to Simon Mitton and Cambridge University Press for publishing the work. Special thanks to Lee Coombs, Martin Germano, Chuck Vaughn, and George Viscome for their stunning astrophotographs. I greatly appreciated the expert assistance of Brent A. Archinal, Kevin Krisciunas, Larry Mitchell, Brian Skiff, and Barbara Wilson, who reviewed drafts of the text and made necessary corrections and welcome suggestions. Thanks to Steve and Tom Bisque of Software Bisque for their generous donation of *The Sky* astronomy software, which was used in creating the finder charts, and to master astronomical cartographer Wil Tirion for the wide-field map of the Messier objects at the back of the book. Thanks also to my good friend Storm Dunlop for his excellent translation of Messier's catalogue from the original French.

Kudos to Nina Barron for her expert, sensible, and sensitive copyediting of the manuscript. Heartfelt appreciation to the Dillingham family – Ken, Lina, Serena, and Karen – for the use of their ranch in the saddle of Kilauea and Mauna Loa volcanoes, where clear skies, steady seeing, and the feeling of home helped me to finish the observations on time. Thanks also to Nina Barron for proofreading the manuscript.

And no words can express the love and devotion I have for my wife, Donna, who, despite having been "husbandless" for a year, found solace in exploring erupting volcanoes, practicing her "free dives" to unknown depths, and peering inquisitively into the eyes of the night.

Finally, I would like to thank our "children," Pele-Hiiaka of Volcano, Milky Way, and Miranda-Pywacket, for keeping their digits off the keyboard when I wasn't looking. Alas, I cannot blame them if any errors have sneaked unannounced into this book; I take full responsibility.

CHAPTER 1

A glimpse into the life of Charles Messier

1

Charles Messier

On the evening of April 13, 1781, French astronomer Charles Messier (1730–1817) made his final observation for what was to become the most extensive catalogue of nebulae and star clusters of the time. The list, which took more than two decades to create, contained the positions and descriptions of 100 objects visible above his Paris horizon (48°51′ north). Just before Messier submitted this list for publication, he received from his contemporary Pierre Méchain (1744–1804) a note that included information on three more objects discovered by him. Although Messier had no time to check the positions of Méchain's additions, he added them to his list, bringing to 103 the total number of nebulae and star clusters in it. This final compendium (the last to be created by Messier) appeared in the French almanac *Connaissance des Temps* for 1784 (published in 1781).

The number of objects in Messier's catalogue has since increased to 110: the original 103 objects, plus 7 more added posthumously by other astronomers for various reasons. (I will discuss these latter objects in more detail later.) In recent times, we have also learned that object number 102 in Messier's 1781 list may be a duplicate observation of object number 101; but this matter remains a topic of much interesting debate. Regardless, the 103 objects in Messier's catalogue and the seven subsequent additions have endured the test of time to become the most popular listing of deep-sky objects targeted by Northern Hemisphere amateur astronomers (both visual and astro-imagers alike), especially by those just beginning in the hobby.

Ironically, this was not Messier's intent. "What caused me to undertake the catalogue," Messier explained in the *Connaissance des Temps* for 1801, "was the nebula I discovered above the southern horn of Taurus on September 12, 1758, while observing the comet for that year.... This nebula had such a resemblance to a comet, in its form and brightness, that I endeavored to find others, so that astronomers would not confuse these same nebulae with comets just beginning to shine ... and this is the purpose I had in forming the catalogue."

Messier, the first astronomer to devote himself to the systematic search for comets, had obviously suffered the frustration of wasting time on "false comets" – diffuse deep-sky objects that could be mistaken for comets during a sweep of the heavens, especially with a small telescope at low power; every minute spent on a "false" comet was time spent away from the potential discovery of a real one in this highly competitive pursuit.

Unlike the skywatchers of today, Messier had no star charts with known deep-sky objects plotted on them, except for a few that included some of the brightest naked-eye wonders, such as the Beehive (M44) or Pleiades (M45) star clusters. To further his success, one could say Messier employed the intelligent strategy of "knowing thy enemy": the objects Messier included in his catalogue are those he wanted to know existed, so that he could ignore them and move along in his attempt to visually capture his ultimate prey.

The magnitude to which Messier was perceived to dedicate his thoughts to the hunt is reflected in a colorful story that has been passed down since the beginning of the nineteenth century. Here is the tale, as told by French astronomer François Arago (1786–1853) in his *Popular Lectures on Astronomy: Delivered at the Royal Observatory of Paris*, published in 1845:

> An interesting memoir of Messier may be found in the *Histoire de l'Astronomie au dixhuitième Siècle* [History of Astronomy in the Eighteenth Century], by Delambre. La Harpe (*Correspondence Litteraire*, Paris, 1801, tom. i, p. 97) says that "he passed his life in search of comets. The *ne plus ultra* of his ambition was to be made a member of the Academy of Petersburgh. He was an excellent man, but had the simplicity of a child. At a time when he was in expectation of discovering a comet, his wife took ill and died. While attending upon her, being withdrawn from his observatory, Montaigne de Limoges anticipated him by discovering the comet. Messier was in despair. A friend visiting him began to offer some consolation for the recent affliction he had suffered. Messier, thinking only of his comet, exclaimed: – *'I had discovered twelve. Alas, that I should be robbed of the thirteenth by Montaigne!'* and his eyes filled with tears. Then, remembering that it was necessary to mourn for his wife, whose remains were still in the house, he exclaimed, – *'Ah! cette pauvre femme,'* [Ah! This poor woman] and again wept for his comet."

THE "BIRD-NESTER" OF COMETS

Given the incredible and efficient technologies used to discover comets in the twenty-first century, both on Earth and in space, many of today's telescopic observers pursue comets mostly as objects of passing interest – especially when they blaze forth to naked-eye splendor or threaten to hit the Earth ... or other planets! (Note, however, that owing to the same leaps in technology, some amateur astronomers across the globe also conduct extremely serious studies of, and searches for, comets, and have contributed greatly to the science.) Nevertheless, to the mid-eighteenth-century observer, comets were among the most mystifying sights in the sky. And the astonishing appearance of six-tailed C/1743 XI (Chéseaux's comet of 1744) – one of the greatest since the dawn of modern astronomy – may have inspired Messier's lifelong passion for comets.

Comet C/2006 P1 (McNaught).

With history having recorded only some 50 comets known by Messier's time, these celestial itinerants presented the burgeoning telescopic astronomers of the day not only with a fascinating challenge (namely to find them) but also the promise of some fame and notoriety. And Messier was the man who first thrust this challenge to the forefront of desire.

Messier was born on June 26, 1730, in the small village of Badouvillier, Lorraine (about 200 miles east of Paris, near the German border). He was the tenth of 12 children born to Nicolas and Francoise Messier, and one of only six to survive into adulthood. His father was a type of mayor/administrator, who, among other duties, collected taxes throughout the serfdom and served as a judge for local misdemeanors; Nicolas passed away in 1741, when Charles was only 11.

When Charles's older brother Hyacinthe took over the Messier household, he began to prep young Charles as a manager of finances – Hyacinthe's particular field of expertise. As his brother's personal clerk, Charles obtained the skills of clean handwriting, draftsmanship, and journal keeping – all traits that would soon help him secure his future employment at the Paris Observatory.

Charles led a safe and secure country life, immersing himself in his spare time in the natural wonders of the local countryside. He also grew up during the height of Rococo development in France, the start of the classical period of music, and the time of the cultural Enlightenment in Europe. These important shifts in scientific and cultural trends may have swept up young Messier's thoughts in their currents, leading him into new areas of imagining. And while it is uncertain what exactly inspired Messier's interest in astronomy, the remarkable sight of Chéseaux's comet in 1744 and the annular solar eclipse of June 25, 1748 (visible from his hometown), must have been major influences.

When a changing political climate in France in 1751 forced his brother Hyacinthe to take a position as a tax collector in his mother's hometown, the orphaned Charles found himself in need of a job. With the help of a family friend, Charles traveled to Paris that October to meet with astronomer Joseph Nicolas Delisle, who then hired Messier to be his apprentice at the Marine Observatory at the Hotel de Cluny in Paris – even though, as Delisle said, the young man had "hardly any other recommendation than that of a neat and legible handwriting and some little ability in draughtsmanship."

Nevertheless, Delisle assigned Messier the task of mapmaking and the recording of astronomical observations. Under the tutelage of Delisle's secretary, Libour, Messier also learned how to use the observatory's astronomical instruments. Charles found them well suited to the purpose of comet hunting, a field of research he not only invented in 1758 but would soon dominate.

Within three years, Messier had become a proficient observer and was promoted to clerk. Then, on January 21, 1759, Messier tasted the bittersweet flavor of success. After 18 months of searching, he recovered the long-sought return of Halley's comet predicted by the late Edmond Halley. As Messier reports in the *Connaissance des Temps* for 1810:

> At about six o'clock I discovered a faint glow resembling that of the comet I had observed in the previous year [the one that led Messier to his discovery of M1 in Taurus]: It was the Comet itself, appearing 52 days before perihelion!

Alas, Messier would eventually learn that he was not the first to catch the comet's return. A Saxon farmer named Johann Georg Palitzsch had spied it telescopically nearly a month earlier, on Christmas night, 1758. For Messier, the news must have come as quite a blow. That recovery observation was monumental, as it confirmed the late Edmond Halley's suspicion that comets were not omens or signs of Divine Providence but celestial bodies governed by gravity and destined to orbit the Sun. This loss of opportunity and a certain degree of fame may have ignited Messier's competitive flame, for he went on to make another independent comet discovery on January 7, 1760, followed by his first comet discovery only 19 days later.

Messier's comet pursuits blossomed after Delisle retired in 1761, the year Messier took charge of the observatory, observed the transit of Venus, and made other acute observations. But Messier's passion clearly was the pursuit of comets. Indeed, of the four new ones discovered between 1762 and 1766, Messier found three of them! Also during that time, in 1764, Messier published in the *Philosophical Transactions* a table of the positions of the Comet of 1764, which he "discovered at the Observatory of the Marine at Paris, the 3rd of January, about 8 o'clock in the evening, in the constellation of the Dragon."

Messier's successes and honors then began to flourish. In 1764, for his contributions to astronomy, the elite Royal Society of London made him a foreign member. In 1769, after his discovery of another comet, the Berlin Academy of Sciences bestowed on him his second foreign membership. In 1770, after he had found yet another comet, he became a member of his beloved French Royal Academy of Sciences in Paris and the official Astronomer of the Navy; in that same year, the Paris Academy accepted his original catalogue of 45 nebulae and star clusters for publication in its *Memoirs* for 1771 (which was to be published in 1774).

THE ORIGINAL CATALOGUE BY MESSIER

Messier did not discover all of the objects in his catalogue. Nor were all the objects in it discovered during comet searches. His is a list of nebulae and clusters compiled from previously existing catalogues of such objects, as well as from the discovery of new objects by him and his contemporaries. As previously mentioned, the catalogue had its origins in 1758, after Messier encountered the Crab Nebula (which we now call the Crab Supernova Remnant, M1) while following the

Messier.

Messier's first original deep-sky discovery (globular cluster M3 in Canes Venatici) came on May 3, 1764. That year, he had begun to conduct, in earnest, a systematic search for these "comet masqueraders," as the late twentieth-century American comet hunter Leslie C. Peltier called them. But it's uncertain as to whether Messier's discovery of M3 sparked the idea for this dedicated search or if it was the first product of it. No matter, we know at least that he swept up M3 while *not* following a known comet, and that he continued to find others systematically after it.

comet of that year; he later magnanimously credited English amateur astronomer John Bevis (1695–1771) with the nebula's discovery in 1731.

Likewise, on September 11, 1760, Messier chanced upon the globular star cluster M2 while following Halley's comet through Aquarius; he later credited its discovery to Jean-Dominique Maraldi (1709–1788), who found it first while looking for Chéseaux's comet in 1746. Again, the comet's remarkable multitailed appearance to the unaided eye may have inspired Charles's passion for comets at an early age.

By January 1765, Messier had compiled a list of 41 previously known or newly discovered objects he had observed and for which he had determined positions. Prior to submitting this list for publication in the Paris Academy's *Memoirs* for 1771, Messier in March 1769 had observed four additional objects – the Orion Nebula (M42 and M43), the Praesepe star cluster (M44), and the Pleiades star cluster (M45)[1] –

[1] Messier probably included these four objects to "round off" the list. The author believes that Messier did not originally include them because they are all bright enough, and were well known enough, not to be confused with a comet when seen under a dark sky. Later, he probably realized that these objects can be mistaken for comets when seen telescopically in a twilit sky. See Chapter 6 for a possible explanation as to why Messier did not include the Double Cluster in Perseus in this list.

bringing to 45 the total number of nebulae and star clusters in his list.

MESSIER'S FIRST SUPPLEMENT

After presenting his original catalogue to the French Academy in 1771, Messier no longer systematically searched the sky for nebulae and clusters. He did, however, continue to collect additional objects that he either encountered during his comet searches and observations or learned about from his contemporaries or other sources. One of these, M49, was the first object to be discovered in the now famous Virgo cluster of galaxies.

Other discoveries of deep-sky objects in the region would follow, especially during the passage of the Great Comet of 1779, discovered by Johann Elert Bode (1747–1826) and independently discovered by Messier, which passed through the Virgo–Coma Berenices region of the sky in that year. The multitude of objects in this area, coupled with the passage of a comet through it, must have caused a sensation and some confusion. In fact, on the evenings of May 5 and 6, Messier tells us he mistook the galaxy M61 for the comet, and that he did not realize his mistake until May 11, when he listed M61 as a new discovery. Unfortunately for Messier, Italian astronomer Barnabus Oriani had beaten him to the discovery six days earlier.

Of the 23 nebulae and star clusters Messier ultimately compiled and published in 1780 as a supplement to his original catalogue in the French almanac *Connaissance des Temps* for 1783 (bringing the total to 68), 13 of them were found by chance owing to the passage of a comet nearby. On the flip side, consider his catalogue entry for M66, which he discovered along with M65 on the evening of March 1, 1780. In that entry, Messier attributes his missing the object earlier to the fact that the comet of 1773–74 passed between it and neighboring M65 on the evening of November 1–2, 1773, when "Doubtless M. Messier, did not see it then because of the comet's light."

The Comet of 1779's passage through Virgo and Coma Berenices also appears to have triggered an avalanche of interest in comets and the deep sky. As we shall see, it also appears to have inspired Messier's greatest rival, Pierre Méchain, who not only began discovering deep-sky objects in 1779 but also began a successful search for comets.

THE SECOND SUPPLEMENT

The contributions of Pierre Méchain to Messier's second supplement are great. Born in Laon on August 16, 1744, Méchain was the son of an architect and a student of mathematics, physics, and architecture. He learned astronomy under the tutelage of Joseph-Jerome le Francais de Lalande (1732–1807), professor of astronomy at the College de France, who found young Méchain at ease with difficult astronomical mathematics.

In 1772, Lalande obtained for him a temporary position of astronomer-hydrographer at the Dépot de la Marine at Versailles. Two years later, the Dépot moved Méchain and its entire department to Paris, where he had his first acquaintance with Messier, who was then well established as the Navy astronomer.

It's hard not to consider the impact Messier must have had on the younger Méchain, for although Méchain's post had him traveling frequently to chart the French coastline, he apparently turned his attention to the night sky in 1774, starting with the occultation of Aldebaran by the Moon.

Aside from his well-earned reputation as a comet fanatic, Messier had an interest in other Solar System objects. For instance, the facsimile, reproduced here, of a page written by Messier in 1754 details some of his solar

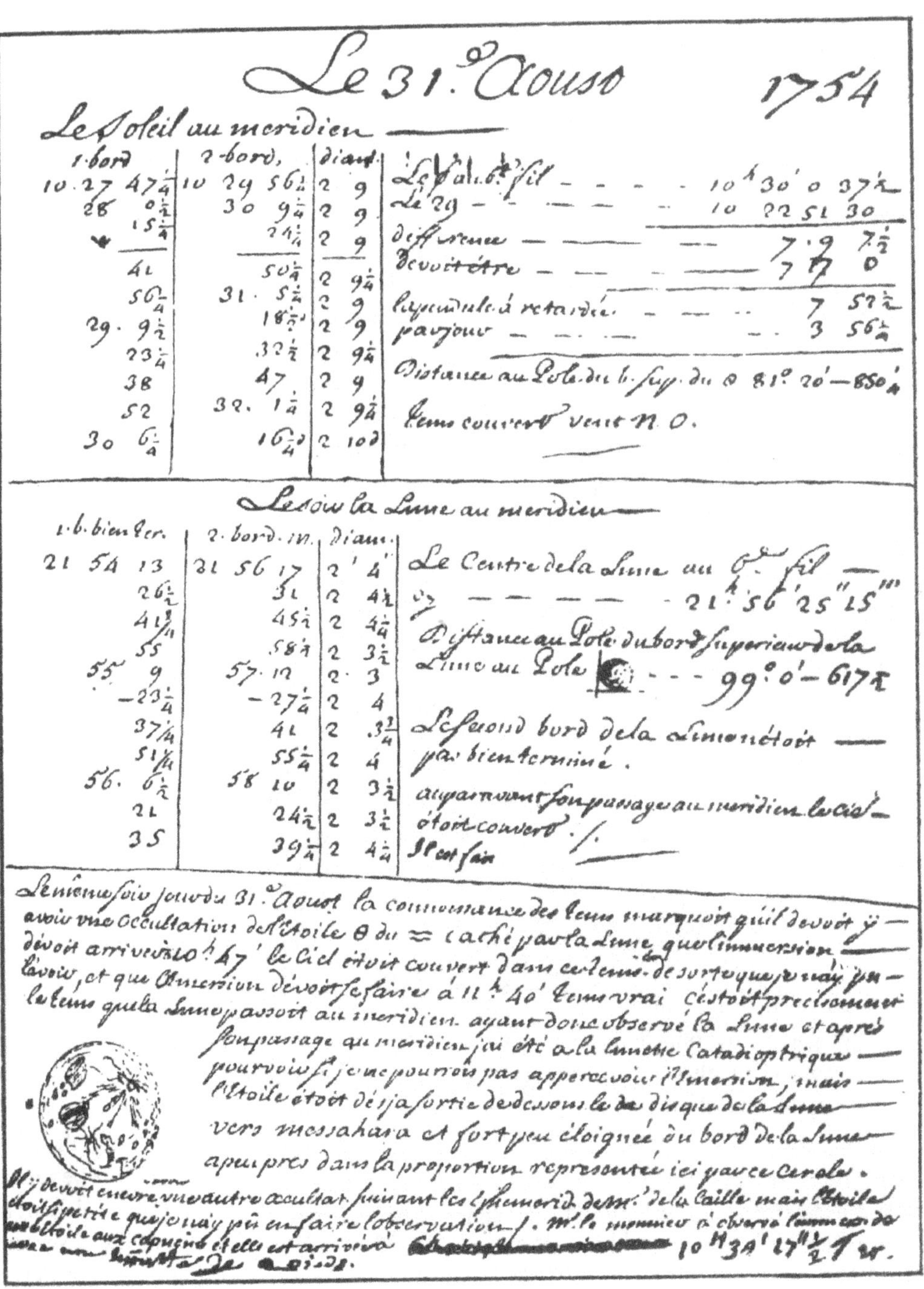
Le 31.e Aoust 1754

Le Soleil au meridien

Le soir la Lune au meridien

Facsimile of a page written by Messier in 1754, detailing solar and lunar observations. *Bulletin de la Société Astronomique de France*, 1929.

and lunar observations, including a drawing of the Moon as it appeared in his simple inverting refractor.

And, in 1776, one month after the American Revolution, Messier made the first observation of Saturn's globe since 1750, discovering a new belt; the following is from an extract of a letter to Mr. Magellan, Fellow of the Royal Astronomical Society, dated "Paris, May 29, 1776":

> June 6, 1776. I have observed, since the 14th of May, a belt of a fainter light on the body of Saturn, opposite to the part of the ring behind the planet. It is pretty broad, and almost as distinct as those of Jupiter. It was with a very good achromatic of three foot and a half, made by Mr. Dolland, that I discovered this appearance. I wish you would communicate it to the astronomers, because those who are furnished with better instruments may, perhaps, see some inequalities in this belt of Saturn, and so the time of the planet's revolution on its axis may be better ascertained than it is at present. Mess. John and James Cassini seem to have been the only astronomers who discovered this phenomenon about the end of the last century.

As mentioned, however, a turning point for many skywatchers occurred in 1779, when the comet of that year sailed through the rich Virgo and Coma clusters of galaxies. That year, Méchain literally burst onto the scene, being adept not only at the visual discovery of comets but also in observing them and calculating their orbits (the latter something for which Messier had no mental facility). One can imagine Messier's chagrin; the "bird-nester" or "ferret" of comets had found a notable competitor. "It would not be surprising to discover a strong jealousy between these rivals," admits Harvard astronomical historian Owen Gingerich in John Mallas and Evered Kreimer's *The Messier Album*, "Yet, if it existed, biographers have remained silent. We can only deduce that the two were professional acquaintances, freely sharing their data."

Méchain made no apparent effort to catalogue new nebulae and clusters, though he did communicate new discoveries to Messier, who, in turn, checked their positions. Messier also continued to add new objects to his list until April 13, 1781 (a month after William Herschel discovered Uranus), when it grew from 68 objects (his original catalogue plus the first supplement) to 100 only a year later; 75 percent of these new discoveries belonged to Méchain.

Although Messier had in his possession three additional discoveries by Méchain, he had no time to check their positions before his publication deadline. Nevertheless, he decided to append them to the list (the latter two without positions) as objects 101, 102, and 103, with a note attesting that "M. Messier has not yet seen." For the final publication, Messier also added more recent observations of the earlier 68 objects.

THE "MISSING MESSIER" OBJECTS AND OTHER CURIOSITIES

Two of the objects in Messier's original catalogue have come under question: M40 in Ursa Major and M24 in Sagittarius. Of the 23 objects in Messier's first supplement (published in 1780), two are of particular note: open star cluster M47 in Puppis and open star cluster M48 in Hydra. And, in Messier's second supplement, we find two of the most controversial listings: M91 in Coma Berenices and M102 in Draco. Some brief explanations of the curiosities follow; all are discussed in more detail under the object entries in Chapter 2.

M40 is a pair of close stars of nearly equal magnitude in Ursa Major that John Hevelius (1611–1687) had described as a "nebulous

Portrait of Charles Messier in 1801, 20 years after his last catalogue observation on April 13, 1781. *Bulletin de la Société Astronomique de France*, 1929.

star" in his *Uranographia* atlas. Although some have disputed the necessity of including this object in the catalogue, Messier clearly notes why he decided to do so: "[E]ven though Hevelius mistook these two stars for a nebula, they are difficult to distinguish with an ordinary telescope of six feet."

M24, the Small Sagittarius Star Cloud, is part of the naked-eye fabric of the Milky Way. Messier, however, appears to have been the first to resolve its light into a cluster-like patch of stars divided into several regions. Many early references credit Messier with discovering not the large Star Cloud but the little 11th-magnitude open star cluster NGC 6603 within this rich segment of the Milky Way's Sagittarius-Carina spiral arm, but John Herschel first discovered the latter object on July 15, 1830.

M47, an open star cluster in Puppis, proved problematical to identify with Messier's catalogued position, which places it in a region of sky devoid of any deep-sky objects. German astronomy popularizer Oswald Thomas (1882–1963) offered a viable solution to this "missing" Messier object. In his 1934 book *Astronomie*, he identifies M47 with the bright and obvious open star cluster NGC 2422. Twenty-five years later, T. F. Morris of the Royal Astronomical Society of Canada's Montreal Centre arrived at the same conclusion, noting that Messier appears to have made an error in computation: if you simply reverse Messier's offset positions from his comparison star 2 Navis, you arrive at NGC 2422, Messier's M47.

M48 was another "missing" Messier object, as no cluster exists at Messier's published position. But, as with M47, Morris proposed that Messier had, once again, made an error in offset. An open star cluster does exist at the correct right ascension, but nearly 5° south of Messier's position: NGC 2548 in Puppis, Messier's M48.

M91 presents us with a respectable challenge because the "missing" object Messier had sighted lies in the rich Coma–Virgo cluster of galaxies, leaving us with several possibilities (which, again, will be explored more deeply in Chapter 2). These include a passing comet, the nearby spiral galaxy NGC 4571, a duplicate observation of M58, and the barred spiral galaxy NGC 4548.

When Texas amateur astronomer W. C. Williams investigated the problem in 1969, he found it possible that Messier had made yet another error in plotting. Williams argues that Messier had determined M91's position by offsetting from M89, but had mistakenly applied the difference in right ascension and declination to M58 (the bright "nebula" he used as a reference to determine the positions of other nebulae in the field). Although somewhat contentious, his conclusion (and the most generally accepted one) is that NGC 4548 is M91.

M102's identity has caused some heated, though enlightening, debates among amateur astronomers. First, it must be made clear that Messier never observed this object. It is one of three objects discovered by Méchain that Messier did not have time to observe and hastily included in his final catalogue. Of those three objects, Messier provided a position (from Méchain) for only one object, 101, which has been identified as the spiral galaxy NGC 5457 in Ursa Major. And while Messier listed no position for 103, it, too, has been definitely identified by description as open star cluster NGC 581 in Cassiopeia.

Object 102, however, is problematical. Messier provides no position, and his description of the object is confusing: "A very faint nebula between the stars o Bootis and ι Draconis, close to a 6th-magnitude star."

The problem is that Omicron (ο) Boötis lies some 40° south of Iota (ι) Draconis, leaving much confusion as to which of several objects in the region Méchain had discovered. If Omicron was a misprint of Theta (θ) Boötis, it's been argued, then either NGC 5879 or NGC 5866 could be M102. On the other hand, if Iota (ι) Draconis is a misprint for Iota Serpentis, as others have suggested, then M102 might be NGC 5928 in Serpens. Which is correct?

In a 1947 paper in *Journal of the Royal Astronomical Society of Canada* (vol. 41, p. 265), Helen Sawyer Hogg writes that, "The question, however, was settled by Méchain himself in 1783, when he announced flatly that Nebula No. 102 was an error, and the same object as No. 101. I translate Méchain's letter as follows:

> On page 267 of the Connaissance des temps for 1784, M. Messier lists under No. 102 a nebula which I have discovered between *o* Boötis and *i* Draconis: this is, however, an error. This nebula is one and the same as the preceding No. 101. M. Messier confused the same as the result of an error in the sky-chart, in the list of my nebulous stars communicated to him.

"Therefore," she continues, "Messier 102 may now be stricken from the records as a non-existing object."

In *The Search for the Nebulae* (Alpha Academic, Chalfont St. Giles, 1975), Kenneth Glyn Jones agreed, writing, "Despite the tempting field for conjecture which Méchain's original description exposes, his later explanation must be accepted and M102 declared as non-existent." But in his *Messier's Nebulae and Star Clusters* (Cambridge University Press, 2nd ed., Cambridge, 1991), Glyn Jones quipped, "Nevertheless, a lingering doubt remains."

M104–M110: AN ADDITIONAL "MÉCHAIN-MESSIER" SUPPLEMENT

As we have seen, Messier's final published catalogue included 100 objects plus a supplement of three additional objects that he did not have the chance to observe. Two years later, in 1783, Bode published a new version of Messier's catalogue (from numbers 46 to 103) in his *Berliner Astronomiches Jahrbuch* (for 1786); this new version included not only Bode's own comments on the objects but also corrections by Méchain (including a copy of his letter clearing up the mystery of M102).

In the same 1783 letter in which Méchain attempts to clear up the M102 muddle, however, he also notes six further nebulae he discovered. Interestingly, while Bode published Méchain's letter toward the back of the volume, he did not include the positions of these new objects in his final Messier list.

Of Méchain's six new nebulae, Sawyer Hogg noticed that two were too vague for her to identify with certainty. She found the four others worthy of addition to Messier's accepted list and opined they should appear as numbers 104, 105, 106, and 107.

Sawyer Hogg also knew that, in 1921, French astronomy popularizer Camille Flammarion had already noticed number 104 listed in Messier's own copy of his catalogue (penned in Messier's own hand) as being discovered by Méchain. Sawyer Hogg suggested this object (NGC 4594, the Sombrero Galaxy) be catalogued as M104, and it has been known as such ever since. Here is Méchain's description of it in his 1783 letter to Bernoulli:

> On 11th May 1781 I discovered a nebula above Corvus which did not appear to contain a single star. It emits a weak light and is difficult to find if the micrometer wires are illuminated. I compared it then on the following day with Spica and obtained its right ascension as 187°9′42″ and South declination as 10°24′49″ This does not appear in the Connaissance des Temps.

Sawyer Hogg adds that "although this object is the second mentioned in Méchain's letter, we will keep Flammarion's numbering, and assign No. 105 to the first nebula mentioned"; she identified this latter object as NGC 3379, one of a group of three spirals in Leo. Méchain's description of it to Bernoulli makes its identity certain:

> [Near M95 and M96] there is a third, somewhat more northerly and even brighter than the others. I discovered this on 24th March 1781, four or five days after the other two.

The last two objects Sawyer Hogg identified from Méchain's descriptions were NGC 4258 (M106), a large spiral galaxy in Canes Venatici, and globular cluster NGC 6171 (M107) in Ophiuchus:

> [M106] In July 1781 I found another nebula close to Ursa Major, near star No. 3 of Canes Venatici and 1° farther south. I estimated its Right Ascension as 181° 40′ and northerly declination as about 49°. I shall shortly try to confirm its position.
>
> [M107] In April 1782 I discovered a small nebula on the left flank of Ophiuchus between the stars z and f whose position I have not yet determined more precisely.

In 1960, Gingerich identified the two Méchain objects that Sawyer Hogg had omitted as the galaxies NGC 3556 and NGC 3992 in Ursa Major, suggesting that they be included in Messier's list as M108 and M109, respectively. These Méchain had included as additional notes to his discovery of M97:

> With reference to [M97], a nebula close to b in Ursa Major, M. Messier mentions, in pointing out its position, two others which I likewise discovered; – one close to [M97] and the other close to g in Ursa Major. I have, however, not yet been able to confirm their positions.

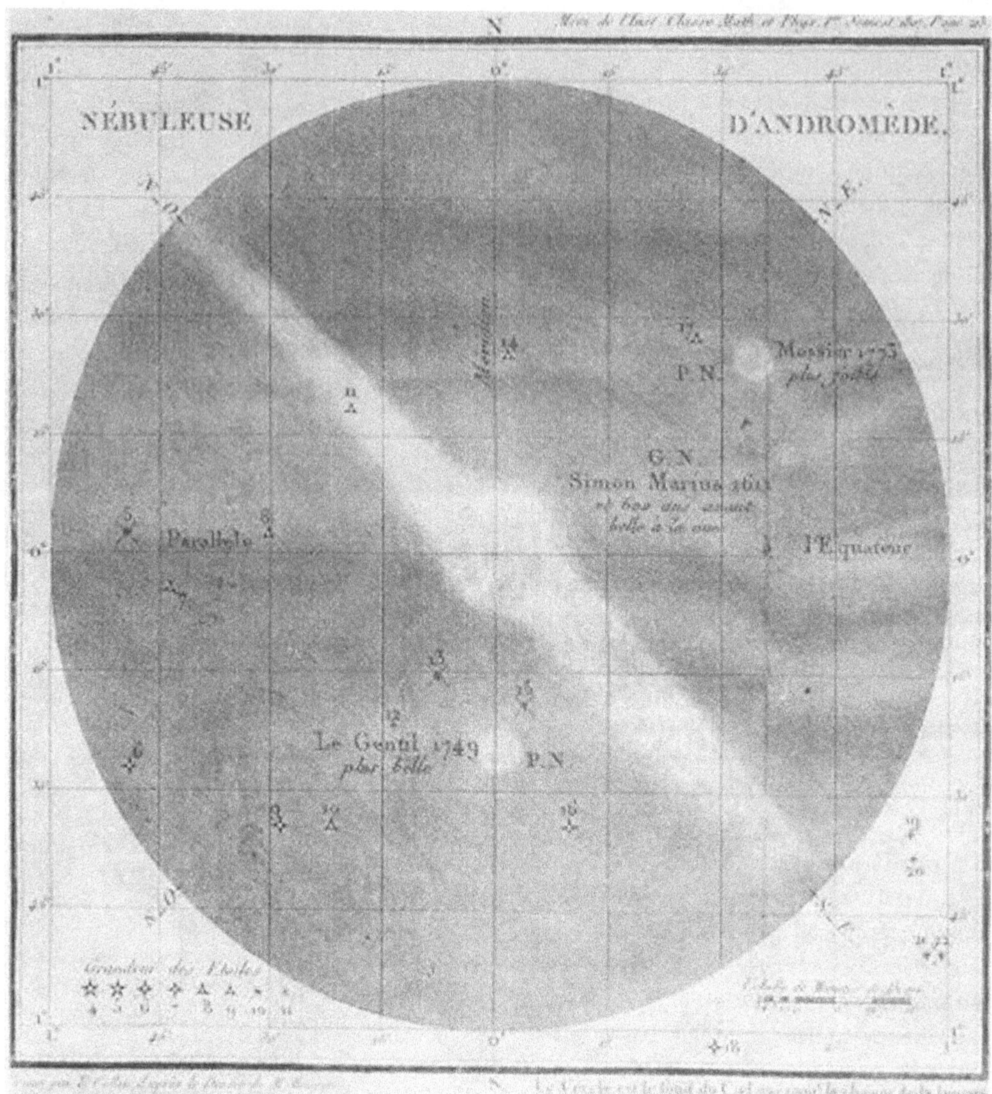

As mentioned, however, the identity of M109 remains somewhat contentious.

Kenneth Glyn Jones suggested the last addition to the Messier catalogue, in 1966. He found a drawing by Messier of the Andromeda "Nebula" (M31) and its two companions. Messier had published the drawing in the *Memoirs* of the French National Institute for 1807. On the drawing, Messier annotated that Simon Marius had discovered the Great Nebula (M31) in 1749, Guillaume Le Gentil had found the small companion (M32) in 1749, and that Messier himself had discovered the second companion (NGC 205) in 1773. Messier, in fact, first spied NGC 205 on August 10, 1773, 10 years before Caroline Herschel independently discovered it from England. For reasons unknown, Messier never mentioned this new object in the supplements to his catalogue. Thus, Glyn Jones suggested that NGC 205 become M110.

Both Gingerich and Glyn Jones emphasize that Messier's final catalogue includes only 103 objects and that only circumstantial evidence exists that Messier ever observed any of Méchain's final catalogue contributions; nor did Messier sanction these objects in print. Furthermore, contemporary observers following in Messier's footsteps knew only of Messier's final list of 103 objects (again, Bode never published positions for Méchain's additional objects in his final Messier list). And Messier himself failed to include NGC 205 in his own supplements.

In fact, one could argue that Messier's final catalogue contains only 100 objects: the 100 nebulae and star clusters whose positions Messier had determined and took the time to observe and visually describe.

OBJECTS MESSIER FAILED TO LOCATE

At the end of his catalogue in the *Connaissance des Temps* for 1784, Messier included a list of objects reported by other astronomers that he had "searched for in vain." What follows is a brief description of each of these "mysteries" and the most likely candidates identifying them.

In his 1690 *Prodronus Astronomiae*, John Hevelius (1611–1687) gives the positions of 14 apparent nebulous stars, eight of which Messier could not locate; note that M40 (a close telescopic double star) was identified by Hevelius as a nebulous star.

1. A nebula in the forehead of Hercules, the brightest of three nebulous stars. Messier searched for this object on June 20, 1764, under good skies, but failed to find it at Hevelius's position. It is most likely an asterism formed by the stars 32, 33, and 34 Ophiuchi.

2–5. Four nebulous stars in Capricornus: one in the Seagoat's forehead, another to the west of its eye (Rho [ρ]), and a third following the eye. Messier notes that, "These nebulae are also found on several planispheres and celestial globes." Messier searched for these four nebulae in July, August, and October of 1764 but failed to find them. They most likely are Sigma (σ), Pi (π), and Omicron (o) Capricorni, respectively. The latter two make a tight naked-eye triangle with Rho Capricorni and undoubtedly appeared nebulous to Hevelius.

6–7. Hevelius gives the position of two other nebulous stars in Cygnus: one north and west of Deneb in the Swan's tail, the other flowing away from that star and the constellation. Messier "carefully searched for these two nebulae," on October 24 and 28, 1764, and did not find them. However, while searching for these objects, Messier encountered a new cluster of faint stars near π^2 Cygni (M39), noting that the position of this cluster differed from the one Hevelius published in his work. Hevelius's two nebulous stars are most likely the close naked-eye double star Omega^1 (ω^1) and Omega^2 (ω^2) Cygni.

8. The brightest of three nebulous stars in the ear of Pegasus. Messier searched for them in vain on the night of October 24–25, 1764. It is most likely 35 Pegasi, which forms a tight naked-eye acute triangle with 37 and 34 Pegasi, near Theta (θ) Pegasi.

9. In his 1740 work *Elements of Astronomy*, Jacques Cassini (1677–1756) describes a nebula his father, Giovanni Domenico Cassini, discovered "in the region between Canis Major and Canis Minor," and which was one of the finest seen through a telescope. When Messier looked for this nebula on several occasions without success, he presumed that it may have been a comet that either was just beginning to shine or in the process of fading.

Nevertheless, in 1772, Messier did in fact discover a cluster of stars (M50) in this region. M50 is close enough to the position of Cassini's object that Kenneth Glyn Jones argues these are "almost certainly the same object." Uncharacteristically, Messier failed to mention Cassini's father's possible discovery in the *Connaissance des Temps* for 1783.

10. In later editions of John Flamsteed's *Historia Coelestis Britannica,* he gives the position of a nebula in the lower leg of Andromeda. Messier searched for this object on October 21, 1780, but couldn't find it. It appears to be the star 55 Andromedae, with a close 6th-magnitude naked-eye companion to the east, just southwest of Gamma (γ) Andromedae.

AN ERA ENDS

After Messier published his list of 103 nebulae and clusters in 1781, we hear no more about his deep-sky discoveries other than that, in the *Connaissance des Temps* for 1801, he writes, "Since the publication of my catalogue I have observed still others: I will publish them in the future." But he never did. One reason may have been his recognition of the outstanding accomplishments of William Herschel over the next few years. Two years after Herschel discovered Uranus in 1781 (which Messier assiduously observed), the German-born English astronomer, inspired by Messier's accomplishments, began a systematic survey of the heavens with telescopes far superior to any available to Messier. Herschel graciously did not include any objects catalogued by Messier in his growing collection of new nebulae and clusters.

The political climate in France had also turned to one of revolution. During the "Reign of Terror" in France (1793–1794), the Navy stopped paying Messier's salary, and he also lost his pension. But, by 1795, the 65-year-old Messier was able to finance his own career. Three years later, he discovered the last comet to bear his name. One can also imagine the rekindled flame in the man's aging heart when he learned that Joseph Leland (1732–1807) had offered a one-time prize of 600 francs for the first comet discovery of the nineteenth century.

How ecstatic Messier must have felt when, on the evening of August 8, 1801, he encountered a magnitude 6.5 comet in Ursa Major – the first new comet of the nineteenth century! And how utterly distraught he must have been to learn not only that his rivals Méchain and Alexis Bouvard (1767–1844) had also discovered the comet on the same night but that Jean Louis Pons (1761–1831) – a doorkeeper at the Observatory of Marseilles and an absolute newcomer to the field of comet hunting – had discovered the comet the night before. Thus Leland's reward of 600 francs went to Pons.

The thirst for monetary compensation for a comet discovery must have been extreme. Consider, for instance, that a German astronomer named Reissig claimed he had sighted Pons's comet earlier on the evening of June 30 through a break in some clouds near the Ursa Major–Camelopardalis border. Thus, he argued, he deserved the reward instead of Pons. This situation remained a historical curiosity, however, until 1969, when the late Joseph Ashbrook calculated the comet's position and determined that it was actually in Andromeda on the night Reissig claimed to have seen it in the far north, so it could not have been the same object.

Regardless, Pons would go on to become the most successful telescopic discoverer of comets in history. "From [1801] until 1827, it was a rare year that failed to register the discovery of at least one comet by Pons," writes Elizabeth Roemer in her 1960 biography of Pons. "Once he found five within eight months (1808 February–September) and later he found five more within twelve months (1826 August–1827 August 2)."

Before his death in 1831, Pons had discovered 37 comets, compared with Messier's roughly 18 finds in 40 years of searching and Méchain's 14 discoveries in 20 years of searching. Roemer goes on to point out that, "By some quirk of fate, the first comet found by Pons, of which Messier was the co-discoverer, was also the last of Messier's discoveries. And the second comet found by Pons, of which Méchain and Olbers were co-discoverers, was the last of Méchain's discoveries."

We also have evidence that Messier may have been in poor physical condition during this time. In 1802, Herschel visited the 72-year-old Messier in Paris, writing in his diary that Messier had "complained of having suffered much from his accident of falling into an ice-cellar" – an accident that had occurred two decades earlier. Messier had also lost his wife and suffered from failing eyesight. Although he continued to make the occasional observation, the climb up the octagonal tower to his telescopes must have also become arduous to his weary bones, especially on cold, damp nights.

Messier's long career began to wane after he suffered a stroke in 1815, which left him partially paralyzed. Two years later, he died at age 87. He left a wonderful legacy, not of comets but of magnificent deep-sky objects that he sought to avoid. The love of the Messier catalogue by modern observers is a lavish testament to Messier, one of the most perceptive and enthusiastic observers of the night sky.

REFERENCES

Alexander, A. F. O'D. *The Planet Saturn: A History of Observation, Theory, and Discovery.* New York: Dover Publications, 1962.

Arago, François. *Popular Lectures on Astronomy: Delivered at the Royal Observatory of Paris.* New York: Greeley & McElrath, 1845.

Brown, Peter Lancaster. *Comets, Meteorites, & Men.* New York: Taplinger Publishing Company, 1974.

Flammarion, Camille. *Bulletin de la Societe Astronomique de France* (1929).

Gingerich, Owen. "Messier and His Catalogue," in John H. Mallas and Evered Kreimer, *The Messier Album.* Cambridge, MA: Sky Publishing Corporation, 1978.

Glyn Jones, Kenneth. *The Search for the Nebulae.* Chalfont St. Giles: Alpha Academic, 1975.

Glyn Jones, Kenneth. *Messier's Nebulae and Star Clusters*, 2nd ed. Cambridge: Cambridge University Press, 1991.

Graun, Ken. *The Next Step: Finding and Viewing Messier's Objects.* Tucson, AZ: Ken Press, 2005.

Kronk, Gary W. *Comets: A Descriptive Catalog.* Hillside, NJ: Enslow Publishers, Inc., 1984.

Levy, David. "Charles Messier and His Catalogue," in Stephen James O'Meara, *Deep-Sky Companions: The Messier Objects.* Cambridge: Cambridge University Press, 1998.

Messier, Charles. "A Table of the Places of the Comet of 1764." *Philosophical Transactions* (Vol. **54**:68 (1764).

Messier, Charles. "A Belt on the Disk of Saturn." *Philosophical Transactions* **66**:543 (1776).

Messier, Charles. *Memoirs de l'Academie des Sciences* (1771).

Meyer, Maik. *International Comet Quarterly* **29**:3 (2007).

Roemer, Elizabeth. "Jean Louis Pons, Discoverer of Comets." *Astronomical Society of the Pacific Leaflets* **8**:159 (1960).

Sawyer Hogg, Helen. "Catalogues of Nebulous Objects in the Eighteenth Century." *Journal of the Royal Astronomical Society of Canada* **41**:265 (1947).

Smyth, Adm. William H. *A Cycle of Celestial Objects.* Richmond, VA: Willmann-Bell, Inc., 1986.

Students for the Exploration and Development of Space (SEDS). http://messier.seds.org.

CHAPTER 2

How to observe the Messier objects

2

The purpose of this chapter is to help new observers get started. (If you're a seasoned observer, you may want to skip to the next chapter, though you should read the section on "Observing Tips" before you do.) It contains enough essential information to help you find a few Messier objects. There is no avoiding the need to learn sky directions, recognize the major constellations and their brightest stars, know how to use a star map, and understand how the sky changes with the seasons. You can teach yourself just about all these things with a star wheel, or planisphere. These special sky maps are inexpensive, fun and easy to use, and very educational. (Star wheels can be purchased at any nature store or planetarium gift shop.) To purchase the proper star wheel, you will need to know your latitude on Earth so you can determine which stars are visible from your location. Consult a world atlas at your local library. You could also use one of the more popular Apps available online, such as *Star Walk*™.

Although anyone can enjoy this book, the following section on sky orientation and many references to sky directions throughout the book are targeted primarily for observers at mid-northern latitudes. Observers in the far south will not be able to see several of the most northerly Messier objects. Chapter 6, however, describes many southerly objects either missed or ignored by Messier. A good supplementary guidebook for southern observers is my *Deep-Sky Companions: Southern Gems*. Otherwise, the material in this book, especially the descriptive notes on the Messier objects, can be used by all.

NAVIGATING THE SKY

Assuming you live in the Northern Hemisphere and have a star wheel and a red flashlight (white light ruins your night vision), the next step is to go outside and orient yourself. One easy way is to use a compass to find the cardinal directions - north, south, east, and west. Do this in the daytime, standing where you expect to set up your telescope. Select an object on the distant horizon - a tree, building, smokestack, or mountain - to mark the main cardinal points. For example, if your star wheel displays a bright star or constellation rising in the east, you know to look in the direction of, say, a willow tree, your eastern landmark.

When night falls, your first mission is to find the North Star, or Polaris. It will be your unfaltering guide. The North Star is very close to where Earth's imaginary axis of rotation intersects the dome of the sky. Unlike other bright stars, it remains essentially in the same position every night, all night, as the Earth turns. Despite its public reputation, the North Star is not the brightest star in the night sky; Sirius, a southern star, holds that honor. In fact, there are 48 stars brighter than Polaris.

The height of the North Star above your horizon (its altitude) is the same as your latitude on Earth. If you lived at the North Pole (latitude 90°), the North Star would be directly overhead (altitude 90°). If you lived on the equator (latitude 0°), the North Star would be on the north horizon (altitude 0°). I live on Hawaii's Big Island (latitude 20°), so I look for a solitary yellowish star of moderate brightness 20° above my north horizon.

How high is 20°? Draw an imaginary line from the north horizon to the point directly overhead (the sky's zenith). That line spans an angular distance of 90°. Twenty degrees is almost one-quarter of the way from the horizon to the zenith. To measure angular distance, hold an upright fist at arm's length and look at it with one eye closed. The amount of sky covered by the fist is about 10°. For me to find the North Star from Hawaii, I would face north, place the base of my upright fist on the horizon line, make a fist with my other hand, and place it on top of the first. Two fists equal about 20°. The North Star should be sitting on the top fist. If you live in New York City (latitude 40°), the North Star will be four fists above your north horizon.

Now you can find the Big Dipper. Using your red flashlight, turn your star wheel until the time of night lines up with the current date. You will be looking northward, so hold the star wheel with the north horizon down; the word "north" or "northern horizon" should be upright. Looking at your star wheel, read the position of the Big Dipper as you would the hand of a clock. For example, on July 1 at 8 p.m., the Big Dipper is to the upper left of the North Star, at the 10 o'clock position. The star closest to Polaris in the bowl of the Big Dipper is Dubhe. It lies 30° (three fists) away. If you open your hands and stretch out your fingers, the angular distance between the tip of your thumb and the tip of your little finger is approximately 20°. Stargazers have long used Dubhe and Merak (the other star marking the bowl's outer side) to point to Polaris, hence the term "pointer stars."

Compare the size of the Big Dipper in the sky with its size on the star wheel. This is an important exercise because, to find less obvious constellations, you will need to scale what you see in the sky with what you see on your planisphere, or vice versa. (Constellations appear larger when they are at or near the horizon than when overhead. It is the same optical illusion that makes the Moon look bigger when it is on the horizon than when overhead.) Scaling star patterns is also necessary when you want to use the more detailed star charts in this book to find Messier objects with your telescope.

To find other star patterns, first locate the brightest stars, get a feeling for the scale, and then look for the fainter stars that make up the constellations. What is wonderful about hunting Messier objects is that you do not have to know the entire sky to find or enjoy them, just like you don't have to know everything about New York City if you want to visit only the Empire State Building, the Statue of Liberty, and Central Park. It is the same way with this book. You can be selective and enjoy the sights you want to see. In fact, you can learn the constellations as you seek out the different Messier objects. At least one Messier object each season is bright enough to be seen with the naked eye. Furthermore, these objects are plotted on good star wheels.

TYPES OF MESSIER OBJECTS

There are three principal categories of objects in the Messier catalogue: star clusters, nebulae, and galaxies. Each is divided into subclasses: open and globular clusters; diffuse and planetary nebulae; and spiral, elliptical, and irregular galaxies. Many of the Messier objects inhabit our own galaxy, the Milky Way.

What does our Galaxy look like? That is hard to know because we live within it. It's like trying to determine the shape of a forest from a spot deep inside it. But astronomers believe it looks something like the accompanying drawing.

Our Galaxy's nucleus measures about 10 light-years across (1 light-year is the distance that light travels in one year – 6 trillion miles!). The nucleus is surrounded by an egg-shaped bulge spanning about 32,000 light-years and containing millions of stars. Encircling the bulge is a flat, pancake-like disk of dust, gas, and stars that measures about 100,000 light-years in diameter. The disk material is probably arranged in spiral arms that emanate from the tips of a bar of dust and gas extending on either side of the nucleus. This entire system is contained in a spherical galactic halo of older

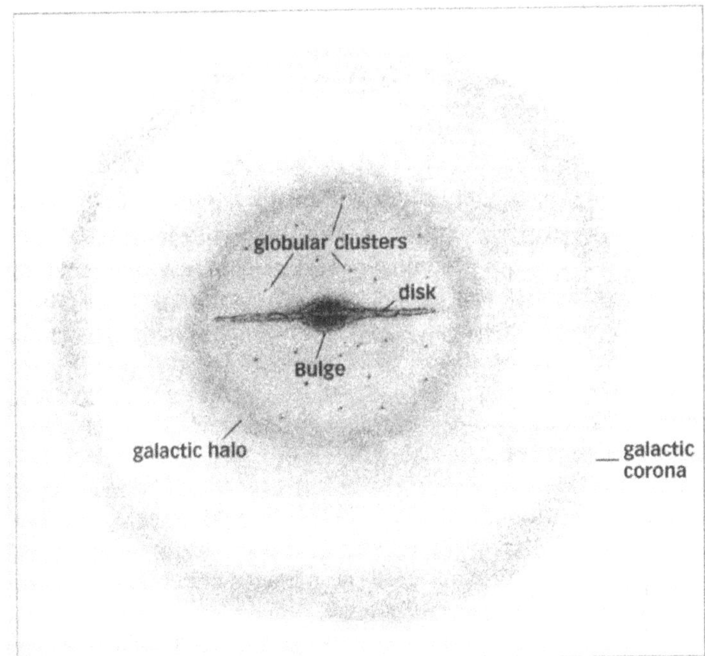

are created through the magic of color photography and do not represent the way they look through a telescope, which is fuzzy and dim. But if you stop to consider how far their light has traveled before it reaches your eye, you will better appreciate their subtle, ghostly glows.

stars and interstellar matter; the diameter of the halo is about 130,000 light-years. Finally, there is the galactic corona, extending some 200,000 to 300,000 light-years. Researchers believe that the mysterious dark matter (material that escapes visual detection; its presence is inferred from gravitational effects) inhabits this region.

The Milky Way Galaxy is our home in the universe. The universe is replete with billions of galaxies, each of which is home to billions of hundreds of billions of stars. The Messier catalogue contains 39 galaxies, all of which are millions of light-years away. At a distance of 2.5 million light-years, M31, the Great Andromeda Galaxy, is the Messier galaxy closest to our own; the farthest is M87, a giant elliptical system in Virgo that is a dizzying 55 million light-years distant.

Note: Don't expect to see the Messier objects as bold, sharply defined masses radiating complex pastel hues. Those images

CATEGORIES OF MESSIER OBJECTS

The Messier catalogue contains five major classes of deep-sky objects: open clusters, globular clusters, diffuse nebulae, planetary nebulae, and galaxies. Open, or galactic, clusters are irregular conglomerations of young stars (a dozen to thousands of them) that travel in a thin disk of stars, dust, and gas along the plane of our Galaxy. Their stars are loosely associated and will probably disperse after making a few trips around the center of our Galaxy. Of the approximately 1,200 known open clusters in our Milky Way, the Messier catalogue contains 30, including those associated with nebulosity.

Globular clusters, on the other hand, are spherical swarms of tens of thousands to perhaps a million ancient stars situated far above and below the galactic disk. These tightly packed bundles of stars move in giant elliptical orbits in our Galaxy's spherical halo. Of the 150 or so globulars known, the Messier catalogue contains 29.

Diffuse nebulae, clouds of dust and gas that lie in the plane of our Galaxy, come in three varieties: emission, dark, and reflection.

Emission nebulae are glowing (ionized) clouds of dust and gas that shine in the visible part of the spectrum. Dark nebulae are dense clouds of interstellar dust that obscure stars and parts of bright nebulae lying behind them, so we see them in silhouette. And reflection nebulae have no emissions of their own but scatter the incoming light from nearby stars. There are seven diffuse nebulae in the Messier catalogue.

Planetary nebulae are expanding shells of gas blown off dying stars about the same size as, or up to a few times larger than, the Sun. The planetary shells last for about 50,000 years before they become too large and faint to be seen from Earth. To astronomers looking through small telescopes a century ago (William Herschel specifically), these nebulae looked like the planet Uranus (tiny round disks with a pale green tinge), and that's why they're called planetary nebulae. Of the 1,000 or so known planetary nebulae, the Messier catalogue contains four.

Finally, there are the galaxies. To see them, we must peer through the Milky Way's dust, gas, and stars into the vastness of space, where other island universes are traveling on mysterious courses. In the grand scope of the known universe, it appears our young Milky Way spiral is in the minority. Most galaxies are older ellipticals – armless galactic nuclei. These ellipticals range in shape from round to very elongated. The Messier catalogue contains 9 ellipticals, 29 spirals, and 1 galaxy classified as irregular.

You can see a representation of nearly all of these categories without optical aid. The following text spotlights some shining examples for each season. These objects are visible at different times of night at different times of the year. Use your planisphere to determine when you can see them, and plan your nights accordingly.

SOME EASY-TO-FIND MESSIER OBJECTS

Winter

Assume it is January 1 at 9:00 p.m. Pick up your planisphere, go outside, and dial in that time and date. Turn the planisphere so that the southern horizon is down and look due south. About two-thirds up from the southern horizon, you should see a tight grouping of six or more stars glittering in a dipper-shaped pattern. This is not the Little Dipper but the Pleiades, Messier 45 (M45), an open cluster. The cluster should be clearly marked on your star wheel. Now you can turn to the page dedicated to M45 in this book and start your adventure! Under dark skies, you might use binoculars or a telescope to search for hints of the faint reflection nebula that envelops the Pleiades.

There's another winter splendor visible at the same time. Look high in the southeast to the constellation Orion, conspicuous by its row of three equally bright stars that form Orion's belt. Less than two finger widths south of (i.e., below) the belt is another chain of three stars, Orion's sword. The middle star in the sword will appear fuzzy to the naked eye. Turn a pair of binoculars to this patch of light and enjoy the magnificent emission nebula (and star cluster) M42, the Orion Nebula. Your telescope will reveal another emission nebula, tiny M43, just north of it.

Spring

If you go outside on April 1 at 9:30 p.m. and face due south, you'll see the bright, blue-white star Regulus about halfway up the sky. Far to its west (right) are the twin stars Castor and Pollux (confirm these with your star wheel). Draw a line between Pollux and Regulus and look a little below the halfway mark. You should see a large, fuzzy glow. This is M44, the Beehive Cluster, another open cluster celebrity.

The keystone (or square) of Hercules is rising at the same time in the northeast. Draw a line between the two western stars in the constellation's "keystone." About a third of the way along that line from the northernmost star is M13, the great Hercules globular cluster! If you do not live under dark skies, you may need binoculars to see it. Finding Messier objects with your binoculars is good practice before using your telescope.

Summer

On August 1 at 9:00 p.m., the curved tail of Scorpius is wonderfully placed for viewing. Look about a fist width above the southern horizon. Two finger widths to the east (left) of the Scorpion's stinger (two stars very close together at the end of the tail) is a large hazy patch, M7, a fine open cluster. Binoculars resolve it into dozens of stars. Another open cluster, M6, lies about one binocular field to the upper right of M7.

The teapot asterism in Sagittarius stands just to the east (left) of the Scorpion's tail. Under dark skies, the Milky Way band seems to steam out of the teapot's spout and waft across the night sky. Use your binoculars to scan that dense Milky Way region because it contains many star clusters and nebulae, about a dozen of which are Messier objects. Here you will also find a network of dark nebulae weaving through the brilliant star clouds.

Fall

The great square of Pegasus dominates the high southern sky on November 1 at 9 p.m. Again, look due south, about two-thirds of the way up the sky, and try to locate the four bright corner stars in the square. They mark the body of the mythical winged horse. Think big because the square averages about 15° (a fist and two fingers) on a side. Using the great square as a landmark, you can find a spiral galaxy with your unaided eye. The journey is a little more involved than the previous ones.

First locate the star that marks the northeast corner (upper left) of the great square. This star is called Alpheratz. Using your hand and fingers, look 15° to the northeast of Alpheratz for a yellowish star of similar brightness. This is Mirach, in the constellation Andromeda. M31, the Great Andromeda Galaxy, is 10° (one fist width) to the northwest (upper right). Under dark skies, it should look like an oval glow to the unaided eye. Binoculars will show it much more clearly. Binoculars will also reveal its companion, M110, an elliptical galaxy, just northwest of it.

Only the tiny, dim glows of the planetary nebulae escape naked-eye detection. Actually, one planetary nebula, the Dumbbell (M27), is visible to the unaided eye, but only from the darkest sites and to those with exceptional eyesight or observing experience. As described later in this book, some planetaries can be detected with binoculars and a bit of perseverance.

Admittedly, not all the Messier objects are easy to locate and identify. But once you master navigating the sky and recognizing star patterns, you will be better prepared to track down the rest of them.

SOME TERMS

Turn to the beginning of Chapter 4, where I describe M1, the Crab Nebula. There you will find a listing of the object's NGC number, coordinates, magnitude, and dimensions. What do these terms mean?

NGC is short for *New General Catalogue of Nebulae and Clusters of Stars.* Compiled by Danish astronomer J. L. E. Dreyer and published in 1888, this work lists 7,840 deep-sky objects in order of right ascension (described subsequently). The NGC includes most of the Messier objects. The Crab Nebula (M1) is entry number 1,952 in the NGC; thus M1 is also known as NGC 1952. Some objects have an IC number, for the *Index Catalogues*, which were supplements to the NGC. For each object, I have included the description from the NGC or IC, as well as a translation of Messier's own descriptions. Together they will give you a historical perspective and a basis for comparison with your own observations.

The Crab Nebula's coordinates ($05^h34.5^m$, +22°01′) represent its right ascension and declination, respectively. Think of these terms as celestial longitude and latitude. If someone gave you the longitude and latitude of New York City, you could locate it on a map of the United States that contains a longitude and latitude grid. Likewise, you can locate the Crab Nebula on a star map by using its celestial coordinates.

Turn to the inside back cover of this book, where you will find a star map covering nearly the whole sky. Right ascension runs from right to left along the top and bottom of the map. Declination runs up and down the map's sides. Note that right ascension is written in hours (from 0 to 24 hours) and that the hours increase to the left (east). These lines of celestial longitude reflect the 24-hour rotation of the Earth, which makes the entire celestial sphere appear to turn. Lines of declination mimic Earth's latitude lines. They increase from 0° at the celestial equator (imagine the plane of Earth's equator reaching out to the dome of the sky) to + 90° at the North Celestial Pole. They decrease from 0° to -90° at the South Celestial Pole. To locate M1, trace the lines of right ascension eastward (to the left) until you find "5 hours" and then continue until you're about halfway between "5 hours" and "6 hours." Next, move down until you reach a declination of +22°. You should see M1 plotted there. A closer view of this region appears in the accompanying finder chart.

Table 1. The Greek alphabet (lowercase)

α	Alpha	ι	Iota	ρ	Rho
β	Beta	κ	Kappa	σ	Sigma
γ	Gamma	λ	Lambda	τ	Tau
δ	Delta	μ	Mu	υ	Upsilon
ε	Epsilon	ν	Nu	φ	Phi
ζ	Zeta	ξ	Xi	χ	Chi
η	Eta	ο	Omicron	ψ	Psi
θ	Theta	π	Pi	ω	Omega

Unless otherwise noted, the coordinates used in this book are precise for "equinox 2000.0." The coordinate system is in constant change because gravitational tugs by the Sun, Moon, and planets cause Earth's axis to wobble like a top. It takes about 26,000 years for the axis to complete a wobble. Although this sounds like a long time (and it is), the gradual shift adds up, so every 50 years or so star charts are revised to incorporate this shift, or precession, of the coordinate system against the backdrop of stars. For this book, the coordinates given correspond exactly to the year 2000, hence equinox 2000.0. Bright stars near the "M" objects are labeled with lowercase Greek letters (see Table 1). This system of stellar nomenclature was introduced in 1603 by Bavarian astronomer Johann Bayer, who labeled stars in each constellation according to their brightness. The most prominent star was given the letter Alpha (α); the faintest became Omega (ω). The brightest star near M1, for instance, is Zeta (ζ), in the constellation Taurus. Astronomers condense it all by saying "Zeta Tauri," which is the Greek letter followed by the Latin genitive of the constellation (see Table 2). There are exceptions, however, such as with the stars in the Big Dipper, which are labeled in order of right ascension, from west to east, not by brightness.

Table 2. Constellations and their Latin genitive forms

Abbrev.	Constellation	Latin genitive
And	Andromeda	Andromedae
Ant	Antlia	Antliae
Aps	Apus	Apodis
Aqr	Aquarius	Aquarii
Aql	Aquila	Aquilae
Ara	Ara	Arae
Ari	Aries	Arietis
Aur	Auriga	Aurigae
Boö	Boötes	Boötis
Cae	Caelum	Caeli
Cam	Camelopardalis	Camelopardalis
Cnc	Cancer	Cancri
CVn	Canes Venatici	Canum Venaticorum
CMa	Canis Major	Canis Majoris
CMi	Canis Minor	Canis Minoris
Cap	Capricornus	Capricorni
Car	Carina	Carinae
Cas	Cassiopeia	Cassiopeiae
Cen	Centaurus	Centauri
Cep	Cepheus	Cephei
Cet	Cetus	Ceti
Cha	Chamaeleon	Chamaeleontis
Cir	Circinus	Circini
Col	Columba	Columbae
Com	Coma Berenices	Comae Berenices
CrA	Corona Australis	Coronae Australis
CrB	Corona Borealis	Coronae Borealis
Crv	Corvus	Corvi
Crt	Crater	Crateris
Cru	Crux	Crucis
Cyg	Cygnus	Cygni
Del	Delphinus	Delphini
Dor	Dorado	Doradus
Dra	Draco	Draconis
Equ	Equuleus	Equulei
Eri	Eridanus	Eridani
For	Fornax	Fornacis
Gem	Gemini	Geminorum
Gru	Grus	Gruis
Her	Hercules	Herculis
Hor	Horologium	Horologii
Hya	Hydra	Hydrae
Hyi	Hydrus	Hydri
Ind	Indus	Indi
Lac	Lacerta	Lacertae
Leo	Leo	Leonis
LMi	Leo Minor	Leonis Minoris
Lep	Lepus	Leporis
Lib	Libra	Librae
Lup	Lupus	Lupi
Lyn	Lynx	Lyncis
Lyr	Lyra	Lyrae
Men	Mensa	Mensae
Mic	Microscopium	Microscopii
Mon	Monoceros	Monocerotis
Mus	Musca	Muscae
Nor	Norma	Normae
Oct	Octans	Octantis
Oph	Ophiuchus	Ophiuchi
Ori	Orion	Orionis
Pav	Pavo	Pavonis
Peg	Pegasus	Pegasi
Per	Perseus	Persei
Phe	Phoenix	Phoenicis
Pic	Pictor	Pictoris
Psc	Pisces	Piscium
PsA	Piscis Austrinus	Piscis Austrini
Pup	Puppis	Puppis
Pyx	Pyxis	Pyxidis
Ret	Reticulum	Reticuli
Sge	Sagitta	Sagittae
Sgr	Sagittarius	Sagittarii
Sco	Scorpius	Scorpii
Scl	Sculptor	Sculptoris
Sct	Scutum	Scuti
Ser	Serpens	Serpentis
Sex	Sextans	Sextantis
Tau	Taurus	Tauri
Tel	Telescopium	Telescopii
Tri	Triangulum	Trianguli
TrA	Triangulum Australe	Trianguli Australis
Tuc	Tucana	Tucanae
UMa	Ursa Major	Ursae Majoris
UMi	Ursa Minor	Urase Minoris
Vel	Vela	Velorum
Vir	Virgo	Virginis
Vol	Volans	Volantis
Vul	Vulpecula	Vulpeculae

Other stars have number identifications. These are Flamsteed numbers. Like the Greek letters, a Flamsteed number precedes the Latin genitive of the constellation: 27 Tauri, for example. John Flamsteed was a prodigious eighteenth-century observer who dedicated 30 years of his life to measuring star positions, which he dutifully catalogued in his *Historia Coelestis Britannica* (1725) in order of right ascension. In his history of celestial cartography, which appears in *Uranometria 2000.0*, the late George Lovi claims that Flamsteed did not number the stars on his star charts; he used only the Greek letter designations. Joseph Jerome de Lalande later assigned the numbers to stars in a French edition of Flamsteed's 1780 star catalogue. But if a Cambridge University Library researcher is correct, Flamsteed may have originated the numbering system after all. In a letter to the editor of *Sky & Telescope* magazine (November 1991), Adam Perkins describes how he found evidence that Flamsteed had used the numbering system as early as 1707.

Magnitude refers to an object's apparent brightness. (There is an involved discussion of limiting magnitude – how faint one can see – in Chapter 3.) Think of magnitude as "class." If something is "first class," it's great; anything else is lower. And that's how the Greek astronomer Hipparchus must have viewed the stars in the second century B.C. when he designated the brightest naked-eye stars as 1st magnitude and the faintest ones as 6th. Mathematically, a 1st-magnitude star is 2.512 times brighter than a 2nd-magnitude star, which is 2.512 times brighter than a 3rd-magnitude star, and so on. The math works out nicely so that a star of 1st magnitude is exactly 100 times brighter than a star of 6th magnitude (because 2.512 is the 5th root of 100). On the brighter end of the magnitude scale, the values soar into the negative numbers: the Sun is magnitude –27, the full Moon is magnitude –12.5, Venus can reach magnitude –4.9, and Sirius, the brightest star in the night sky, is magnitude –1.6.

Most astronomy books state, as a general rule, that the faintest star visible to the unaided eye is 6th magnitude; the faintest star visible with 7 × 50 binoculars is 9th magnitude, and 12th magnitude is the faintest you will see with a 4-inch telescope. However, these numbers are very conservative and should only be used when making generalities about average skies. For some reason, these magnitude limits have become chiseled in stone and have misled beginners for decades! This is a very important point. Be sure to read the section on magnitude limits in Chapter 3 for more discussion.

The apparent size of a deep-sky object is an angular measure of its dimensions against the celestial sphere. The units of angular measure are degrees (°), arcminutes (′), and arcseconds (″): 1° = 1/360 of a circle; 1′ = 1/60 of a degree; and 1″ = 1/60 of an arcminute. The angular separation between the two pointer stars in the Big Dipper is 5°. Both the Sun and Moon are about 1/2° in diameter, or 30′. Most of the Messier objects have much smaller angular measures. The Crab Nebula, for example, is an irregularly shaped haze 6′ long and 4′ wide, or 6′ × 4′, or about one-fifth the Moon's diameter. (It is important to note, however, that although the Moon appears large, this is an optical illusion. It's much smaller than the width of your little finger held at arm's length.)

The light of each Messier object is spread over a specific area of sky. If M1 shines at 8th magnitude, its light output is the same as an 8th-magnitude star. But its light also covers a larger area of sky than the star. Have you ever used a flashlight with an adjustable beam? Think of how bright the light appears when the beam is concentrated and how much weaker it looks when the beam is diffused. Likewise, a diffuse 8th-magnitude Messier object will appear dimmer than an 8th-magnitude star. This dimming effect is intensified under less than perfect sky conditions.

An object's magnitude alone, then, can be deceiving, especially with faint, diffuse objects such as a galaxy. Therefore, to help provide you with a better measure of a galaxy's visibility, the data lists for galaxies in this book include a value for surface brightness. Think of it as dividing the object's magnitude by its area, though the math is more complicated than this. Take the galaxy M65 in Leo, for example. It shines at magnitude 9.3 and occupies an area measuring 8.7′ × 2.2′. Its surface brightness is 12.4, meaning that each arcminute of the galaxy shines roughly with the brightness of a magnitude 12.4 star.

That raises another key issue – extinction. Your latitude determines how far south on the celestial sphere you can see. For instance, if you live at the North Pole (+90°), you cannot see any Messier objects with declinations south of the equator (0°). Although all the Messier objects in this book are visible from northern latitudes, some will be closer to the horizon, where the atmosphere is densest. Generally, stars near the horizon appear dimmer by a few tenths of a magnitude than stars shining overhead. If the air is very polluted, a star's light could be diminished by one or more magnitudes! Because the Messier objects are large and diffuse, their visibility is greatly affected by such pollution.

STAR COLOR

The descriptions of the Messier objects also spotlight bright, nearby stars, listing their magnitude and spectral classification. The spectral classification reveals many physical characteristics of a star, including its surface temperature, size, and density.

Spectral classes are designated by the letters O, B, A, F, G, K, and M, which correspond to surface temperature. The hottest stars have the letter designation O, while the coolest stars have the designation M. You can remember this sequence by the mnemonic "Oh, Be A Fine Girl (Guy), Kiss Me." A star's apparent color is directly related to the temperature of the gas at its surface and to its surface area. Small blue stars, for example, are very hot (some 40,000 K [Kelvin]), while red giant stars are relatively cool (about 3,000 K). (The Kelvin scale begins at absolute zero – about -273.16 °C [degrees Celsius], the coldest temperature that can be approached. Water freezes at 273 K and boils at 373 K.)

Each spectral type is divided further into 10 subclasses denoted by the numbers 0 through 9. The higher the number, the cooler the star. The Sun, for example, is classified as a G2 star, which is slightly cooler than a G1 star and slightly hotter than a G3 star.

Finally, a star's luminosity class is indicated by a roman numeral from I to VI. Supergiant stars are I; bright giants, II; giants, III; subgiants, IV; main-sequence stars/dwarfs, V; and subdwarfs, VI. Thus, Zeta (ζ) Tauri, designated as a B2 IV star, is a hot, blue, subgiant star.

Table 3 lists the spectral types, apparent color, approximate surface temperature, and an example of a familiar member of each class.

FINDING YOUR TARGET

When planning your night's observing, select an object you would like to see, read the translated Messier catalogue and NGC

Table 3. Star classifications

Spectral type	Apparent color	Surface temperature (K)	Example star
O	blue	25,000–40,000	Zeta Orionis (O9)
B	blue	21,000–25,000	Spica (B1)
A	blue to white	27,500–11,000	Vega (A0)
F	white	26,500–7,500	Polaris (F8)
G	white to yellow	25,000–6,000	Sun (G2)
K	orange to red	23,500–5,000	Arcturus (K2)
M	red	23,000–3,500	Antares (M1)

descriptions of its telescopic appearance, and examine the accompanying photograph and drawing. Place a bookmark on the page(s) for that object because you will refer to it often in the field. Next, scan the all-sky map in the back of the book to find the object's general location, the constellation it is in, and a nearby bright star or two to use as guideposts. (The brighter stars and other reference stars are labeled with their identifying Greek letters or Flamsteed numbers.) Now turn to the smaller-scale finder chart in the section of the book describing that object to zero in on your destination. Some of the finder charts show stars to 6th magnitude, others to 7th or 8th magnitude, depending on the chart's scale. Once you locate and identify your target, use your red flashlight to read the more extensive notes, which will steer you to several visual attractions in and around that particular object.

One thing you can start doing now is to practice aiming your telescope on one of the bright seasonal Messier objects mentioned earlier. Do so until it becomes second nature, because you cannot expect to hit your celestial target if your telescope and finder scope are not carefully aligned. Nothing is more frustrating than hunting for a faint Messier object when you're starting with the wrong guide star. It is amazing how similar the star fields appear when you are looking through a telescope. Take the time now to practice aiming. You can also do this during the day using a distant object on the horizon.

Pointing with accuracy is impossible unless you know how to scale what you see through the eyepiece with what you see on the finder chart. The maps show more sky than you can see through your telescope, and your telescope will reveal many more stars than are plotted on the maps. Therefore you have to determine what area of sky you can see with your low-, medium-, and high-power eyepieces. To determine the field diameter, simply choose any star near the celestial equator (such as one of the stars in Orion's belt) and time how long it takes for the star to drift across the entire field. When multiplied by 15, this number will give you the field diameter in minutes and seconds of arc. For instance, if it takes five minutes for the star to drift across, then your field diameter is 75′, or 1 1/4°. Most telescopes at low power will show at least 1° of sky.

There are electronic devices now that can effortlessly guide you (or your telescope) automatically to thousands of celestial objects, virtually eliminating the time spent searching for them. But you should save that wizardry for star parties, when you are showing the sky to friends, or on nights when time is of the essence. For now, think in terms of "exploring the sky." Besides, you do not want to be so dependent on machinery that you are helpless without it.

How do you find the fainter Messier objects? First, use the wide-field map in the back of the book to get a ballpark location for the "M" object you want, noting its position among the brighter stars. Now look at the detailed finder chart on the object page and see which naked-eye stars make a pattern with the object – simple patterns like a triangle, a straight line, or a square. Next, find those stars in the sky (I always use binoculars to confirm the star field) and place the crosshairs of the finder scope where the Messier object fits into the geometrical pattern, even if you cannot see the object. When using a finder scope, I keep both eyes open; one eye is focused on the crosshairs, and the other eye is looking at the sky. If your telescope is properly aligned, and if you have practiced your aiming, you should be very near, if not right on, the object. Use the lowest magnification (the one with the widest field of view) and look for a faint, diffuse patch of light.

If the object is not immediately recognizable, try to identify the stars in the telescope's field with those on the finder chart. You might have to hop from star to star to reach your destination. To do this, a knowledge of field diameter and sky direction is invaluable. If you get confused, return to a bright, easy-to-find Messier object, like the Pleiades or the Orion Nebula, and practice moving your telescope up and down, left and right, to determine the field's diameter and orientation.

Note that the finder charts in this book are all oriented with north up and west to the right. The unnerving part about field orientation is that a telescope inverts the image, so south is at the top and east is to the right. To match the view through your finder scope, you will have to rotate the book. Furthermore, if your telescope is on an equatorial mount, the field will often be rotated with south to the upper right or upper left. Nudge the telescope tube toward Polaris and notice where stars enter the field – that is north. For telescopes on an altazimuth mount, as mine is, just tilt the star map until it matches your view.

If your telescope's finder scope is equipped with a diagonal, remove it and replace it with a regular eyepiece (or invest in a new finder); also avoid using a diagonal with your telescope when you are using a star chart, unless you want to retrace every star chart so that it represents a mirrored view. Finally, if you are not having any success in your search, take a break and then start over. It takes practice to hit your mark.

OBSERVING TIPS

Have you ever left a theater in the middle of a movie to get some popcorn? You leave confident that you can find the way back to your seat because you can see fairly well in the low light. You enter the brightly lit foyer, get the popcorn, and return to the theater. Suddenly, you can't see a thing except the screen. You walk down the aisle with arms extended like Frankenstein's monster, groping for the right seat. Several minutes later, your night vision returns and everything appears normal. The same thing happens every time you step outside from a lighted house and stand under the stars.

The retinas of our eyes are packed with light-sensitive neural receptors called rod cells and cone cells. The cones are less sensitive to light than the rods and therefore work best in daylight. When we go from a very bright to a very dark environment, our cones essentially shut down, but our light-sensitive rods do not kick in instantly, so we become temporarily blinded. The visual pigment rhodopsin in the rods bleaches out in intense light and takes about 30 minutes to regenerate. It takes about 30 minutes for our eyes to fully dark-adapt. You can start seeing things a lot sooner than that, but don't expect to discern really faint details in a Messier object until you are fully dark-adapted.

For instance, tests have shown that eyes dark-adapted for 30 minutes are six times more sensitive to light than eyes dark-adapted for 15 minutes. My experience in the field has been that this time varies with the individual. I don't know if there is any way to validate what I do to hasten the regeneration of rhodopsin in my eyes, but the following seems to work. I pull my coat, jacket, or sweater over my head and stare intently into the darkness for a minute or two. Other times I pick the darkest spot around and just stare at it, occasionally squinting real hard, until more and more detail becomes visible.

Rod cells are also peculiar in that they do not lie directly in the center of vision (that is where the cones are) but are on the perimeter. Faint celestial objects will be best seen using averted, or peripheral, vision. In other words, do not center a faint galaxy in the eyepiece and stare directly at it – the galaxy will probably vanish because you are trying to see it with your day-sensitive cone cells. Look off to one side a little, but focus your attention on the object.

I tend to favor a particular spot in my right eye when looking for faint details. To see a knot in a galaxy's faint spiral arm, for example, I have a tendency to position the galaxy to the upper left of the field, in the direction of my temple. So, if the galactic knot is at the 11 o'clock position, I direct my line of vision to the 4 o'clock position. Since our eyes interact with specific parts of the brain, this action might reveal a subconscious knowledge of the most sensitive region in my field of rods. Perhaps a bunch of rods work together at that spot to enhance my vision. In any case, I've developed an awareness of how my own mind and eyes react to detecting faint objects. Try to discover that magic spot in your own eye or eyes (if there really is one), or just find the most comfortable position for looking. Your magic spot may be totally different from mine.

The late Carolyn Hurless, a prolific observer of variable stars, shared the following observing secret with me when I was about 16 years old. She jokingly called it "heavy breathing" and said it was a tactic that the late variable star observer and comet discoverer Leslie Peltier employed when trying to detect very faint variable stars from his observatory in Ohio. The trick is to hyperventilate, taking several very deep breaths (actually through your nose with your mouth closed) before you put your eye to the eyepiece. This sends a fresh supply of oxygen to the brain and eyes and increases your alertness.

Take a few minutes to scan the field with this fresh oxygen supply, then begin inhaling and exhaling slowly through your nose. (Actually, I now do this through puckered lips. Luckily I'm in the dark!) Increase the frequency (so the breaths are shorter and deeper) as you zero in on a target. This reminds me of the way a bat hunts using echolocation. It sends out pulses of sound at a certain frequency until it finds a meal. Then it zeros in by increasing the frequency of the pulses. You have to be careful, however, not to overdo the heavy breathing. On a few occasions, I have nearly passed out trying to see faint stars!

Hyperventilating is great for detecting starlike objects and tiny, faint patches of nebulosity. But very large and diffuse objects present a different challenge. Take something like the Andromeda Galaxy, Orion Nebula, or nebulosity in the Pleiades. These are very extensive

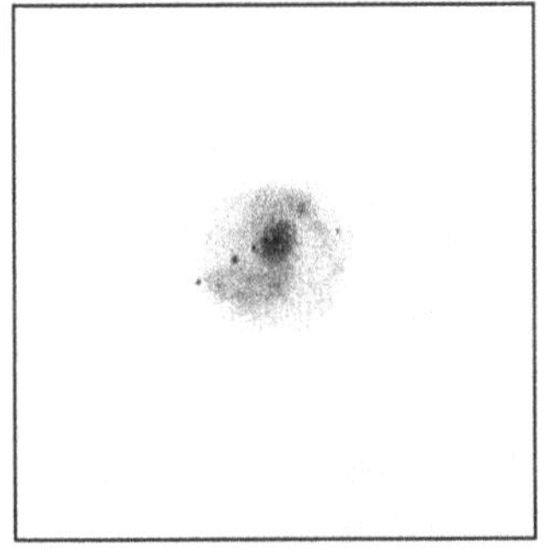
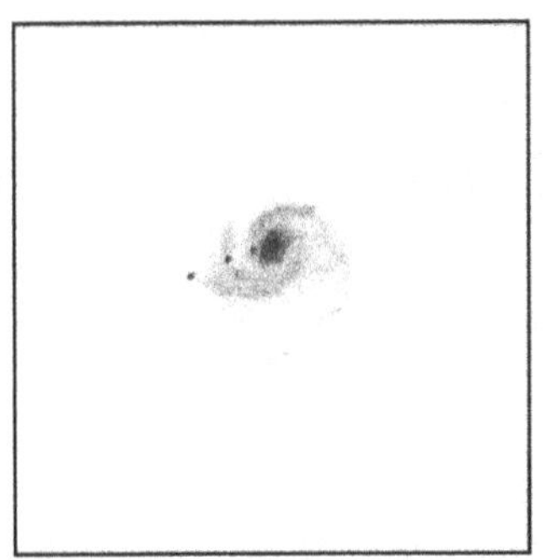
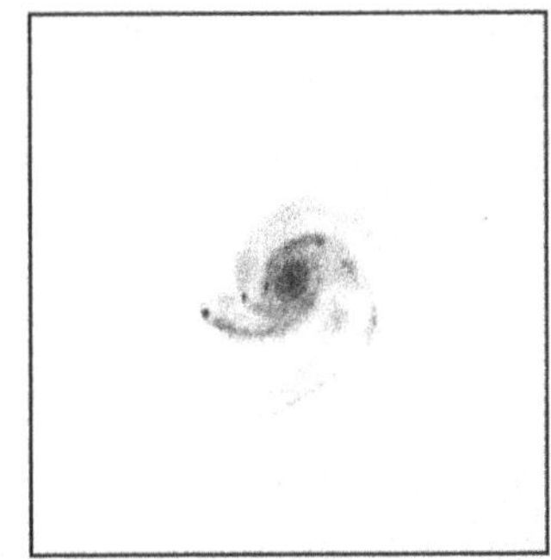

These sketches demonstrate how increased time behind the eyepiece leads to seeing more detail.

glows. The question is, how do you determine how far the object extends before it fades into the bright, starry background? The nineteenth-century Harvard astronomer George Bond would place the object well out of the field of view of the great Harvard 15-inch refractor and let the object drift back into view. When he suspected seeing a change in the sky background, he would make a note. This is essentially what I have done for the larger Messier objects. But you can also manually move the telescope far afield and gradually bring the object back in (without waiting for the Earth to rotate) and achieve the same result.

If I am not sure whether some faint detail is real or not, I will jiggle the telescope ever so slightly to set the sky in motion. The eye is better at detecting tiny moving objects than tiny stationary ones. For example, it is easier to detect a very faint satellite moving among a myriad of stars than it is to detect an equally faint stationary nova.

Although the effects of diet and other factors on visual observing are still being studied, being in good physical condition helps night vision and seems to quicken dark adaptation. It is best to be well rested; I can see one full magnitude fainter when I'm rested than when fatigued. On the moonless nights during which I made the observations for this book, I exercised regularly, avoided alcohol, wore warm, comfortable clothing, and observed from a relaxed position.

Finally, the most important factor in detecting faint details is the amount of time you spend observing. Don't rush through the Messier objects like you are at a pie-eating contest (see my comments on "Messier marathons" in Appendix B) but rather savor each one. The more time you spend looking at a particular object over the course of a night or several nights, the more detail you will see. That is how I approached the observations in this book. The accompanying drawings show how, over time, I could see more detail in the galaxy M101 in Ursa Major.

For most of the Messier objects, I observed each for an average of six or more hours over three nights. Some of the more complex objects were observed for three hours per night for several nights. Don't expect to see all the detail in my drawings with just a glance. Challenge yourself to spend the time to really study these objects, which are some of the most splendid deep-sky wonders visible from our unique perspective in space, and to strive to see even more detail than I have shown. Once you have located a Messier object, let's say a galaxy, spend about a half hour just enjoying it.

Next, try to sketch as much detail as you can without referring to the book. (Even if you can't draw, make a sketch; you will be surprised at how your artistic talents will increase by trying.) Then sit back for a moment, take a sip of hot tea, relax your eyes, examine the photograph and drawing, and compare your sketch with the photograph. Before you return to the eyepiece, read the descriptive information on that page and pick out a specific detail you want to find; maybe this particular galaxy has a faint outer arm, which you overlooked. When you return to the eyepiece, you now focus your attention only on trying to see that particular detail. If you do this for each feature within that particular object, using different magnifications, your drawing will ultimately come together into a coherent whole! By training your eye in this way, you will be able to pick up these subtle features more easily on subsequent observations of the object.

CHAPTER 3

The making of this book

3

This book is your companion under the stars. Similar to a field guide to birds, insects, or flowers, this book will help you locate and identify the Messier objects – the most famous deep-sky splendors visible from the Northern Hemisphere. It contains, among other things, finder charts to help you locate them, photographs, pencil drawings of their telescopic appearance, detailed descriptions, and descriptions of other interesting celestial sights near them. I have also included a lineup of some of my favorite non-Messier objects.

Each Messier object section opens with a photograph and some essential data, including the object's coordinates, magnitude, apparent size, and distance. Descriptions of the object from a translation of Charles Messier's original catalogue and from the *New General Catalogue of Nebulae and Clusters of Stars* (1888) or the two subsequent *Index Catalogues* (1895 and 1908) follow. These data and descriptions should help you realize what to expect when you first glance at a Messier object through the eyepiece.

The objects are ordered as Messier catalogued them. However, you can view them in any order you wish, as long as a particular object is above the horizon when you plan to observe it. Use a star wheel or an astronomy app to determine which constellations, and therefore which Messier objects, are visible on the date and time of night you want to look.

THE TELESCOPE

This book was conceived at the May 1994 Texas Star Party, where I spent several nights observing the Whirlpool Galaxy (M51) with a Tele Vue Genesis 4-inch refractor. Until that time, I had spent most of my life observing the Moon and planets with permanently mounted, long-focus, observatory-class refractors with old, high-quality glass. Any views of the deep sky were confined to high powers and narrow fields. For instance, the lowest magnification I normally used with the 9-inch f/12 Clark refractor at Harvard College Observatory was 137×, which offered me a 1/2° field of view. Only one-ninth of the Andromeda Galaxy (M31) would fit in that field! And when I used the 18-inch f/16 Clark refractor at Amherst College, a typical low power was 365× and the field of view was even smaller.

Even as a young Messier-object hunter in the 1960s, I used long-focus reflectors and refractors and never saw the objects in a field greater than 1°. The Genesis refractor, when coupled with a 22-millimeter (2-inch) Nagler eyepiece, will show the Andromeda Galaxy at 23× in a field of view close to 3°! Nearly the entire galaxy and its companions fill the field. Furthermore, changing eyepieces and adding a Barlow lens allows me to study the galaxy's nuclear region with 10 times that magnification.

When it came time to decide which telescope to use for this project, I had no hesitation in choosing the 4-inch f/5 (500-millimeter) Genesis refractor. The unobstructed optics in the Genesis are of unquestionable quality. Al Nagler (formerly a NASA optical engineer, who designed the wide-field optics for the Apollo Lunar Landing Simulator) created a special four-element optical primary for the Genesis refractor. When coupled with his Tele Vue eyepieces (which have an equally revolutionary optical design), the Genesis transcends the traditional limitations of both the long-focus and rich-field refractors by combining the best qualities of both.

For this study, I used only two eyepieces: a 22-millimeter Panoptic (adaptable 1 1/4-inch and 2-inch) and a 7-millimeter Nagler (1 1/4-inch); these provided magnifications of 2× and 72×, respectively. A 1.8× Barlow

and holds the telescope rock steady. Sometimes when I am out observing on the summit of Kilauea volcano, the winds will suddenly gust to gale force, yet the telescope hardly jiggles. The mount has also performed quite heroically during several minor earthquakes (a frequent occurrence when you live on an active volcano!); I never lost my field of view during these events, and sometimes the small volcanic quivers helped me to confirm the existence of some faint details in a nebula.

lens with the 7-millimeter eyepiece gave me a "high" power of 130×, which is a magnification of about 32× per inch of aperture. As a rule, 50× per inch of aperture is considered the maximum useful limit under ideal atmospheric conditions. Although the Genesis can perform beyond that theoretical limit (especially under very dark and stable skies at high altitudes), I restricted myself to those magnifications because the vast majority of us constantly deal with less than perfect atmospheric conditions.

I did not use any filters when making the observations for this book. Observers who live under light-polluted skies, however, should consider using light-pollution or skyglow-reduction filters to sharpen the contrast between the diffuse Messier objects and the background sky. Hydrogen-beta and oxygen-III filters work wonders on emission nebulae, but be aware that they also dim starlight by about one magnitude.

The Genesis is cradled in the altazimuth yoke of a Gibraltar mount. This heavy, ash-wood tripod certainly lives up to its name

I should add that no one twisted my arm, dangled a carrot, or offered me the key to Manhattan Island to use this particular telescope for this project. The Genesis and I met serendipitously under dark Texas skies, where I realized its quality and potential. The rest is history. Naturally, you can enjoy the Messier objects through any telescope, large or small. And most of the objects are even visible in 7 × 35 binoculars.

Since the writing of the first edition of this book, I acquired a 5-inch Tele Vue refractor and then downsized to a 3-inch. A few observations from these telescopes have been added to the essays in Chapter 4.

OBSERVING LOCATION

After the 1994 Texas Star Party, I returned home to Boston, Massachusetts, where city lights wash out all but the brightest stars from the night sky. Unquestionably, if this project were to go forward, I would have to travel to complete the book. My plan was

altitude of 3,600 feet. It is some three miles east of the summit of Kilauea volcano, which rises 4,200 feet above sea level. The streetlight nearest to our house is 1 mile to the north. Beyond that, some 300,000 acres of Hawaii Volcanoes National Park and reserved forest border our subdivision to the north, south, and west. Hilo, the nearest city to the east, is about 45 miles away. The massive, swollen back of Mauna Loa volcano, which rises nearly 14,000 feet above sea level, blocks our view of Kona, 100 miles to the west. All around us for 3,000 miles is the vast Pacific Ocean. Adding to the visual pleasure, the entire island has a lighting ordinance to help keep the sky dark for astronomers at the nearly 14,000-foot-high Mauna Kea Observatory to the north.

to observe from the darker skies of western Massachusetts on weekends and augment these observations with any I could make at various star parties. By a twist of fate, in the fall of 1994, my wife, Donna, was offered a job on the Big Island of Hawaii, and I followed her there that December.

A better stage could not have been set for performing the observations for this book. We purchased a house in Volcano, Hawaii, at an

Here in Volcano, on clear moonless nights, the Milky Way is bright enough to cast shadows, and Venus can "pollute" the sky with its brilliance. The bright haystack of the zodiacal light, the dim, dusty alley of the zodiacal band, and the much fainter elliptical glow of the gegenschein – a faint brightening of the zodiacal band at the antisolar point; the counterglow to the zodiacal light – are all visible from our front yard. These large, faint features of the night – created by sunlight reflecting off dust-sized particles in the plane of our Solar System – are the hallmarks of a truly dark observing site. In their *Observing Handbook and Catalogue of Deep-Sky Objects*, Brian Skiff of Lowell Observatory and Christian Luginbuhl write, "Frequently the sole requirement for the visibility of a notoriously difficult nebula is not a large telescope or a special eyepiece, but a truly dark sky." Although none of the Messier objects are notoriously faint or difficult, the exotic details of each – a spiral arm, a dark lane, a tenuous wisp of nebulosity – can be extinguished in all but the darkest of skies.

LIMITING MAGNITUDES

That truly great skies are truly dark skies is a myth. What is really meant by dark skies is that the atmosphere is so free of man-made pollutants and nighttime lighting that multitudes of fainter stars can be seen. The more starlight you see, the brighter the night sky appears. From the darkest sites, the zodiacal light and band, the gegenschein, and the Milky Way add illumination to the celestial backdrop. (This dark-sky myth is deeply rooted. Sometimes, visitors atop Mauna Kea – the world's best observing location, where magnitude 8.5 stars are within grasp of the naked eye – complain that the site is substandard because the sky does not look dark!)

Limiting magnitude is itself a very good indicator of a sky's quality. But another myth can blind us from determining that limit accurately. Many popular amateur astronomy books state that 6th magnitude is the limit of naked-eye vision and that if you have good eyes and a good site you can probably see about 0.5 magnitude fainter. Magnitude limits are also attributed to telescopes of given apertures: a 4 1/2-inch aperture will show stars to magnitude 12; an 8-inch will reveal stars to magnitude 14. What these general rules fail to consider is the human contribution in seeing faint objects. Several years ago, I researched the origins of the magnitude formula. Most profound is the classic 1857 work by the nineteenth-century English astronomer Rev. N. R. Pogson (*Monthly Notices of the Royal Astronomical Society*, vol. 17). His words, which reveal the individuality of limiting magnitudes, seem to have been forgotten:

> I selected [the ratio] 2.512 for the convenience of calculation.... If then any observer will determine for *himself* the smallest of Argelander's magnitudes just discernible by fits, on a fine moonless night, with an aperture of one inch, and call this quantity L, or the limit of vision for one inch, the limit l, for any other aperture, will be given by the simple formula, l = L + 5 × log aperture. Numerous comparisons, made with various telescopes and powers, at different seasons of the year, have furnished me with the value L = 9.2 for my own sight, which is, I believe, a very *average one*, and therefore suitable for such a determination.

All emphases are mine. Pogson created a formula that worked for him. He encouraged others to discover their own limits with this formula. Pogson clearly states that his eyesight was average, and he was keenly aware that limiting magnitude varies with aperture, season, and atmospheric conditions. Also, consider that the glass he used for his observations was inferior to that employed by today's amateurs. A better formula is the "brat" equation,

$$N = brAtf(m)C,$$

where N is the number of photons available per unit time to the eye; b is the most effective bandpass for the human eye (100 nanometers); r is the transmission through the atmosphere, optics, and reflective coatings; A is the light-collecting area of the aperture

(πr^2); t is the eye's storage time for collecting photons (0.1 second); $f(m)$ is the decrease in stellar magnitude, m, being discussed ($f(m) = 2.512^{-m}$); and C is the value of incident stellar radiation of a zero-magnitude standard star beyond the Earth's atmosphere. C = 10,000 photons/second/nanometer per square centimeter.

The origin and significance of this formula is explained in more detail in an article entitled "Some Thoughts on Limiting Visual Stellar and Cometary Magnitudes with Various Apertures," by Daniel W. E. Green, in the *International Comet Quarterly* (April 1985). The equation demonstrates that, under ideal conditions, the naked eye can, theoretically, detect enough photons to see a 9th-magnitude star! It is also not impossible to see stars fainter than 15th magnitude with a 6-inch reflector, as the keen-eyed variable star observer E. H. Mayer has done from his dark-sky site in Ohio.

In 1901, Heber Curtis of Lick Observatory wrote a memoir "On the Limits of Unaided Vision" (*Lick Observatory Bulletin* no. 38). He began with the following comment: "It is generally stated that stars of the sixth magnitude are as faint as can readily be seen by the unaided eye, though it is well known that under conditions of exceptional clearness favorably placed stars from a half to a whole magnitude fainter can be made out."

Curtis proved this in a visual experiment conducted at the observatory, in which he detected with the unaided eye stars of magnitude 8.2 without difficulty and stars of magnitude 8.3 with difficulty. He also glimpsed a star of magnitude 8.5. (Brian Skiff notes that Curtis's data are not on the new standard magnitude system. It might be interesting, Skiff suggests, to refer to the original paper and do this study again with new magnitudes for the stars Curtis saw.)

Regardless, Curtis's observations support my own limiting-magnitude studies at the 9,000-foot level of Mauna Kea, where I consistently detected stars as faint as magnitude 8.4 with the unaided eye. In another study performed at the 1994 Texas Star Party, Florida amateur astronomer Jeannie Clark reported seeing a magnitude 7.9 star with her unaided eye. (For another discussion on the power of vision, see "Telescopic Limiting Magnitudes" by Bradley E. Schaefer, a pioneer in the study of human perception in astronomy, which appeared in the *Publications of the Astronomical Society of the Pacific*, vol. 102, February 1990.)

Although "conventional wisdom" says you need an 8-inch telescope to resolve 14th-magnitude stars in globular clusters, in fact it is possible to reach that magnitude with a 4-inch telescope. The same misconceptions apply to naked-eye and binocular magnitude limits as well. Of course, there are factors that influence the limit of our vision, the most debilitating of which is light pollution – that ugly glow over our cities and suburbs that robs us of our views of the stars. Light pollution is amateur astronomy's greatest nemesis. If writing this book has done anything for me, it has made me appreciate how beautiful the night really can be. I fear that if we do not act now to preserve the few untainted sites we have left for amateur astronomy, night itself may become a myth.

At the 1994 Texas Star Party, Brian Skiff and I used a 7-inch refractor to do some visual tests of limiting magnitudes. When we turned it on the globular cluster M3 and compared what we saw to a magnitude sequence published in the *Observing Handbook and Catalogue of Deep-Sky Objects* (see photo), we independently recorded a magnitude 15.1 star and concluded we could have gone fainter. That same magnitude sequence is reproduced

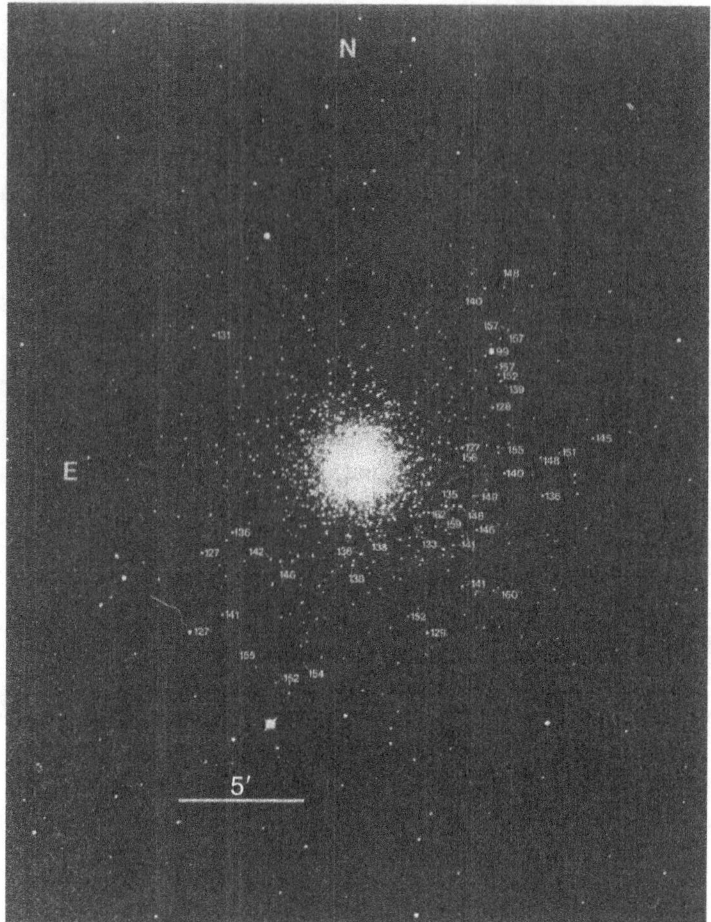

here to help you determine your own limits. And do not forget that using averted vision – looking slightly off to the side of the object of interest – can be very helpful when viewing faint objects. Skiff concludes from naked-eye tests that he can see three magnitudes fainter with optimally averted vision than he can with direct vision when his eyes are fully dark-adapted.

Altitude also has a significant effect on how faint you can see. The higher you are, the fainter you can see, up to a point. I gain roughly 1/2 to 1 magnitude for every 3,000 feet above sea level. The improvement levels off at about 9,000 feet, for two reasons. First, you're above the densest layers of the Earth's atmosphere (which greatly affects the visual limit), and second, the relative lack of oxygen in the air at that altitude means less oxygen gets to the eyes, thus adversely affecting vision.

MAGNITUDE ESTIMATES AND OTHER DATA

Estimating the brightness of diffuse objects is not easy. The traditional method, first employed by J. B. Sidgwick, is to compare the size and brightness of the diffuse object with a selection of similarly bright stars that have been defocused until they appear the same size as the diffuse object. In another method, both the star and the diffuse object are simultaneously defocused and their brightnesses and sizes compared. The results of these methods can be quite accurate, but it has been my experience, particularly in comet studies, that they are inadequate for objects with strong central condensations surrounded by faint, extended halos. Many Messier objects, especially the galaxies and globular clusters, fall into this troublesome category.

Simple in-and-out focusing methods lead one to underestimate the intensity of these awkwardly diffuse objects because the contribution of light from the extended halo is either negated (in the first method) or lost (in the second).

Twenty years ago, I devised my own method for estimating the magnitude of diffuse objects, which factors in the light loss of this outer envelope. This is accomplished by racking the diffuse object out of focus until it is a uniform glow. I then defocus my eyes to select candidate comparison stars. The selected stars are racked well out of focus, and I compare the glow from each stellar disk with that of the out-of-focus diffuse object. I then move back and forth between the diffuse object and the comparison stars until a reasonable match is obtained in both intensity and size. The entire procedure can take more than an hour! Try this method and see if you come up with magnitudes similar to those in this book.

Charles Morris, a well-known comet observer, has described a similar method in which the entire diffuse glow of the defocused comet is compared with a star image racked out to the same size (meaning the star is defocused even more than the comet). His method was published in 1980 in the *International Comet Quarterly*, volume 2.

Nearly all my magnitude estimates for the Messier objects were made with the unaided eye or with binoculars. I included them in the list of data for each object only if I found a marked disagreement with published values. Otherwise, the object data provided in the book are from the sources listed in the sections that follow.

SOURCES OF DATA AND INFORMATION

The data and information in this book were drawn from a variety of modern sources. Many of these sources were used in the first edition of this book as well as the other fours books in my *Deep-Sky Companions* series, so you can compare the properties of these objects with confidence. Generally speaking, I gleaned recent research findings on the physical nature of each object from the *Astronomical Journal* or the *Astrophysical Journal*, and citations are given. From each object's apparent diameter and distance, I calculated its physical dimensions using the formulas that appear on page 35 of the first edition of this book. Other information, such as constellation lore, properties of stars, and each object's position, apparent magnitude, angular size, and surface brightness, comes from the following excellent sources (primary sources are listed first).

Star names, constellations, and mythology

Allen, Richard Hinckley. *Star Names: Their Lore and Meaning*. New York: Dover Publications, 1963.

Ridpath, Ian. *Star Tales*. New York: Universe Books, 1988. See also http://www.ianridpath.com/startales/contents.htm.

Staal, Julius D. W. *The New Patterns in the Sky: Myths and Legends of the Stars*. Blacksburg, VA: McDonald and Woodward, 1988.

Stellar magnitudes and spectra

Stars. http://stars.astro.illinois.edu/sow/sowlist.html.

Hirshfeld, Alan, Roger W. Sinnott, and Francois Ochsenbein, eds. *Sky Catalogue 2000.0*, Vol. 1, 2nd ed. Cambridge, England: Cambridge University Press; and Cambridge, MA: Sky Publishing, 1991.

Stellar distances

Stars. http://stars.astro.illinois.edu/sow/sowlist.html.

ESA. *The Hipparcos and Tycho Catalogues*. Noordwijk: European Space Agency, 1997.

Double stars

Luginbuhl, Christian B., and Brian A. Skiff. *Observing Handbook and Catalogue of*

Deep-Sky Objects. Cambridge: Cambridge University Press, 1989.

U.S. Naval Observatory (USNO). *The Washington Double Star Catalog.* Washington, DC: Astrometry Department, U.S. Naval Observatory. http:// ad.usno.navy.mil/ad/wds/wds.html.

Variable stars

American Association of Variable Star Observers (AAVSO). http://www.aavso.org/.

Open star clusters

Archinal, Brent A., and Steven J. Hynes. *Star Clusters.* Richmond, VA: Willmann-Bell, 2000.

Open star cluster distances generally were gleaned from the professional literature.

Globular star clusters

Harris, William E. *Catalog of Parameters for Milky Way Globular Clusters.* Hamilton: McMaster University. http://www.physics.mcmaster.ca/~harris/mwgc.dat.

Skiff, Brian A. "Observational Data for Galactic Globular Clusters." *Webb Society Quarterly Journal* **99**:7 (1995), updated May 2, 1999.

Globular star cluster distances generally were obtained from the professional literature.

Planetary nebulae

Cragin, Murray, James Lucyk, and Barry Rappaport. *The Deep-Sky Field Guide to Uranometria 2000.0.* Richmond, VA: Willmann-Bell, 1993.

Luginbuhl, Christian B., and Brian A. Skiff. *Observing Handbook and Catalogue of Deep-Sky Objects.* Cambridge: Cambridge University Press, 1989. (See dimensions and central star magnitudes.)

Skiff, Brian A. "Precise Positions for the NGC/IC Planetary Nebulae." *Webb Society Quarterly Journal* **105**:15 (1996). (See positions.)

Strasbourg-ESO Catalogue of Galactic Planetary Nebulae.

Planetary-nebula distances generally were taken from the professional literature or from the World Wide Web page of the Space Telescope Science Institute (www.stsci.edu).

Diffuse nebulae

Cragin, Murray, James Lucyk, and Barry Rappaport. *The Deep-Sky Field Guide to Uranometria 2000.0.* Richmond, VA: Willmann-Bell, 1993.

Distances of diffuse nebulae were gleaned from the professional literature.

Galaxies

Cragin, Murray, James Lucyk, and Barry Rappaport. *The Deep-Sky Field Guide to Uranometria 2000.0.* Richmond, VA: Willmann-Bell, 1993. (See positions, angular size, apparent magnitude, and surface brightness.)

National Aeronautics and Space Administration (NASA). *The Extragalactic Database.* Pasadena, CA: Infrared Processing and Analysis Center. http://nedwww.ipac.caltech.edu/. (See types, mean distance, radial velocity, and notes.)

Tully, R. Brent. *Nearby Galaxies Catalog.* Cambridge: Cambridge University Press, 1988. (See inclination, total mass, and total luminosity.)

Some galaxy distances were obtained from the professional literature.

Extragalactic supernovae

"List of Supernovae." Cambridge, MA: Central Bureau for Astronomical Telegrams. http://cfa-www.harvard.edu/iau/lists/Supernovae.html.

Other details of discoveries were taken from individual *IAU Circulars.*

Historical objects

Glyn Jones, Kenneth. *The Search for the Nebulae.* Bucks: Alpha Academic, 1975.

Smyth, Admiral William Henry. *A Cycle of Celestial Objects.* Richmond, VA: Willmann-Bell, 1986.

Other historical anecdotes in this book were obtained from various individual and professional papers from the nineteenth and early twentieth centuries.

General Notes

Note that the World Wide Web Uniform Resource Locators, or URLs, are subject to change. The dimensions, magnitudes, and positions of all other additional deep-sky objects in this book were taken from *The Deep-Sky Field Guide to Uranometria 2000.0*.

In most cases, the physical diameters of the galaxies were calculated using the following formula:

$$\text{diameter} = 0.292 \times D \times R,$$

where D is the object's diameter in arcminutes and R is the object's distance in megaparsecs. The physical diameters of all the other Messier objects were calculated using a similar formula,

$$\text{diameter} = 0.000292 \times D \times R,$$

where D is the object's diameter in arcminutes and R is the object's distance in kiloparsecs.

This book contains the most up-to-date astronomical data for the Messier objects of any book in the popular literature.

THE IMAGES

All the object images in this book are reproduced in black and white, with north up and east to the left. Most are roughly 15 × 15 arcminutes. And, unless otherwise noted, all were taken by my longtime friend and colleague Mario Motta with an STL 1001E SBIG camera on his homemade 32-inch telescope described later in the book. (Detailed credits for these and additional photos appear at the end of the book.)

I've also reproduced images from the Hubble Space Telescope (HST) and other large telescopes. These images were used for the sole purpose of inspiring you to use your imagination. You certainly will not see anything like these images when you look through your telescope, but how else can you fully appreciate what it is you are seeing? So do not be discouraged; be enlightened.

An advanced amateur astronomer, Mario Motta has built several observatories and telescopes. His "Wingaersheek Observatory" – which sits atop his study in his home in Gloucester, Massachusetts – houses his homemade 32-inch reflector under a 20-foot dome that overlooks Ipswich Bay to the north and a large saltwater marsh to the south.

One of the largest telescopes in New England, Mario's Monster (as I congenially call it) is a state-of-the-art, robotic telescope that weighs 1,200 pounds and sports a special optical system of the Relay design. It was the brainchild of Mario, his optical-designer friend Scott Milligan, and other members of the Amateur Telescope Makers of Boston (ATMOB), of which Mario has been a longtime member and past president. It is this same telescope that Mario used to take the images in this book and the ones in *Deep-Sky Companions: The Secret Deep*.

The images appear in black and white because while some stars, clusters, and nebulae do display subtle hues (though some astronomers would argue otherwise), for the most part the objects appear as grayish-white hazes or clusters of stars projected against a black background. Color images, while often beautiful, may mislead beginners into thinking they can expect to see vivid reds, greens, and blues; such expectations would only lead to disappointment when the real thing is seen. Because my mission is to excite you about what you can see through a telescope, I decided to stick with black-and-white images only.

The images are meant not only to inspire you to look at these objects but also to help you confirm your visual impressions of them – especially when you think you're seeing an elusive detail in a nebula, or a pattern of faint stars in a cluster. For example, if you study the globular cluster M13 at high magnification, you might suspect several dark lanes that slice across the stars near the cluster's

Of course, no image can record emotion. When I finally pick out a faint blur in the sky and realize that it is a galaxy 40 million light-years away, well, the visual image might not be all that impressive, but I am in awe of it nevertheless.

THE DRAWINGS

All the drawings in this book are composites based on several observations with magnifications of 23×, 72×, and 130×. They are shown with north up to match the photographic view; the finder charts also have north up. I spent a minimum of six hours on each object over three nights. The observations were made only on the clearest, moonless nights, at altitudes no lower than 3,500 feet. The drawings represent what good observers from dark, sea-level sites might expect to see with an 8- to 10-inch Schmidt-Cassegrain, a popular backyard telescope.

I did not try to accurately position every resolvable star in a globular cluster. I did, however, try to plot, to the best of my ability, the locations of any particularly bright members, especially if they would help you orient the drawing to the photograph for comparison. Otherwise, the drawings of the globular and open clusters reveal the patterns of major star streams, dark lanes, and irregularities in shape. Diffuse objects, such as the Orion Nebula, were extremely challenging and required longer dedicated efforts. I treated the Orion Nebula, for example, as if it were a dozen individual objects, and focused on one area at a time over the

center. By orienting the photograph to match your visual impression, you will discover that the dark lanes really do exist. These features were first seen by Lord Rosse with his 72-inch speculum-mirror telescope in the 1800s, but they're within reach of moderate-sized amateur telescopes today.

The photographic appearance of stars and nebulosity depends on the length of the exposure, the sensitivity of the film, and how the film is developed. Many of the images in this book show more stars and nebulosity than you can expect to see with your telescope. Years ago there was an unfortunate trade-off in deep-sky photography. To expose the faint outer arms of a spiral galaxy, the photographer had to overexpose the brighter core or nuclear region. But that is no longer a problem with today's CCD imagers who can manipulate images – not only of galaxies, but also of globular star clusters and other objects with a high dynamic range in visual contrast – with specialized software to show all these details. All images of deep-sky objects in this book have north up and east to the left.

course of several weeks. (For a revealing look at how increased time spent behind the eyepiece can enhance the amount of detail you can see, refer to the sequence of drawings of the galaxy M101.)

Although I strived for accuracy in the drawings, my renditions are not perfect, and my interpretations of what I see may be different from those of others. For instance, my eye and mind might work together to follow a particular pattern of stars in a globular cluster - say, a counterclockwise spiral of stars - but other observers might see a perfectly reasonable clockwise spiral of stars in the same region.

For some objects, open clusters in particular, I took the liberty of drawing the whimsical creatures (e.g., bats, alligators, fireflies) that I visualized in the patterns of the stars during moments of fancy. An example is M41, whose stars form an outline of a fruit bat reaching for a bite to eat. I have also highlighted certain geometrical patterns in the drawings for emphasis. The Spanish surrealist artist Salvador Dali referred to drawing as the honesty of art, saying "there is no possibility of cheating. It is either good or bad." With that in mind, I hope you enjoy my renditions of the Messier objects and find them useful in lending perspective to your observations.

THE FINDER CHARTS

The constellation map at the back of the book is sufficient for locating most, if not all, of the Messier objects, which are fairly bright and easy to spot once you're looking in the general vicinity. But the smaller-scale finder charts provide additional detail. They can be particularly helpful in identifying other objects that lie near the Messier objects. The scales and magnitude limits of the finder charts vary. For some, a smaller scale (and higher magnitude limit) may have been needed to distinguish several objects that are close together, whereas a wider scale (and lower magnitude limit) was appropriate to show objects separated by greater distances or to show an object's location relative to a large pattern of stars in a constellation.

The following symbols are used to represent the different object types on the finder charts:

MESSIER'S OBJECT DESCRIPTIONS

Messier's own descriptions of the objects he catalogued are included in Chapter 4. The translation of the French catalogue in the *Connaissance des Temps* for 1874 was expertly done by Storm Dunlop, author and translator of numerous books on astronomy and a fellow of the Royal Astronomical Society. Storm gained access to the catalogue in the library of the RAS in London, where it is preserved. His is the most precisely interpreted and smoothest-reading English translation of the catalogue that I have seen, and I am grateful to him for bringing his linguistic gifts

Diffuse nebula	Open cluster	Galaxy
Dark nebula	Globular cluster	Variable star
Planetary nebula	Supernova Remnant	Double star

and knowledge of astronomy to bear on this important task.

There are a few things to note about Messier's catalogue and its translation. His object descriptions appeared on right-hand pages, while the facing left-hand pages contained columns listing the object number, the right ascension and declination, his estimate of the object's angular diameter, and the date of his observation (which I have put in brackets preceding each description). This edition of the catalogue, the last one published before Messier's death, included 103 objects; numbers 104 through 110 were added later.

Messier used the third-person grammatical form when referring to himself, saying of M32, for example, "M. Messier saw it for the first time in 1757, and has not noted any change in its appearance." This style was not imposed by the translator. Bracketed [] words or phrases within the descriptions were added in translation for purposes of clarification. Messier also made frequent use of the term "ordinary telescope," which has been translated literally by others in the past. The more correct meaning, "simple refractor," is used here. The implication is that the telescope used was not a compound (achromatic) refractor. When Messier describes an observation made with a "one-foot telescope" or a "simple three-foot telescope," he is referring to the scope's length, not its aperture. Messier's use of the word "parallel" was translated literally; for example, "the cluster is close to Antares and on the same parallel." Dunlop suggests that Messier might have meant something more like "zone of declination," because he often referred to an object being on the same parallel but slightly above (or below) it. Messier occasionally refers to an object being plotted on the English *Atlas Céleste*, by which he means an original English edition of Flamsteed's *Atlas Coelestis*. A French-language edition, *Atlas Céleste*, was published in 1776, several years before Messier's catalogue.

The Latin names of constellations are given throughout, rather than the vernacular names that Messier used. Punctuation has been modernized.

M1

Crab Nebula
NGC 1952
Type: Supernova Remnant
Con: Taurus

RA: $5^h34.5^m$
Dec: + 22°01′
Mag: 8.4; 8.0 (O'Meara)
Dim: 6′ × 4′
Dist: ~ 6,500 light-years
Disc: John Bevis, 1731

MESSIER: [Observed September 12, 1758] Nebula above the southern horn of Taurus, which does not contain any stars. Its light is whitish and elongated like a candle flame. Discovered when observing the comet of 1758. See the chart of this comet, *Mémoires de l'Académie* 1759, page 188; observed by Dr. Bevis around 1731. It is plotted on the English *Atlas Céleste.*

NGC: Very bright, very large, extended roughly along position angle 135°; very gradually a little brighter in the middle, mottled.

MESSIER HAD NO IDEA HIS FIRST catalogued object would be among the most intriguing in the heavens. M1, the Crab Nebula, is one of 100 or more known supernova remnants in our Galaxy – a corpse of a star that experienced a fast life and a violent death. A supernova explosion is the final stage in the life of a star several times more massive than the Sun. Such a red supergiant star (like Betelgeuse in the shoulder of Orion) voraciously consumes its nuclear fuel in about 10 million years (100 times faster than the Sun). When the star's thermonuclear energy is exhausted, its Earth-sized core collapses under the force of gravity and, within seconds, shrinks until the core's density equals that of an atomic nucleus. Unable to contract further, infalling gas rebounds off the resistant core. A quarter-second later, the star ends its life in a fantastic explosion, the peak energy of which can rival that of its host galaxy.

The Crab is what remains of a cataclysmic stellar explosion that occurred in our own Milky Way Galaxy. So powerful and so close (~ 6,500 light-years) was the blast that Chinese skywatchers described it as a "guest star" in the annals of the Sung dynasty. They first recorded it on July 4, 1054, during the day. It rivaled the brightness of Venus, appeared reddish-white, and was observed for 23 days during the day and 30 times longer during the night.

The mysterious "guest star" then vanished from the annals of history until British physician and amateur astronomer John Bevis (1695–1771) encountered its glow telescopically around 1731. Messier independently swept it up while searching for the Comet of 1758 and included it as the first object in his 1771 catalogue of nebulae and clusters. In a letter dated June 10, 1771, Bevis informed Messier of his discovery, and Messier duly gave credit to Bevis in later editions of his work.

Images of M1 taken throughout 11 years during the early twentieth century ultimately led to the discovery in 1921 that the object was a supernova remnant expanding about 0.2" per year. Further evidence came when astronomers identified strong radio energy radiating from it in 1948 and then bright x-rays in 1964. In 1968, observations with the Arecibo Radio Telescope in Puerto Rico revealed a rapidly pulsing source (a pulsar) at the nebula's center – for the first time linking this strange new class of objects to the collapsed core of the supernova's progenitor star. Called a neutron star (because it is comprised entirely of neutrons), the Crab's pulsing heart is the closest thing to a black hole that astronomers can observe directly. It measures only about six miles across, yet has a mass of a half million Earths squashed into its sphere; a teaspoonful of its matter would weigh as much as a majestic mountain.

The Crab's pulsar is also known as a "millisecond pulsar," as it rotates at a rate of 30 times per second. While one would expect a pulsar's rotation rate to "spin down" over time, M1's doesn't, because material transferred to it periodically by an unseen companion gradually "spins up" its rotation.

An image (shown here) released from the Chandra X-ray Observatory in 2000 shows an "inner ring" around the pulsar, possibly a shock wave of high-energy particles, that has pushed outward over a light-year from the neutron star. The image also captures jets of high-energy particles shooting off perpendicular to the ring. Although the ring appears elliptical in the Chandra image, the feature is presumably a circle tilted toward our line of sight.

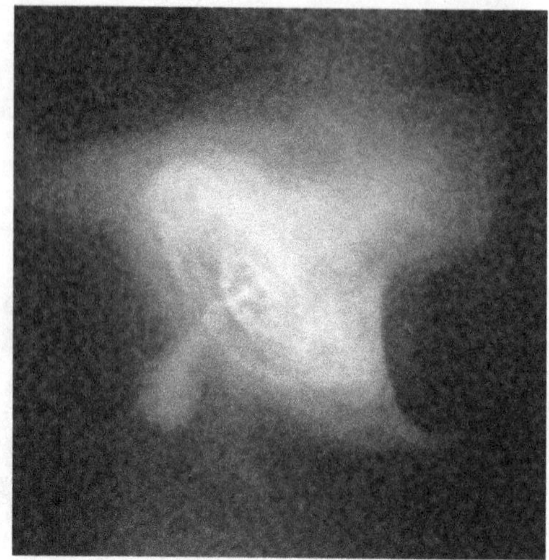

The 2005 Hubble Space Telescope image release presented on the next page shows the entire nebula in unprecedented detail. It is in fact the largest mosaic of the Crab Nebula ever made. The 24 individual Wide Field and Planetary Camera 2 exposures – taken in October 1999, January 2000, and December 2000 – have proven that the inner part of the Crab Nebula is "far more dynamic than previously understood." Every few days, its wisps stream away from the pulsar at half the speed of light.

As the only supernova remnant in Messier's catalogue, M1 warrants special attention. In small telescopes, it is a 6′ × 4′ irregular patch of nebulosity situated a little more than 1° northwest of 3rd-magnitude Zeta (ζ) Tauri, a hot, blue subgiant star. First use the accompanying chart to look for a trapezoid of stars with Zeta Tauri as the southern corner and magnitude 6.5 Star *a* to the northwest. M1 lies only a little less than 30′ west of Star *a*.

Under excellent skies, the nebula can be spied with binoculars (amazing, if you consider that nearly 1,000 years have passed since the explosion). Visual-magnitude estimates

M1

N

Taurus

a

ΔM1

E

W

ζ

1°

S

for this very famous object vary from roughly 9th magnitude to 8th (the latter being my own visual estimate). The difficulty is that any degradation in sky transparency will cause the nebula's outer regions to fade significantly and affect its apparent brightness.

If you can see the nebula in a 2° or larger field, it's worth the look, as the Crab Nebula shares the field with Zeta Tauri, and its glow looks much like the ghost image of that star. But M1, which measures roughly 10 × 7 light-years in true extent, is so much more enormous. I once watched it materialize in the twilight through the 5-inch and was amazed at how elusive its form appeared until nearly the true beginning of nightfall. So light pollution will hinder its visibility. If you have difficulty spotting it, try fixing the telescope on its position and gently tapping the tube while using averted vision. Patience is key.

The nebula is composed of three parts: the 16th-magnitude pulsar (the pulsating neutron star), an inner bubble of material (a powerful wind of radiating particles bound to the object's magnetic field), and an outer shell of dense material released in the supernova explosion. How much you can see depends greatly on aperture.

With a glance at 23× through the 4-inch, the nebula's midsection appears pinched. A longer look reveals two halves slightly askew or misaligned, as if two plates along a tectonic fault had suddenly slipped. At higher magnifications, the nebula looks patchy, with three distinct sections running southeast to northwest. The southern and middle sections are of similar brightness, while the northern one is smaller and much fainter. When photographed in polarized light, the nebula reveals a similar trilobate or serrated aspect, indicating the existence of very strong magnetic fields.

Has anyone with a large telescope ever tried to view M1 with polarized glass, like the rotating polarizer used in terrestrial photography?

The Crab's eastern edge contains a prominent notch, or bay, accentuated by a long filament flowing to the southeast. This filament meanders through the nebula's midsection to the western side, where it extends beyond the main body. Look carefully and see if you can detect a gray river adjoining the filament, which visually separates the southern and middle portions. An enhancement at the northern boundary of the southern section abuts the gray river. It looks like an elongated patch of nebulosity (running east to west) with a possible dual nature.

When I first saw this patch, I wondered if this might be the blended image of the Crab's rapidly spinning neutron star and its equally bright neighbor star. Alas, at magnitude 16, the neutron star is too faint for such a small instrument. The enhancement, however, is just south of the Crab's true double heart. Interestingly, this patchy feature also shows well in polarized-light images.

I did view the neutron star through a 20-inch Dobsonian at the 1990 Winter Star Party in the Florida Keys. From the best sites, the pulsar can be seen in a 10-inch telescope with high-quality, unobstructed glass. That night in Florida, I also noticed that the main nebulous body is surrounded by a fainter glow composed of a network of fine filaments. William Parsons, the 3rd Earl of Rosse (1800–1867), first noted the Crab's filamentary structure in 1844. And though it is commonly stated that large telescopes are needed to bring out these delicate features, they can be glimpsed in a 4-inch glass. The problem is not one of aperture but of sky background and contrast.

Data from Hubble Space Telescope observations of the Crab Nebula reveal that these filaments are cloaked in a glowing plasma, which

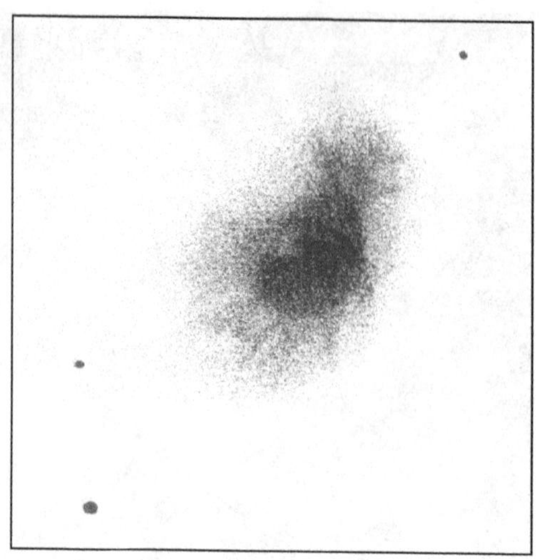

may develop where fast-moving plasma from the Crab's inner bubble of material pushes on the dense outer shell. HST data also show regions of magnetic instability, where fingers of plasma pour back into the inner bubble as it pushes outward. Some astronomers have suggested that there is also an invisible element to M1 – more material beyond the Crab's visible extent – through which this inner bubble is expanding. The bubble, then, might be sweeping up this unseen material and channeling it into the visible, finger-like structures resolved by the HST. If so, this finger formation in the filaments is an ongoing process.

Finally, have you ever wondered how M1 derived its nickname, "the Crab"? The name stems from a drawing made in 1844 based on observations by Lord Rosse with his 36-inch reflector at Birr Castle in Ireland. The drawing bears some resemblance to a horseshoe crab, though, as the late Kenneth Glyn Jones pointed out, it really looks more like a pineapple. Lord Rosse virtually repudiated that sketch, yet the moniker remains. Interestingly, in the 4-inch, the nebula's main body looks very much like a crab or lobster claw, so I derive some pleasure in having found a way to preserve this intriguing historical interpretation.

M2

NGC 7089
Type: Globular Cluster
Con: Aquarius

RA: $21^{h}33.5^{m}$
Dec: -00°49′
Mag: 6.5; 6.3 (O'Meara)
Diam: 16′
Dist: 37,500 light-years
Disc: Jean-Dominique Maraldi II, 1746

MESSIER: [Observed September 11, 1760] Nebula without a star in the head of Aquarius. The center is bright, surrounded by circular luminosity; it resembles the beautiful nebula that lies between the bow and the head of Sagittarius. It is clearly visible with a two-foot telescope, set on the same parallel as α Aquarii. M. Messier plotted it on the chart showing the path of the comet observed in 1759, *Mémoires de l'Académie* 1760, page 464. M. Maraldi saw this nebula in 1746, while observing the comet that appeared in that year.

NGC: Very remarkable globular cluster, bright, very large, gradually pretty much brighter toward the middle, well resolved into extremely faint stars.

IN FALL, THE VAST SUMMER MILKY Way slips drowsily into the western horizon after sundown. Hours will pass before mighty Orion and other bright winter constellations rise in the east. Looking overhead, we now peer straight out of the Galaxy - away from the crowded stellar cities of the Milky Way's arms - into a suburb of stars whose residents include some of the most inconspicuous constellations in the night sky; chief among them Aquarius, the Water Bearer. It is largely indefinable, and its faint stars must compete with light pollution. Nevertheless, Aquarius contains a secret treasure well worth hunting for - the spectacular globular cluster M2.

Jean-Dominique Maraldi (Maraldi II; 1709–1788) first spied this "nebulous star" from Paris on September 11, 1746, while searching for Chéseaux's comet. Maraldi called it "very singular" in that he could not resolve any star within it, nor within the entire telescopic field. He thought this peculiar given that "most of the stars which are called 'nebulous' are surrounded by a large number of stars ... if my judgment is correct." His reasoning was clearly hypothetical, and seems to reflect at least one common belief of the time: that all nebulae were comprised of stars, the resolution of which depended on the aperture of the telescope.

Messier independently chanced upon this object on September 11, 1760, and plotted it on a chart he made for the Comet of 1759. The popular nineteenth-century British observer Admiral William Henry Smyth (1788–1865) seemed especially fond of Messier's second catalogue entry: "This magnificent ball of stars condenses to the centre and presents so fine a spherical form that imagination cannot but picture the inconceivable brilliance of their visible heavens to its animated myriads."

M2 is certainly a marvel. This 175-light-year-wide swarm of 150,000 stars is replete with yellow and red giant stars about 13 billion years old. Photometry has also shown that M2's inner region becomes bluer toward the center, while the reverse holds true for its outer extents. The inner color gradient appears to be caused by color mixing between the cluster's red giants and its blue horizontal-branch stars, while the outer region's gradient may result from the cluster's unresolved background (Young-Jon Sohn et al., *Journal of Astronomy and Space Sciences,* vol. 16, p. 91, 1999).

The cluster lies about 37,000 light-years from the Sun and 34,000 light-years from the Galactic center. Its members each contain, on average, about 1/45 as much metal (i.e., elements heavier than helium) per unit hydrogen as the Sun. The most remarkable feature of M2, however, is its clearly elliptical form, elongated northwest to southeast. When C. J. Grillmair (SIRTF Science Center) and M. Irwin (Cambridge University) examined the large-scale distribution of stars in an area over 50 square degrees around the cluster, they found evidence for tidal tails (*The Astrophysical Journal*, 641:L37–L39, 2006). These features can be caused by strong gravitational interactions between the Galaxy and a globular cluster that nears the Galactic center in its elliptical orbit. Such interplay can cause stars to escape the cluster's gravitational field and become tidal debris forming tails in front of and behind the cluster's orbit. In 2006, Adam Laucher and his colleagues (*Astrophysical Journal*, vol. 651, p. L33) found that "clusters with the lowest concentration factors are more inclined to have significant extratidal debris, and that the cluster concentration is a good predictor of the presence of significant tidal tails. In their study, they found, however, that M2, a Class II globular (meaning that it's a halo object in an elongated orbit around the Galactic center, with a high stellar density at the core), shows only some evidence of weak tidal extensions.

To find M2 without using setting circles or automatic devices, you need a proper knowledge of the constellations because the cluster lies in a region relatively barren of stars. First use the chart at the back of this book to locate 2nd-magnitude Epsilon (ε) Pegasi (the nose of the mythical winged horse, Pegasus) and 3rd-magnitude Beta (β) Aquarii. Now use the accompanying chart to identify magnitude 5.5 star 25 Aquarii, which lies roughly midway along a line between those two stars. Note also that 25 Aquarii marks the southeastern vertex of a nearly equilateral triangle with Enif and 5th-magnitude Beta (β) Equulei.

The chart also shows that 25 Aquarii marks the northern tip of a kite-like formation of stars comprised of three roughly 6th-magnitude stars – including 26 Aquarii and Star *a*. M2 lies a little more than 1 1/2° southwest of Star *a*. Under dark skies, several experienced observers have spied M2 with their unaided

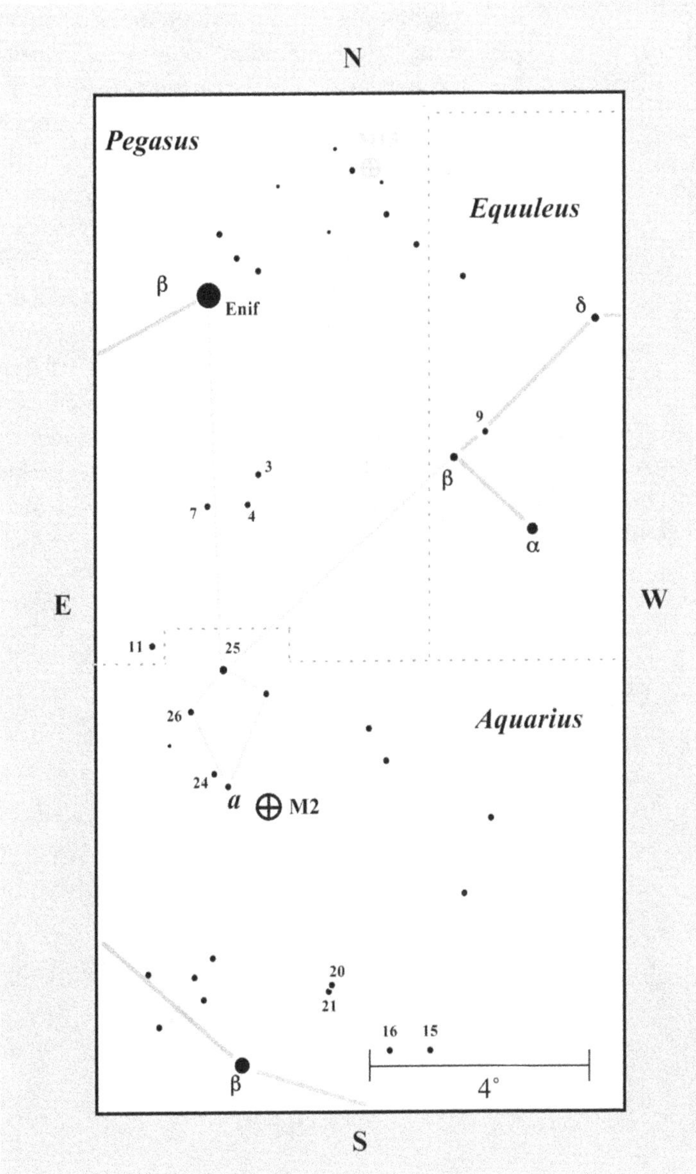

eyes using averted vision. This requires nearly pristine sky conditions, a singular dedicated pursuit, and much patience.

Except for a 10th-magnitude star about 5′ to its northeast, M2 appears rather lonely at low power. The globular lies far from the plane of the Milky Way, so its telescopic surroundings appear relatively devoid of bright background stars. But the visual impact of the cluster itself makes up for that deficiency.

At 23× through the 4-inch, it displays a very tight, starlike center surrounded by a yellow outer core that has a diffuse, pale-blue halo. Increasing the magnification to 72×, resolves the cluster into a sprinkling of stars, the brightest of which shine around 13th magnitude. But double the power and, with averted vision, the dotted haze becomes a multitude of faintly sparkling gems. The strong illusion of a central star at low power vanishes with moderate to high magnifications, when the globular is transformed into a mysterious ball of starlight and shadows.

If you can let your gaze alight a moment on the individual shadows, the illusion of M2's spherical nature will be shattered. The outer envelope takes on a peculiar north–south asymmetry with an explosion of starlight at its fringes. Now the mystery shadows seem to blow out from the center with the star streams, forming spidery

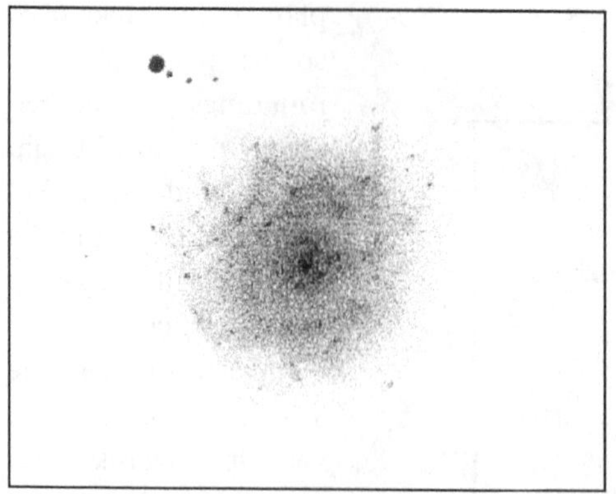

arms. One particularly obvious rogue shadow has been noted by several observers. It slices through the northeast section of the outer halo and runs northwest to southeast. The neighboring 10th-magnitude star is a perfect guide because the shadow lies roughly halfway between it and the cluster's center.

Look at M2 several times over the course of two weeks because it contains a prominent variable star that can greatly impact the cluster's appearance. A French amateur astronomer named A. Chèvremont discovered this pulsating star in 1897. Located on the eastern edge of the cluster, just slightly north of center, Chèvremont's variable ranges in brightness from magnitude 12.5 to 14.0 over about 11 days (that period may fluctuate). At maximum, Chèvremont's variable should be within range of a good 2-inch telescope.

M3

Jellyfish Cluster
NGC 5272
Type: Globular Cluster
Con: Canes Venatici

RA: $13^h42.2^m$
Dec: +28°23′
Mag: 6.2; 5.9 (O'Meara)
Diam: 19′
Dist: ~ 33,300 light-years
Disc: Charles Messier, 1764

MESSIER: [Observed May 3, 1764] Nebula discovered between Boötes and one of Hevelius's Hunting Dogs [Canes Venatici]. It does not contain any stars, the center is bright, and its light decreases imperceptibly [away from the center]; it is circular. Under a good sky it can be seen with a one-foot telescope. It is plotted on the chart of the comet observed in 1779, *Mémoires de l'Académie* for that year. Observed again 29 March 1781, still as fine.

NGC: Very remarkable, globular cluster, extremely bright, very large, very suddenly much brighter in the middle, stars 11th magnitude and fainter.

M3 IS ANOTHER BRIGHT AND STUNNING globular star cluster that presents some visual challenges, including seeing it with the unaided eye, and, as discussed in Chapter 3, as a test of your telescopic limiting magnitude (see the chart). Charles Messier discovered it in 1764; it was his first official discovery (others had spotted M1 and M2 before him). Although uncertain, the discovery of M3 might have spurred Messier to not just catalogue chance observations like M1 and M2 but to start a dedicated and *systematic* search for other uncatalogued nebulous objects.

The new "nebula" remained a "mottled" glow until 1784, when German-born English astronomer William Herschel (1738–1822) resolved M3 into a "beautiful cluster of stars." In 1844, British observer Admiral William Henry Smyth (1788–1865) referred to this "brilliant and beautiful congregation of not less than 1000 small stars" as "one of those balls of compact and wedged stars, whose laws of aggregation it is so impossible to assign; but the rotundity of figure gives full indication of some general attractive bond of union." The late Kenneth Glyn Jones noted in his book *Messier's Nebulae and Star Clusters* that Admiral Smyth's drawing of M3 "does resemble a jellyfish," thus the suggested moniker in the opening data table.

M3

Surpassed only by M13 in the northern skies, M3 certainly is a visual dynamo, with a half-million stars shining in this venerable cluster. In a 1998 paper in the *Monthly Notices of the Royal Astronomical Society* E. Carretta and colleagues referred to it as among the "most important and prominent clusters in the northern hemisphere," noting particularly that it contains the largest number of variable stars known within a single cluster; astronomers have identified at least 260, many of which are RR Lyrae stars. (For comparison, M2 has about 20 known variable stars.)

M3 lies in the halo of our Galaxy, about 33,300 light-years from the Sun and nearly 40,000 light-years from the Galactic center – near the position of the North Galactic Pole as seen projected on the celestial sphere. We see it moving toward us in its highly elliptical orbit at about 90 miles per second. In true physical extent, the cluster is quite large, spanning about 185 light-years, though its gravitational potential reaches out about four times as far. M3 is about 10 billion years old and metal-poor. Its individual members contain, on average, about 1/32 as much metal as the Sun, and the spectral type of the cluster's integrated light is F6.

In the early 1950s, Allan Sandage studied the relationship of the brightness (luminosity) and temperature of individual stars in M3 and discovered the enigmatic stars now known as "blue stragglers" – old stars that burn hot and blue. Since then, three main theories have been proposed to explain their origins: collisions between stars, mergers of stars, and mass transfer between two stars. In 2011, however, Aaron Geller (Northwestern University) and Robert Mathieu (University of Wisconsin, Madison) may have solved the mystery (*Nature*, Vol. 478, p. 356). Using the WIYN Observatory in Tucson, Arizona, to study blue stragglers in the very old globular star cluster NGC 188 in Cepheus, they found that the companion stars involved in blue stragglers are white dwarfs with about half the mass of the Sun. Since collisions and mergers require more massive companion stars, the only viable solution, they say, is that of mass transfer. A blue straggler stays "young," then, by siphoning off material from a close stellar companion.

To find this stunning cluster, use the chart at the back of this book in conjunction with the accompanying one to review the stars of Boötes and Coma Berenices. M3 lies just northeast of an orange 6th-magnitude star about 6° east of Beta (β) Comae Berenices and about 10° (one fist) north and slightly west of Eta (η) Boötis; more appropriately, it marks the northeastern corner of a right triangle with Beta Comae Berenices and magnitude 4.5 Tau (τ) Boötis. Try locating it first in binoculars; look for a little glowing orb of light just 30′ northeast of magnitude 6.5 Star *a*.

Once seen, this cluster is easier to spy with the unaided eye than M2 for several reasons: M3 is slightly brighter, a 6.5-magnitude guide star lies next to it, and the cluster appears much higher in the Northern Hemisphere sky. Here's the catch: M3 lies so close to the guide star that it may be hard to resolve the two! Both Brian Skiff (Lowell Observatory) and I achieved this from the dark skies of southwest Texas.

Although M3 lies in a relatively star-poor field, it is contained in an acute triangle of 9th-magnitude stars, and an obvious topaz star borders the cluster to its southeast. With a glance using 23×, the 15′-wide glow displays a starlike core surrounded by a gradually fading halo. But look closely at the bright central point because it appears bounded by a very tight inner shell with a distinct peach color. (I have noticed that when the cluster is low in the sky, this inner region looks yellow.)

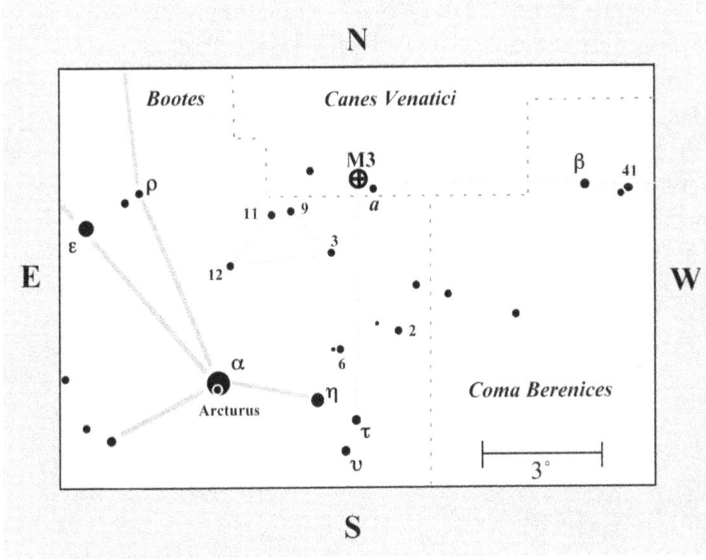

At 72×, the cluster metamorphoses into a finely resolved sphere with an uncanny three-dimensional quality; I feel like I'm looking down on a snowball melting on black ice. The cluster's core appears bulbous and its edges flat and sprawling. Moderate power also reveals a tantalizing string of stars that connects the cluster to the topaz star about 10′ to the southeast. When seen in a simple inverted telescope, M3 seems to dangle from the star like an earring. The brightest stars should be easily resolved at magnitude 12.7, while the cluster's full resolution requires you to be able to see to a limit of magnitude 15.6.

Some observers, such as the late Walter Scott Houston – whose Deep Sky Wonders column in *Sky & Telescope* magazine gave my generation of observers constant food for thought – have noticed that the central region is skewed to the west. I, too, see this, but believe it to be an illusion caused by a curious

The observed color of globulars is open for debate. Some astronomers argue that the colors are merely optical illusions. But I find that at very low magnifications most bright globulars show distinct, albeit faint, colors – mostly shades of yellow and blue. Other observers have independently seen these tints, and true-color images of globulars reveal them as well. Admittedly, green tints, which some globulars appear to have, may be illusory, being a combination of the yellow and blue casts. Not surprisingly, I have also found that the blue and green colors of globulars seem to vanish with notable increases in magnification.

Continuing at low power, M3's peach-tinted inner shell is framed by a larger and fainter mantle, which shines weakly with an aqua luster (more greenish with some blue). Beyond this is an even larger envelope, whose overall color is blue-green. M3 should certainly be rated as one of the most colorful globulars in the Northern Hemisphere for small telescopes. At the least, its color scheme is intriguing and apparent.

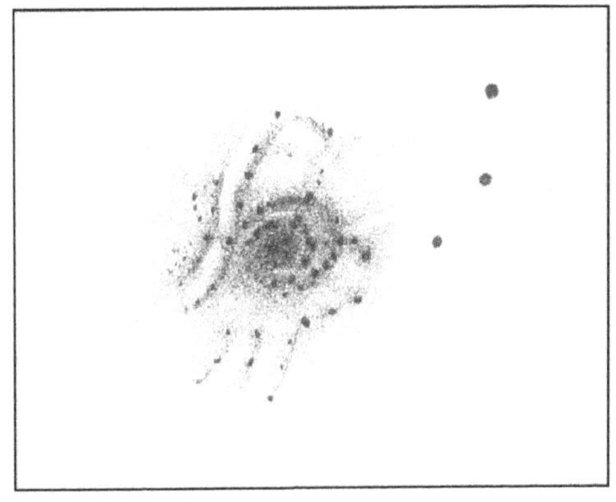

gathering of stellar clumps west of the cluster's true center. A series of westward-extending arms adds to this illusion. Otherwise, medium and high powers resolve the cluster into a myriad of ultrafine stars, like tiny emerald chips scattered on a golden carpet. So uniform and gauze-like is the light from each individual star enveloping the bright core that I imagine it as a candle burning behind a curtain of green lace.

Visual observers should look for the mysterious dark spots that inhabit M3's nuclear region. William Parsons first noted them from his observatory at Birr Castle in Ireland as "small, dark holes"; they show up well on some high-resolution images today.

M4

Cat's Eye
NGC 6121
Type: Globular Cluster
Con: Scorpius

RA: $16^h23.6^m$
Dec: -26°31.5′
Mag: 5.4
Diam: 36′
Dist: ~ 7,200 light-years
Disc: Philippe Loys de Chéseaux, 1746

MESSIER: [Observed May 8, 1764] Cluster of very faint stars. With a small telescope it looks like a nebula. This star cluster is close to Antares and on the same parallel. Observed by M. de la Caille, and included in his catalogue. Observed again 30 January and 22 March 1781.

NGC: Cluster, with 8 or 10 bright stars in a line with 5 stars, well resolved.

RIDING HIGH IN THE SUMMER SKY, globular cluster M4 awaits the inquisitive gaze of amateur astronomers using the smallest of apertures - including the unaided eye. Swiss astronomer Philippe Loys de Chéseaux (1718–1751) discovered the glow "Close to Antares" in 1746, describing it as "white and round." Some five years later, Abbé Nicolas-Louis de Lacaille (1713–1762) swept it up with his 1/2-inch 8× telescope during an expedition to the Cape of Good Hope in South Africa, noting that its form "resembles the small nucleus of a faint comet." Messier first noticed that it was a cluster of very faint stars when he observed it in 1764. William Herschel (1738–1822), the discoverer of Uranus, noticed in 1783 that M4 has a "ridge of 8 or 10 pretty bright stars," which Admiral Smyth saw as "running up to a blaze in the centre."

When you probe the depths of this cluster, keep in mind that you are looking at an object about 75 light-years in diameter and roughly 13 billion years old, a senior member of our Milky Way Galaxy, which formed at an early stage of the Galaxy's evolution, during the collapse of the protogalactic cloud. Its members contain, on average, about 1/14 as much metal per unit hydrogen as our Sun; the spectral type of its integrated light is F6/7. We see it receding from us at 44 miles per second.

M4 once held the distinction of being the globular star cluster closest to the Sun, beating out NGC 6397 in Ara by a mere 300 light-years. But, in 2007, Charles Bonatto of Universidade Federal do Rio Grande do Sul and his colleagues, writing in the *Monthly Notices of the Royal Astronomical Society* (vol. 381, p. L45), announced the discovery of a

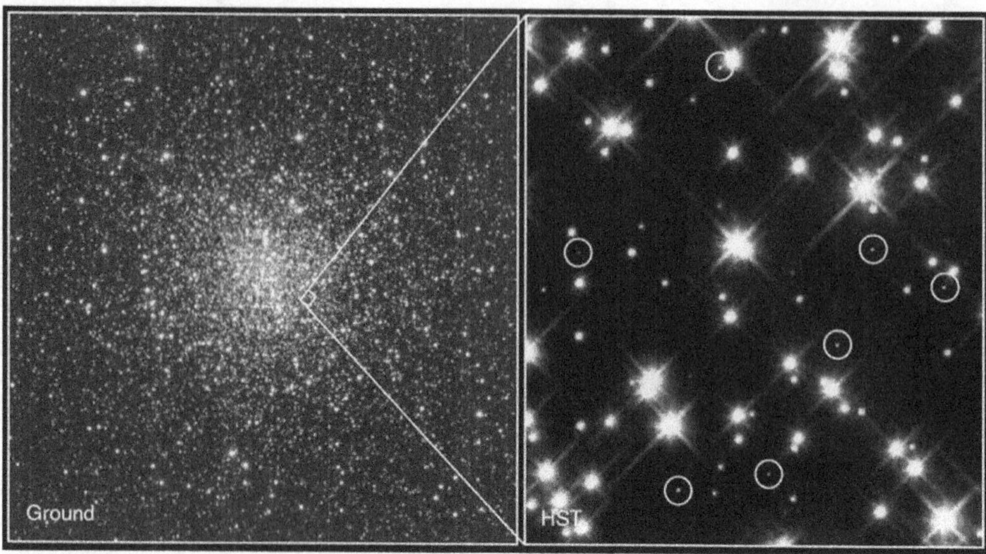

new, low-mass globular cluster, called FSR 1767, only 4,900 light-years distant. FSR 1767 has only about one-tenth the number of stars as M4 and a puny diameter of only 6.5 light-years. Nevertheless, M4 is clearly the closest globular star cluster accessible to amateur astronomers in both hemispheres.

Like M1, the Crab Nebula, M4 harbors an 11-millisecond pulsar (PSR B1620–26), in which an old neutron star is spun by accretion during an x-ray binary phase. In 1993, six years after the pulsar's discovery, it was suggested that the pulsar is a member of a triple system. In 2003, astronomers used the Hubble Space Telescope to confirm that the third object in the system was indeed a planet, with a mass 2.5 times that of Jupiter. It is the oldest and farthest known planet. As the release stated, "Its very existence provides tantalizing evidence the first planets were formed rapidly, within a billion years of the Big Bang, leading astronomers to conclude planets may be very abundant in the universe."

The Hubble Space Telescope (HST) also revealed 75 white dwarf stars in a tiny area within M4 (see the accompanying image); astronomers predict the cluster to contain about 40,000 of these "stellar corpses." "The observation was so sensitive," the release stated, "that even the brightest of the detected white dwarfs was no more luminous than a 100-watt light bulb seen at the moon's distance" (239,000 miles).

To find M4, use the chart at the back of this book and the one here to find 1st-magnitude Alpha (α) Scorpii (Antares). The cluster lies a little more than 1° to the west. Although brighter than either globular star cluster M2 or M3, M4 lies so close to that brilliant red supergiant Antares that glare all but ruins a clear and consistent naked-eye view of it (try blocking the light of Antares with the edge of a roof or some other terrestrial obstruction). With binoculars, the cluster appears as a very bright, round, and diffuse glow.

Telescopically, M4 is dazzling even in the smallest of instruments. It is a Class IX globular, meaning it is a loose system with a more homogeneous concentration of starlight toward the center. At 23× through the 4-inch,

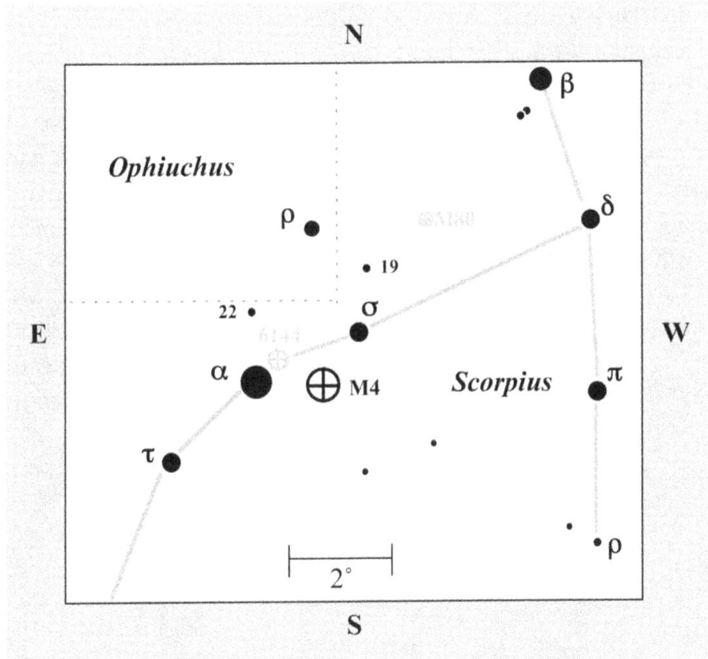

the 35′-wide disk of M4 immediately resolves into a loose throng of 11th-magnitude stars, which betrays the cluster's overall roundness and symmetry; bisecting this sphere is a spindle of about a dozen 10th- to 12th-magnitude stars running north to south. With medium power, this ridge appears more obvious and elongated. Here is a cat's keen eye, with its slit of a pupil, peering at you through the dark vapors of night. Although the Cat's Eye moniker for M4 is my own, do not confuse it with NGC 6543, the Cat's Eye Nebula, an 8th-magnitude planetary in Draco.

Behind that stunning slit of starlight is a swollen haze of unresolved cluster members. Just when I feel I can resolve a faint member, that moment passes and I catch sight of another "spark," until my eye spies another, and so on. One can spend hours being hypnotized by the comings and goings of these fleeting flashes. Burnham had a similar impression of M4 after he observed the cluster through the Yerkes 40-inch refractor. He writes:

> It happened that a few days before, I had obtained a small but fascinating device called a spinthariscope in which the effect of radioactivity is made visible to the eye; ever since I have mentally associated M4 with alpha particles. ... The observer, after a period of dark adaptation, looks into the lens [of the spinthariscope] to see a view resembling M4 "brought to life" with hundreds of microscopic "stars" blazing up and vanishing every second.

Shock waves of dark lanes radiate from the needle-like hub. Defocus the telescope slightly to follow these dark ripples. So many populate the region that with some imagination you can envision a shattered glass photographic plate of M4, or a cobweb silhouetted against an approaching swarm of lightning bugs, or, better yet, diamonds snatched up in a loose black net, with many leaking through the fibers into limitless space.

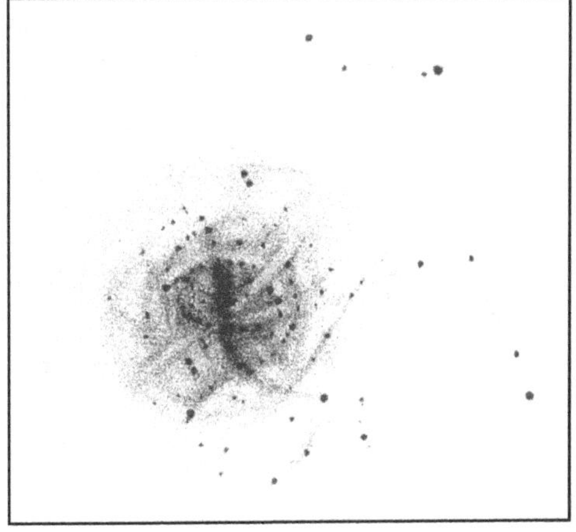

The outlying northern section appears to sink gradually into a black ocean, and if you look hard enough, you can perceive a submerged shoal of faint stars, a celestial reef. Many of the stars outside the core form concentric horseshoe-shaped patterns that loop northward from the south. Stellar arms fly off in various directions. In this way, the cluster looks more like a collision of two pinwheel galaxies such as M33 or M100, with flat, broad, and loose spiral arms.

While in the area, check out the tiny 9th-magnitude globular cluster NGC 6144 only about 40′ northwest of Antares. Its feeble glow reminds me of a ghost image of Sigma Scorpii, a hot, blue giant (B1) star that shines at 3rd magnitude. NGC 6144 is a highly neglected object, hidden in the glare of Antares and overshadowed by M4. It is equal in brightness to some of the fainter Messier globulars in Sagittarius. At 23× through the 4-inch, the cluster is a beautiful, 5′-wide round glow, with a slight and gradual brightening toward the center but no real obvious central concentration. It starts to resolve at 101×, especially around the edges.

M5

NGC 5904
Type: Globular Cluster
Con: Serpens (Caput)

RA: $15^h18.5^m$
Dec: -2°04′
Mag: 5.7
Diam: 23′
Dist: 24,500 light-years
Disc: Gottfried Kirch, 1702

MESSIER: [Observed May 23, 1764] Beautiful nebula discovered between Libra and Serpens, close to the sixth-magnitude star Flamsteed 5 Serpentis. It does not contain any stars; it is round, and it may be seen very well under a good sky with a simple one-foot refractor. M. Messier plotted it on the chart for the comet of 1753, *Mémoires de l'Académie 1774*, page 40. Observed again 5 September 1780, and 10 January and 22 March 1781.

NGC: Very remarkable, globular cluster, very bright, large, extremely compressed in the middle, stars from 11th to 15th magnitude.

THE FIFTH OBJECT IN MESSIER'S catalogue is a powerful and dynamic sight in telescopes of all sizes. Even at low power, M5 is a slightly stellar conflagration with a blazing heart. A wide and loose, slightly elliptical exterior becomes increasingly tight toward a starlike center. The cluster looks as if it is collapsing under the force of gravity, triggering atomic reactions in its core. And with a 7-millimeter eyepiece, the entire cluster seems electric, bursting with fiery sparks.

Now, contrast that with Mary Proctor's musings on the same globular, which she viewed through the world's largest refractor, the 40-inch Clark telescope at Yerkes Observatory. The description is from her 1924 book *Evenings with the Stars*: "Myriads of glistening points shimmering over a soft background of starry mist, illumined as though by moonlight. ... For a few blissful moments, during which the watcher gazed on this scene, it suggested a veritable glimpse of the heavens beyond." Observing is indeed a highly personal experience; sometimes we see what we feel.

The discovery of M5 has an interesting history. Usually credit goes fully to German astronomer Gottfried Kirch (1639–1710). In a delightful 1949 article titled "Out of Old Books," which appeared in the *Journal of the*

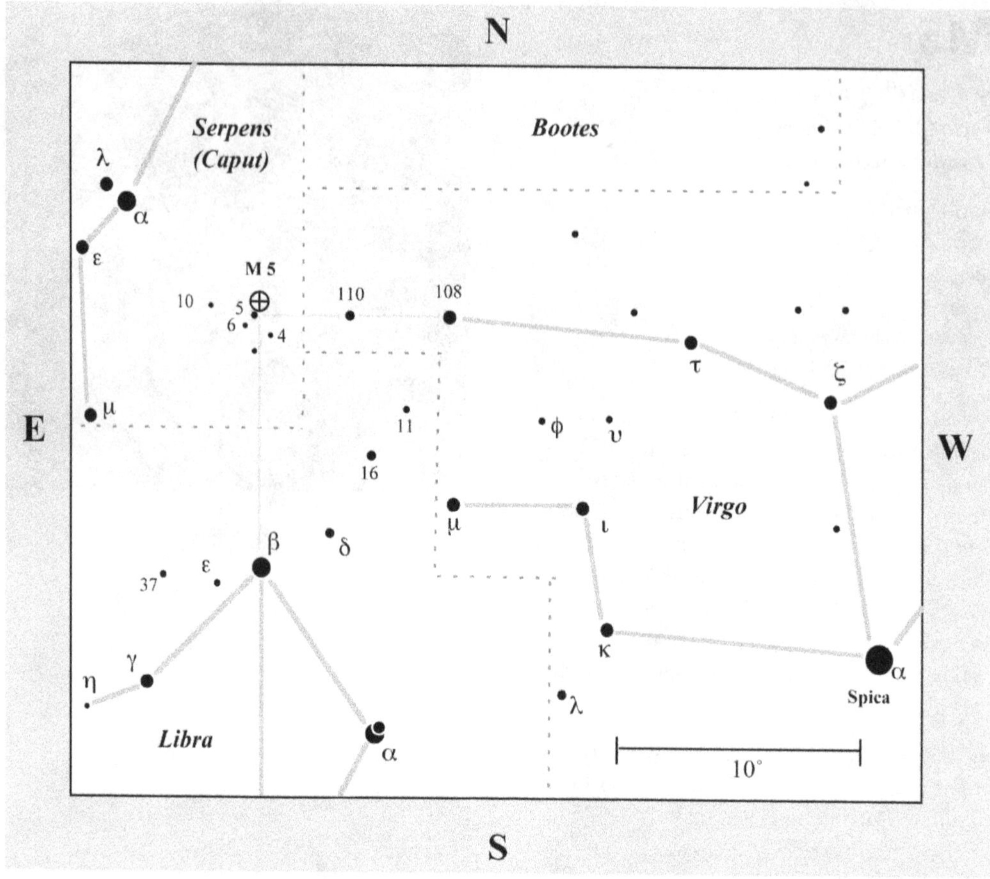

Royal Astronomical Society of Canada (vol. 43, p. 45), Helen Sawyer Hogg notes that, according to the diary of Gottfried's wife, Margarethe (who assisted him in observing, recording, and computing),[1] she had just discovered the Comet of 1702, which Gottfried pursued diligently afterward. On May 5, however, while searching for a new comet, Gottfried encountered a "nebulous star" near 5 Serpentis. Sawyer Hogg continues, "The next night, May 6, he and his wife together confirmed the discovery of the object. It was Margarethe Kirche who drew the chart of its position and made up the record. Almost fifty years later the object became known as Messier 5 when Messier published his first list of nebulae and star clusters in 1771."

M5 lies about 24,500 light-years from the Sun and 21,200 light-years from the Galactic center. In true physical extent it is quite large, spanning about 160 light-years. Next to globular cluster M3 in Canes Venatici, M5 is the richest globular cluster in variable stars in the northern sky, with at least 143 variables and suspected variables. Its integrated spectral type is F7, and each of its members contains, on average, about 1/20 as much metal per unit hydrogen as the Sun. We see it approaching us at 33 miles per second.

[1] J. L. E. Dreyer first noted this in his supplement to John Herschel's *General Catalogue.*

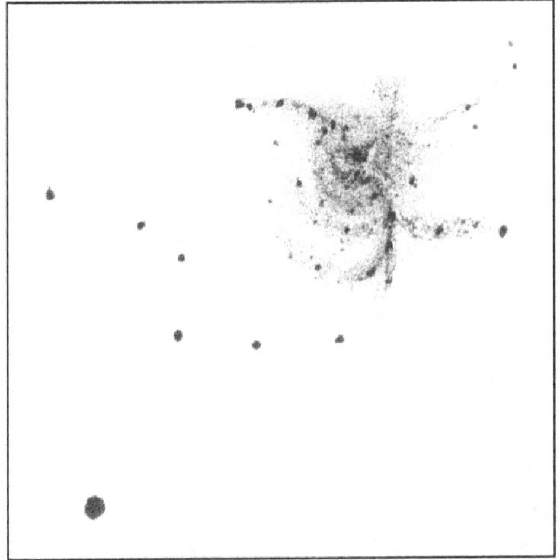

To find M5, use the chart in the back of the book to locate 5 Serpentis (a golden 5th-magnitude star with a 10th-magnitude ashen companion a little more than 2′ away). You can also use the more detailed chart on the previous page to locate it. Note that it is a little more than 10° (about a fist held at arm's length) north of magnitude 2.5 Beta (β) Librae and less than 4° east of the magnitude 4.4 star 110 Virginis. Once you identify 5 Serpentis, you'll find the cluster only 20′ northwest of that fine double star.

Telescopically, especially at low power, color – namely, a straw interior with a powder blue exterior – is immediately obvious in M5. Few accounts I've seen mention the obvious curved wings of 12th-magnitude stars stretching northeast to south from the core. This region reminds me of an airborne owl, whose feathered wings shimmer with reflected light. At medium power, the wings are amazingly well defined. Christian Luginbuhl and Brian Skiff share a different impression in their book *Observing Handbook and Catalogue of Deep-Sky Objects*, one of a "rich open cluster superposed on a bright galaxy."

At 130×, the nuclear region appears brightest to the north, with a faint stubby wing just to its west. Toward the southeast, stars spiral or fan out in long arms – a pattern William Parsons noticed through his 36-inch reflector at Birr Castle in Ireland in 1875. These features seem to originate from stellar kinks along a central bar, the kinks themselves being chance associations of stars. The cluster's brightest stars shine around 12th magnitude, and full resolution can be achieved if your telescope can reach 15th magnitude. Admiral Smyth called it "a superb object ... a noble mass, refreshing to the senses after searching for faint objects; with outliers in all directions, and a bright central blaze, which even exceeds [M3]."

By the way, M5 contains two millisecond pulsars: PSR B1516+02A (M5A) and PSR B1516+02B (M5B). In a 2008 paper in *The Astrophysical Journal* (vol. 679, p. 1433), Paulo C. C. Freire and his colleagues report results of 19 years of Arecibo radio observations of these objects. They found that M5A is an isolated millisecond pulsar (MSP) with a spin period of 5.55 milliseconds and M5B is a 7.95-millisecond pulsar in a binary system with a low-mass companion. Thus, three of the first five objects in Messier's catalogue contain millisecond pulsars: M1, M4, and M5.

As you gaze at this incredible, 13-billion-year-old object – one of the finest globular clusters in the northern sky for small telescopes – try contemplating the following: In the two-dimensional sky, M5 is superposed on the edge of a faint, distant cloud of galaxies, which is composed of several groups of galaxies – a mind-boggling aggregate with some 200 members per square degree. Four degrees (about two finger-widths) due west of M5, at least eight galaxies surround 110 Virginis (not shown on the finder chart). All are visible at low power, but they are best seen at moderate magnifications. Take your time studying this region and confirm the galaxies on your star atlas.

M6

Butterfly Cluster
NGC 6405
Type: Open Cluster
Con: Scorpius

RA: $17^h40.3^m$
Dec: –32°15.5′
Mag: 4.2
Diam: 33′
SB: 11.8
Dist: ~1,600 light-years
Disc: (?); arguably observed throughout history as a naked-eye nebulous star

MESSIER: [May 23, 1764] Cluster of faint stars between the bow of Sagittarius and the tail of Scorpius. To the naked eye this cluster appears to form a starless nebula, but even the smallest instrument shows it to be a cluster of faint stars. (M6)

NGC: Cluster, large, irregularly round, little compressed, one star of magnitude 7, then stars 10th-magnitude and fainter.

M7

NGC 6475
Type: Open Cluster
Con: Scorpius

RA: $17^h53.8^m$
Dec: -34°47.1′
Mag: 3.3; 2.8 (O'Meara)
Diam: 75′
Dist: ~780 light-years
Disc: Claudius Ptolemy, second century A.D.

MESSIER: [Observed May 23, 1764] A larger cluster of stars than the previous one [M6]. This cluster appears to be a nebula to the naked eye; it is not far from the previous one, lying between the bow of Sagittarius and the tail of Scorpius.

NGC: Cluster, very bright, pretty rich, little compressed, stars of 7th to 12th magnitude.

LOCATED ABOUT 5° NORTH-NORTHEAST of the blue subgiant star Lambda (λ) Scorpii, the easternmost of the Scorpion's two stinger stars, open star clusters M6 and M7 are two of the most striking details projected against the hub of our Galaxy – a dramatic naked-eye oval expanse of milky starlight and dark dust. These clusters are among the more brilliant Messier objects in the night sky. (M7, the southernmost Messier object, is arguably the brightest spot in the entire visual Milky Way.) M6 and M7 all but dominate the summer Milky Way and erupt like distant fireworks. Both clusters are visible to the naked eye as puffs of "smoke" even under full moonlight!

Personally, I cannot look upon one of these clusters without immediately trying to contrast it with the other. And though they are separated by 3 1/2° of sky, I view them as a double cluster, fraternal twins. Actually, the two clusters are some 820 light-years apart, but aren't we entitled to enjoy them as we please?

M6 is the fainter of the two clusters. Discovery credit is a bit muddled. In his *Celestial Handbook*, the late Robert Burnham Jr. notes that it "appears to be mentioned, along with M7, in the catalogue of Ptolemy. They seem to be the *Girus ille nebulosus* of the 1551 edition of the *Almagest*, and also appear in Ulug Beg's star catalogue as the *Stella nebulosa que sequitaur aculeum Scorpionis*, 'The Cloudy Ones Which Follow the Sting.'"

The Italian astronomer Giovanni Batista Hodierna (1597–1660) logged it as a star cluster in his *De Admirandi Coeli Caracterbus* of 1654 (Ha. II-4). Swiss astronomer Philippe Loys de Chéseaux included this "fine [cluster of stars]" in his *Catalogue of Truly Nebulous Stars*, compiled in 1745 or 1746. And, a decade later, Abbé Nicolas-Louis de Lacaille listed it as the 6th object in his Class III objects (stars accompanied by nebulosity), calling it a "large group of a great number of small stars, a little compressed and occupying the space of a semi-circle of 15′ or 20′ diameter, with a slight nebulosity spreading in that space."

Of course, the nebulosity all these early observers were seeing was just, most probably, unresolved starlight, as Charles Messier noted in 1764, saying that: "To the naked eye this cluster appears to form a starless nebula, but even the smallest instrument shows it to be a cluster of faint stars."

M6 contains 126 stars to magnitude 14, about half of which are magnitude 11 or brighter; the entire cluster contains more than 330 stars. Most of the stars are dazzlingly crisp blue-white gems. The major exception is orange BM Scorpii, a semiregular variable whose light fluctuates from magnitude 5.5 to 7 in about 850 days. M6 is estimated to be 100 million years old; M7 is more than twice that age. At a distance of ~1,600 light-years, M6 spans ~15 light-years of space; M7 is nearly the same size – though it appears twice as large in our skies (owing to the fact that it is half as far from us as M6).

To find this dazzling wonder, use the chart at the back of this book to find magnitude 1.6 Lambda (λ) Scorpii, the easternmost star in the Scorpion's stinger. M6 is about 5° north-northeast of that star. You can see its relative position in the accompanying chart.

Under a dark sky, M6 can be seen with the unaided eye. From suburban locations, I suggest using binoculars, then placing Lambda in one corner of the field. If you have the binoculars oriented correctly, M6 will be on the other side of the field of view, looking like a

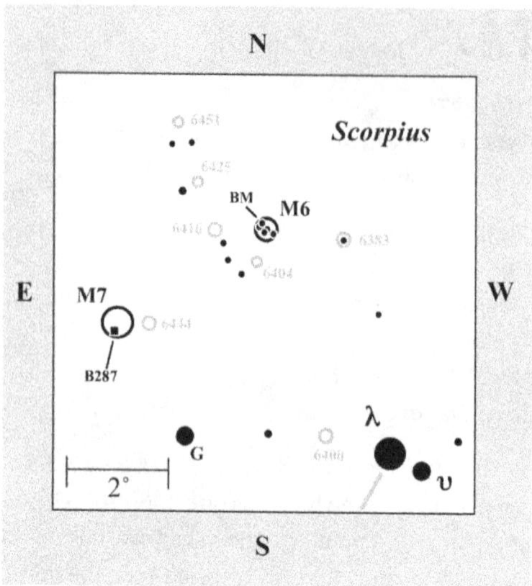

little gathering of sparkling stars in looping patterns.

M6 is famous for its butterfly pattern, which looks more like a 1960s butterfly motif (the kind worn on bell-bottoms) than, say, a swallowtail, but nevertheless, the imagery is quite striking. Through the 4-inch at 23×, the cluster's butterfly pattern is best enjoyed, especially since, as the drawing here shows, the butterfly's antennae add a tad more detail to the design.

Aside from the butterfly, look to the northwest, where a definite V-shaped pattern of ice-blue stars dominates the telescopic view; this grouping shows up especially well with moderate powers. One night in 2011, I had the chance to view M6 through a 12-inch Schmidt-Cassegrain telescope under bright moonlight. I found the butterfly pattern standing out most remarkably. But I could not take my eye away from BM Scorpii, because its stunning orange sheen appeared so captivating when seen against the blue ice-chip stars comprising the rest of the cluster.

See if you can detect BM Scorpii with the unaided eye; it marks the northeast tip of the

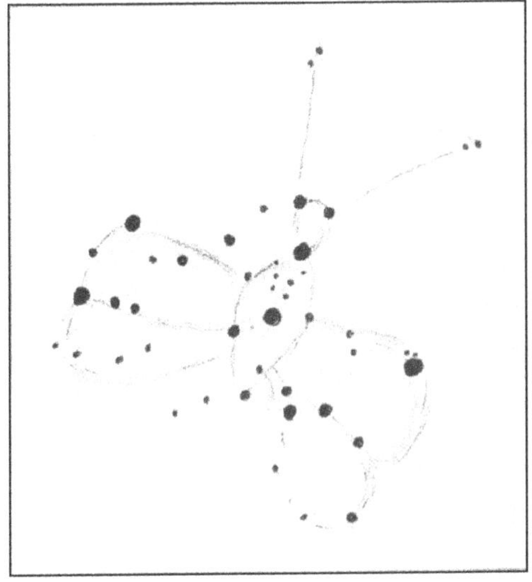

butterfly's wing. There should be no problem doing this when it shines at maximum. Actually, the cluster contains four widely separated 7th-magnitude stars, which under ideal conditions could be detected with the unaided eye.

Now use your binoculars to study the Milky Way region surrounding M6. Then train them on nearby M7 and its surroundings. See anything noticeably different? M6 is surrounded by as much darkness as M7 is by light! This aspect is even more apparent in the 4-inch at 23×. In fact, when sweeping from M7 to M6, I noticed that M7 lies in the fattest part of a bright wedge of Milky Way - a rich star cloud - that points to M6. Tiny M6 sits like an island off the glittering sand of this Milky Way beach.

With your binoculars or your telescope at low power, look at the region immediately surrounding M6; you should see a wall of open clusters running from the northeast to the southwest, including NGC 6451, NGC 6425, NGC 6416, and NGC 6404. A little south of west is another loose sprinkling of stars, NGC 6383.

I am not sure why over the last few centuries more observers didn't sing praises about M7. In the seventeenth century, Giovanni Hodiera wrote merely, "Counted 30 stars." Lacaille described M7 from the Cape of Good Hope as a "group of 15 or 20 stars very close together in a square." And, in his 1864 catalogue, John Herschel characterized it as "a cluster, very bright, pretty rich, little compressed."

I, however, find that from the northern tropics and the Southern Hemisphere, there are few grander naked-eye sights in the deep sky than this magnificent wonder. M7 blazes against the river of the Milky Way like the head of some great lost comet.

Compared with the icy-hued stars of M6, those in M7 appear more golden, like sun-drenched hay. M7 has about 80 stars, all of which shine brighter than 10th magnitude, and a dozen of these are brighter than 7th magnitude. The challenge, then, is to resolve this cluster with the naked eye. Clearly, half of these stars are 6th magnitude and should be discernible from suburban skies. If you live under a very dark sky, try during the quarter moon, which will diminish the cluster's background noise of fainter, unresolved stars. Another trick is to catch M7 in the morning or evening twilight when its light just starts to fade or emerge, respectively.

With 10 × 50 binoculars, M7 looks like a gem-studded cross inside a double halo of similarly bright stars. With imagination, it is possible to perceive the cross and the two halos all appearing on different planes. Telescopically, M7 resembles a cosmic flower opening in the morning mist of the Milky Way, the long axis of the cross being the flower's stamen and the halos its petals. (NGC 6444, a modest open cluster to the west, is a burst of pollen. I made the accompanying drawing at

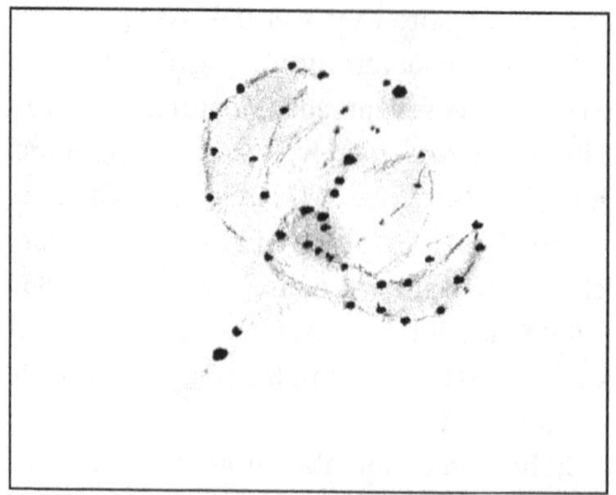

the left using the 4-inch; it shows the flower aspect.

Return to the binocular view. This time, defocus the view slightly and see if you can detect the black slash running along the western side of the cross's long axis. That dark wound in the Milky Way points to Barnard 287, a pond of dark matter south-southeast of the cluster.

Now use your telescope at low power and center M7 in the field of view. Abutting the northern edge of the cluster is a 1°-long and slender snake of dark nebulosity – part of a larger 90′ × 60′ complex of dark nebulosity known as Barnard 283 (B283). James Dunlop (1793–1848) first noticed the region's patchy Milky Way and dark nebulae from Parammatta Observatory in New South Wales, Australia, saying they "very much resemble small cirrocumuli clouds." In the *Barnard Catalogue of Dark Objects* (Part I of his *A Photographic Atlas of Selected Regions of the Milky Way* – a "volume of his pioneer celestial photographs made at the Lick Observatory in the years 1889–1895"), Edward Emerson Barnard gave the photographic appearance of the feature an opacity of 5 (the darkest of the dark).

Through the 5-inch at 33×, B283 is extremely striking. It consists of two parallel streams of dark nebulosity (oriented east to west) connected by a U-shaped "joint" of dark-gray "smoke" centered on a 6th-magnitude star, which I thought had a warm hue. To appreciate the darkness of the parallel streams, I had to relax my gaze while using keen averted vision; in time, the dark bands wafted clearly into view, looking just as Dunlop offered: like the wavy black ribbons of "cirrocumuli" clouds.

I could also see the western edge of the northern black ribbon curling slightly to the north-northwest like an elliptical attachment. All manner of thin wisps extend westward and southward from the southern ribbon of darkness. In fact, the most obvious western wisps form a prominent triangular patch – a sideways D (with the triangle's tip pointing to the east), which could stand for Dunlop; interestingly, the southern wisps branch off to the east in two parallel rays. When seen together with the southern dark stream, they form the letter E – the first and middle initial of Barnard's name.

M8

Lagoon Nebula
NGC 6523 (Nebula) and NGC 6530 (Cluster)
Type: Diffuse Nebula and Cluster
Con: Sagittarius

RA: $18^h03.8^m$
Dec: -24°23′
Mag: 4.6; 3.0 (O'Meara)
Dim: 45′ × 30′ (nebula)
Diam: 14′ (cluster)
Dist: ~5,200 light-years
Disc: Giovanni Batista Hodiera, before 1654; John Flamsteed independently discovered it around 1680

MESSIER [Observed May 23, 1764]: A cluster of stars that appears to be a nebula when observed with a simple three-foot refractor; with an excellent instrument, however, one sees only a large number of faint stars. Near this cluster there is a fairly bright star, which is surrounded by a very faint glow; this is the seventh-magnitude star Flamsteed 9 Sagittarii. The cluster appears to be elongated in shape, extending from northeast to southwest, between the bow of Sagittarius and the right foot of Ophiuchus.

NGC: A magnificent object, very bright, extremely large, extremely irregular in shape, with a large cluster.

USE THE CHART AT THE BACK OF THIS book and the one here to find 3rd-magnitude Lambda (λ) Sagittarii, the orange K-type star that marks the top of the famous teapot asterism in Sagittarius. The emission nebula M8 lies only about 5° west and slightly north of that star. It's very conspicuous to the naked eye under a dark sky, appearing as a large curl of galactic vapor, just beyond the western edge of the Milky Way, that rises from the teapot's spout. The nebula is complemented on its eastern side by NGC 6530 – a 14′-wide loose sprinkling of 113 very young stars, all of which are probably intimately associated with the M8 nebulosity that enshrouds them in loops and swirls. The nebula itself is powered by the radiative energy of the very hot 6th-magnitude star 9 Sagittarii, a 9th-magnitude star (known as Herschel 36), and possibly some obscured stars.

I was surprised at my magnitude estimate of 3.0 for M8, which was derived by defocusing

M8

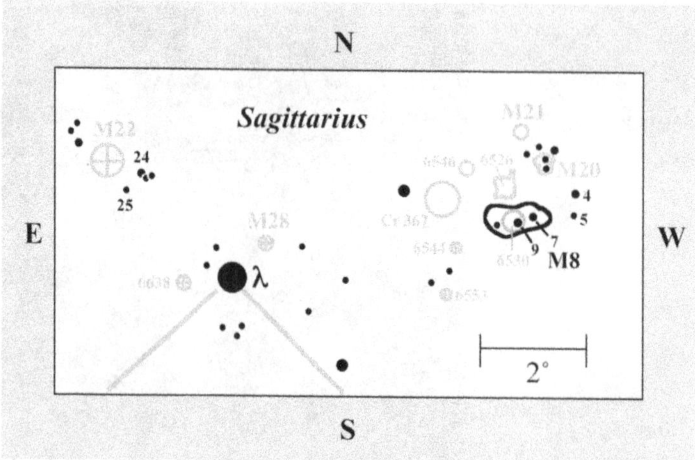

my eyes and comparing the nebula's light to that of Lambda and Gamma (γ) Sagittarii. Other sources rate M8's magnitude anywhere from 4.6 to 6.0. The discrepancy could arise from the object's low altitude and from the dimming effects of light pollution. But Brent Archinal believes these fainter estimates are more accurate for the cluster itself and not the nebulosity.

Here is a simple visual check. Look at M8 and then compare it to M6, which shines at magnitude 4.4. To my eyes, M8 appears brighter. Another check is to see which emerges first from the evening twilight.

The word "Lagoon" was probably first used in association with M8 by Agnes M. Clerke in her 1890 work entitled *The System of the Stars.* The name refers to the striking black furrow running east to west that divides the nebula's brightest regions and looks more like a channel than a lagoon. The channel is but a part of the lagoon, which is really a V-shaped body of darkness that embraces a bright island of nebulosity to the west and a curved "sandbar" of nebulosity to the east. If you look long and hard enough, you should actually see two lagoons: a "deep" and narrow inner one and a wider, "shallower" outer one.

When you look at M8 with your telescope at low power, can you see the skeleton-like fingers of nebulosity to the south of the Lagoon's main dark channel? Anyone familiar with the classic horror film *Creature from the Black Lagoon* might find some similarity between these nebulous fingers and those webbed hands of the Creature.

Through the 4-inch at 23×, I found the dark channel so prominent that it must be one of the most dramatic examples of dark nebulosity in any deep-sky object visible in small telescopes. The entire nebula appears to be caught between worlds of darkness and light. John Herschel described it as a "collection of nebulous folds and matter surrounding and including a number of dark, oval vacancies." Use moderate power and averted vision to see the dark nebulosity that cages the bright nebula.

Particularly delicate at 23×, the nebulosity and its myriad dark lanes look like a frozen flower petal that has fallen to the ground and shattered. If you mentally erase the nebulosity, you should see a crossbow of seven prominent stars – the skeleton of this Messier object.

To the naked eye and in binoculars, the nebula and the staff of the crossbow constitute what I immediately recognize as M8, and this is what Messier saw. The NGC 6530 designation applies to the stars in the eastern part of the bow. Concentrate on the position of NGC 6530 in the nebula and see if it seems to weigh down the southern part of the nebulosity. To me, it is a very

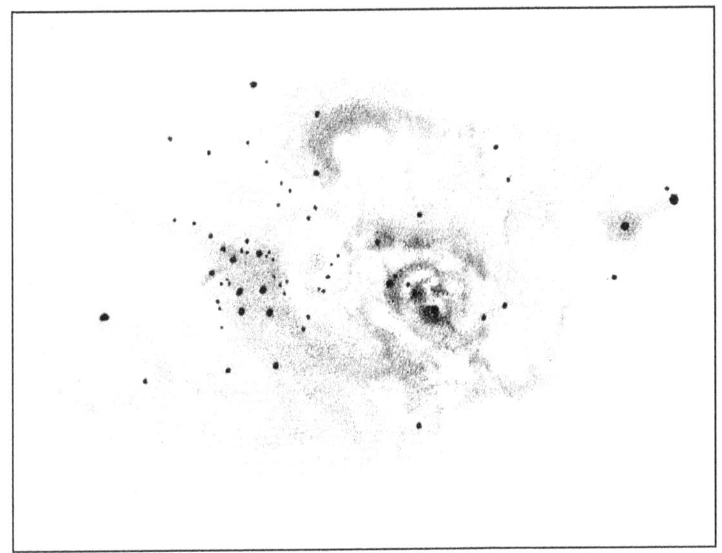

nascent clouds like tornadoes on Earth. These twisters are silhouetted against the brilliant backdrop of ionized gas in the "hourglass." The large difference in temperature between the hot surface and cold interior of the "hourglass" clouds, combined with the pressure of starlight, may produce strong horizontal shear to twist the clouds into their tornado-like appearance. A curious knot of nebulosity appears isolated from the "hourglass" to the north of Herschel 36. Could it be part of the "hourglass" separated by these dark twirling clouds?

incongruous sight. I remedy that illusion by creating another: I like to imagine that the stars of the crossbow and the cluster are foreground objects and that the nebulous matter is far behind them.

Reserve plenty of time to study the heart of M8, just 3′ west-southwest of 9 Sagittarii. It is easily recognizable as a very dense region of nebulosity surrounding the star Herschel 36. With moderate magnification, the region shines with a dusky yellow or straw color. Its main features are two bright knots (southeast and northeast of Herschel 36) and a "wishbone" of dark lanes to the northwest. The bright knots, whose tapered ends join in the middle, make up the famed "hourglass" nebula – a mysterious source of radio emission.

Once in orbit, NASA's Hubble Space Telescope rapidly began solving many deep-sky mysteries. In our own galactic neighborhood, it opened doors to secret cosmic gardens (like the "hourglass" region in M8) and revealed the wonders within. The HST's portrayal of the "hourglass" is nothing short of spectacular. In a region only one-half light-year long, interstellar "twisters" – eerie dark funnels – project from

Just to the north of this region, you will encounter a long, diffuse swath of nebulosity that arcs over all these features. To the west, a faint nebulosity veils the 5th-magnitude F5 star 7 Sagittarii and its fainter neighbor farther to the west-northwest; a dim bridge of cloud connects them to M8.

The June 2011 issue of *Gemini Focus* (the official publication of the Gemini Observatory) spotlighted a composite image of the Lagoon Nebula's "southern cliff" taken with the 8-meter Gemini South telescope on Cerro Pachón in Chile. This stunning cloudscape of dust and gas, which runs along the nebula's south edge, surrounds a nursery of intermediate- and low-mass stars – most of which hide in the tips of bright-rimmed pillars of dust. "As these 'baby stars' grow," Argentine astronomers Julia I. Aria and Rodolfo H. Barbá, explain, "they eject large amounts of fast-moving gas which plow into the surrounding nebula, producing bright shocks known as Herbig-Haro objects." The

Gemini image shows a dozen of these objects spanning sizes from a few thousand astronomical units (about a trillion kilometers) to 1.4 parsecs (4.6 light-years) – a little larger than the distance from the Sun to the nearest star, Proxima Centauri.

Photographs, such as the one that opens the discussion of this object, also reveal several tiny, round globules of dark matter known as Bok globules after the late Harvard University astronomer Bart Bok, who interpreted the mysterious dark nebulae, which were first recorded in photographs by turn-of-the-century astronomer E. E. Barnard. These dense, obscuring clouds have diameters of about one-third of a light-year and larger and are believed to be sites of star formation.

Bok globules are similar to, but different from, the dense star-spawning nebulae called "evaporating gaseous globules" (EGGs) that the Hubble Space Telescope imaged in M16, a nebula and cluster in Serpens. Bok globules are about 50 times larger than EGGs, whose placental clouds of dust and gas are about the size of the Solar System. Like the EGGs, Bok globules undergo photoevaporation, the process in which intense ultraviolet radiation from young, hot O- and B-type stars blows away surrounding dust and gas, uncovering sites of star formation.

Bok globules and EGGs are too faint to be seen in small telescopes, but I did unknowingly record a mock Bok globule. It is actually Barnard 88, a comet-like dark nebula in the northeast section of the outer cloud (see the photograph). Visually, it appeared as a notch or bay in that swirling swath of brightness. But when I compared the drawing to the photograph, I saw that the location of the notch matched that of Barnard's black "comet."

Before you leave this object, use your binoculars to look immediately east-northeast of the Lagoon Nebula, where you'll find a glowing patch of milky starlight that includes the sparse open star cluster Collinder 367 superimposed on yet another dimly glowing nebula that matches its extent. The entire region is awash in dim patches of glowing gas, starlight, and dark clouds.

M9

NGC 6333
Type: Globular Cluster
Con: Ophiuchus

RA: $17^h19.2^m$
Dec: -18°31′
Mag: 7.8
Diam: 12′
Dist: 22,800 light-years
Disc: Charles Messier, 1764

MESSIER: [Observed May 28, 1764] Nebula without a star, in the right foot of Ophiuchus; it is circular and its light faint. Observed again 22 March 1781.

NGC: Globular cluster, bright, large, round, extremely compressed in the middle, well resolved, stars of 14th magnitude.

AS YOU HUNT DOWN THE MESSIER objects, try to forget any preconceived notions you might have about their appearance. For example, it is easy to assume that all Messier globular clusters will at first appear as tight, fuzzy balls of starlight. But M9 is clearly an exception. It immediately looks different, as though someone has tried to erase it from the sky.

In a way, it's amazing that Charles Messier found the nearly 8th-magnitude glow with his inferior instrument in 1764. The cluster remained only a dim patch of light until William Herschel trained his 20-foot reflector on it in 1784 and resolved it into stars; he likened its appearance to that of a miniature M53 in Coma Berenices, which is interesting, because Herschel believed M53 to be one of the most beautiful sights in the heavens. Admiral Smyth seems to have concurred, describing it as a "myriad of minute stars, clustering into a blaze in the centre, and wonderfully aggregated, with numerous outliers seen by glimpses."

The beauty in part is owing to the fact that we see the cluster projected against the central bulge of the Galaxy. The study of globulars is important to the study of the Milky Way's evolution. Astronomers have two main theories of how our Galaxy formed: (1) a single collapse and fragmenting gas cloud, and (2) built up over time from smaller clouds that merged. Either way, the most metal-poor stars should be the oldest in the Galaxy and give us some understanding as to the initial stages of our protogalaxy. M9 is about 25,800 light-years from us and only about 5,500 light-years from the Galactic center – so close that gravitational forces have pulled it slightly out of shape. As suspected, astronomers found it to be metal-poor, with an age about twice that of the Sun. So it likely formed

M9

At 23×, through the 4-inch it remains dwarfish and looks highly compressed. (Its appearance is all the more paltry if you happen to take in the more dazzling Ophiuchus globulars M10 and M12 first.)

Curious as to why M9 looked so unusual, I swept the region at 23× and was surprised to discover that the cluster and its immediate surroundings are cloaked in very impressive dark nebulae. Most obvious is the oily gulf known as Barnard 64 just to the cluster's west-southwest. A larger cloud, whose shape reminds me of a dinosaur footprint (with the toes pointing to the east), encompasses both Barnard 64 and M9. Another larger patch, Barnard 259, lies about 1° to the southeast. I wondered whether these gloomy realms caused this globular to look so gray and dim. In fact, M9 lies so close to the Galactic center that heavy absorption of the cluster's light by interstellar dust dims its intensity by at least one magnitude.

Unfortunately, through a small telescope, the cluster itself reveals little in the way of stunning detail; what's most conspicuous is what's absent. At low power, it has a very strongly condensed center and a tiny unresolved halo, outside of which, to the south, lies a fine double star. Moderate power reveals some structure, namely caps on the northwest and southeast edges studded with stars no brighter than 13th magnitude. Lord Rosse clearly noticed the

during an epoch when the universe was just a fraction of its current age. In a Hubble Space Telescope image released in 2012, shown here, we see M9 in spectacular clarity; it is the most detailed image of the cluster ever taken, revealing its myriad stars spread across 80 light-years of space. The cluster is approaching us at 142 miles per second.

To find M9, use the chart at the back of this book to find magnitude 2.5 Eta (η) Ophiuchi, which is about a quarter of the way from Alpha (α) Scorpii (Antares) to Alpha Aquilae (Altair) – or nearly 15° (a fist and two finger widths) to the northeast of Antares. Then switch to the chart here. With an angular diameter of 12′ (though only half of that is obvious in small telescopes), M9 appears as but a tiny patch of light in 10 × 50 binoculars.

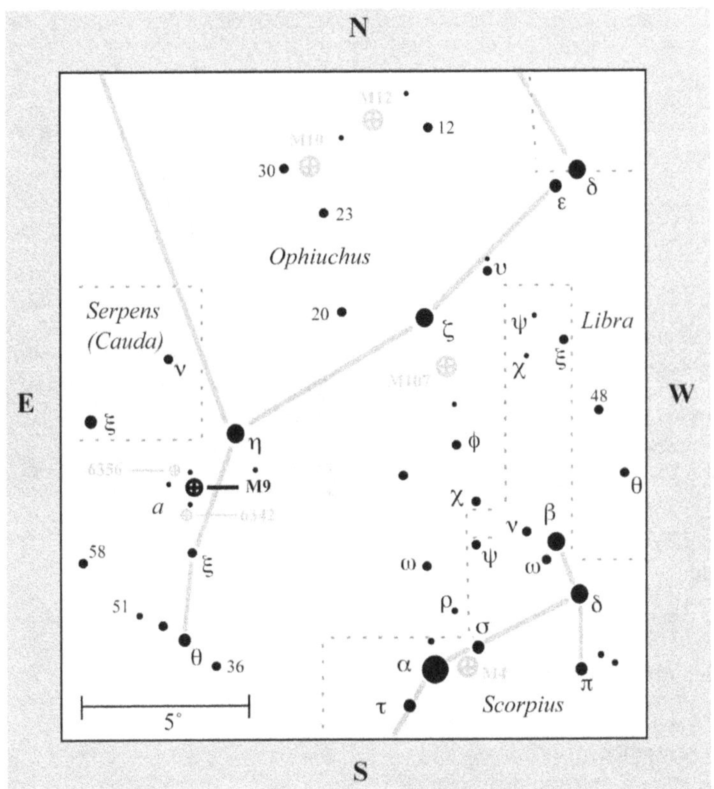

southern cap, seeing an outlying portion separated from the main body by a dark passageway. Thus the globular appears slightly oval, with arms like those in a barred spiral galaxy. High powers start to resolve the cluster, but with difficulty. It requires keen averted vision, so the stars just seem to flitter about. Better resolution comes with larger telescopes that can reach 16th magnitude.

Two smaller, fainter globular clusters lie just outside the boundaries of the dark envelope. NGC 6342 is a dim magnitude 9.5 object a little more than a degree to the southeast (and surrounded by Barnard 259), and magnitude 8.2 NGC 6356 is situated equidistant to the northeast. This triangle of globulars – M9, NGC 6342, and NGC 6356 – "overlaps" a similar triangle of 6th-magnitude stars.

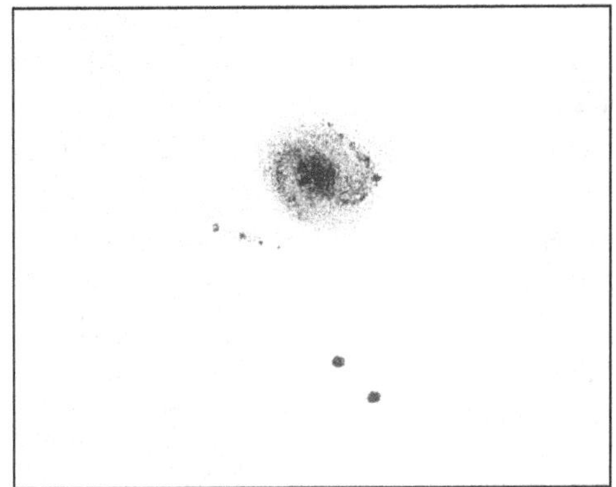

M10

NGC 6254
Type: Globular Cluster
Con: Ophiuchus

RA: $16^h57.1^m$
Dec: –4°06′
Mag: 6.6
Diam: 20′
Dist: 14,350 light-years
Disc: Charles Messier, 1764

MESSIER: [Observed May 29, 1764] Nebula without a star, in the belt of Ophiuchus, close to the thirtieth star in this constellation, which is of magnitude six, according to Flamsteed. This is a beautiful, circular nebula; it can be seen only with difficulty with a simple three-foot refractor. M. Messier plotted it on the second chart of the path of the comet of 1769, *Mémoires de l'Académie 1775*, plate IX. Observed again 6 March 1781.

NGC: Remarkable globular, bright, very large, round, gradually very much brighter in the middle; well resolved, stars of 10th to 15th magnitude.

M10 IS A BRIGHT AND BEAUTIFUL globular star cluster in Ophiuchus, nearly equidistant from us and the Galactic center. It makes a nice pairing with globular cluster M12 just 3° to the northwest. Charles Messier discovered these clusters on successive nights in May 1764. William Herschel resolved M10 in 1784, noting a very compressed core of stars, which is interesting because, while the core is bright, it is not compact.

Indeed, M10 is a relatively loose cluster with an intermediate metallicity; each of its members contains about 1/32 as much iron per unit hydrogen as the Sun. Nevertheless, its bright core has been considered a reasonable candidate for harboring an intermediate-mass black hole. However, in a 2011 paper in *The Astrophysical Journal* (vol. 743, p. 11), E. Dalessandro and colleagues describe how they used the Hubble Space Telescope to study both the core and the external regions of M10 and found that the fraction of binary stars (which, owing to their greater mass, tend to congregate in and around the core) decreases from about 14 percent within the core to about 1.5 percent in M10's outer regions. The researchers concluded that the estimated binary fraction is "sufficient to account for the suppression of mass segregation observed in M10, without any need to invoke the presence of an intermediate-mass black hole in its center." The image here was taken by the Hubble Space Telescope and shows the inner 13 light-years of its core.

M10

Once every 53 million years, M10 crosses the Galactic plane. Tidal attractions between loose clusters like M10 and the strong pull of the Galactic center can rob the cluster of stars and create tidal tails. In a 2000 paper in *Astronomy and Astrophysics* (vol. 359, p. 907), Stephane Leon (Paris Observatory) and colleagues found evidence for a possible 500 light-year-long tidal tail associated with M10, which has a tidal radius of about 85 light-years. The tail appears to lie in the cluster's orbital path, and the velocity diffusion of the stripped stars is similar to that of the cluster velocity dispersion. M10 last crossed the Galactic plane about 20 million years ago, and evidence for some recent gravitational shocking can be seen in the cluster.

From dark skies, this magnitude 6.6 globular cluster in Ophiuchus is an easy binocular object and can be spied with the unaided eye, especially because the 5th-magnitude star 30 Ophiuchi, an orange giant star of spectral type K4 III, is so close to it. If you are under a dark sky, just stare at this star and your averted vision should pick up M10.

To find 30 Ophiuchi, I start at magnitude 2.6 Beta^1 (β^1) Scorpii (which has a blue, 5th-magnitude companion, β^2 Scorpii) and follow a line a little more than 10° (a fist width) to the northeast to an equally bright and blue star, Zeta (ζ) Ophiuchi. Now continue on that line for another 6° (about two finger widths) to a 5th-magnitude star, 23 Ophiuchi. Two and a half degrees farther along is 5th-magnitude 30 Ophiuchi. M10 is just 1° west-northwest of 30 Ophiuchi.

At low power, M10 appears to be very typically detailed, but I could not escape the clear impression that its outer halo of stars has an ice-blue sheen, whereas its interior exudes a

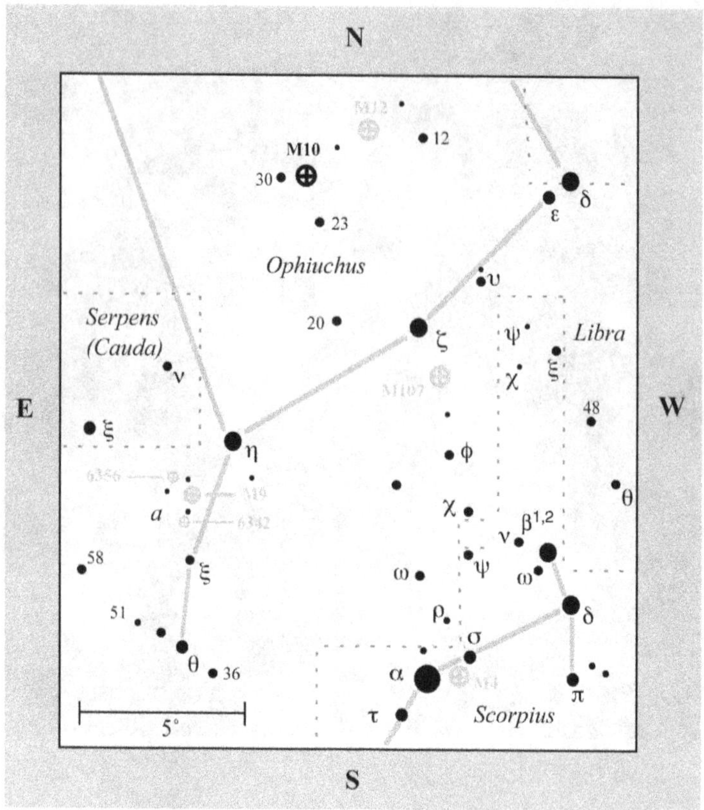

pale salmon light. Look for a yellow spark at the very center. These colors are similar to those of M3, though not as obvious. When I relax my gaze, I see two short, thin arms running through the center of the cluster from north-northeast to south-southwest. Suddenly M10's core looks like Saturn with its rings seen nearly edge-on through a thick fog.

Moderate power shatters this illusion and creates another. Now the cluster is caged – a ball of energy inside a pyramid of 12th-magnitude stars that may or may not belong to the cluster (they're at that discernible fringe in the 4-inch). This pyramid itself is enclosed in another, larger pyramid of stars. The cluster's center shines with a soft, uniform glow. When I use averted vision, the surrounding halo, which otherwise shimmers with a faint light, shoots sparks of starlight across that sheet of "plasma." This is not a flamboyant sight but an elegant one. Its starlight is rather demure and

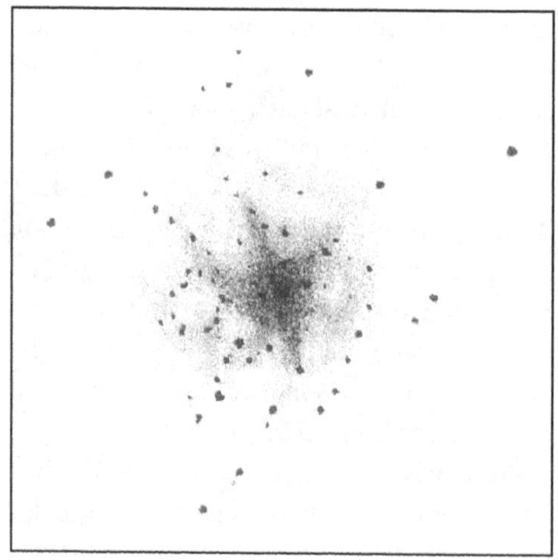

does not vary much in intensity. Most components look like diamond dust. Indeed, the cluster starts to resolve at magnitude 14.7, a tantalizing threshold for moderate apertures under dark skies. Even its sluggish recessional velocity of 63 miles per second speaks of a certain stately galactic reserve.

Although I have noticed several dark patches in the cluster's outer halo, I did not notice, as Lord Rosse did, that the "upper" one-sixth of the cluster is fainter than the rest. Does such a variation in brightness exist? What do you see?

M11

Wild Duck Cluster
NGC 6705
Type: Open Cluster
Con: Scutum

RA: $18^h51.1^m$
Dec: –6°16.2′
Mag: 5.8; 5.3 (O'Meara)
Diam: 13′
Dist: 5,460 light-years
Disc: Gottfried Kirch, 1681

MESSIER: [Observed May 30, 1764] Cluster of a large number of faint stars, near the star *K* in Antinoüs [Aquila], which may be seen only with good instruments. With a simple three-foot refractor it resembles a comet. This cluster is suffused with faint luminosity. There is an eighth-magnitude star in the cluster. Kirch observed it in 1681, *Philosophical Transactions* no. 347, page 390. It is plotted in the large English *Atlas Céleste.*

NGC: Remarkable, cluster, very bright, large, irregularly round, rich, one star of 9th magnitude among stars of 11th magnitude and fainter.

M11 IS A SMALL (13′) BUT EXTREMELY rich open cluster, bristling with the light of at least 680 stars; of these, about 400 shine brighter than 14th magnitude. The cluster measures about 20 light-years in diameter, and its core is very dense. If you lived at the center of M11, you would see several hundred 1st-magnitude stars in the sky, and possibly 40 or so with an apparent brightness ranging from 3 to 50 times the light of Sirius, a scenario Robert J. Trumpler of Lick Observatory calculated in 1932.

Gottfried Kirch, who was associated with Berlin Observatory, discovered M11 in 1681. Upon discovery, he was "at first uncertain the phenomenon was a comet or a nebulous star." Its lack of motion, however, revealed its nature as a "nebulous star." William Derham (1657–1735), an English clergyman and amateur astronomer, first resolved it into stars, noting, "It is not a nebula but a cluster of stars somewhat like that which is in the Milky Way."

In a 2010 paper in *Astronomy & Astrophysics* (vol. 513, p. A29), S. Messina (INAF – Catania Astrophysical Observatory) and colleagues announced the discovery of 75 periodic variables in the ~250-million-year-old cluster, of which 38 are candidate members. These include 6 early-type periodic variables, 2 eclipsing binaries, and 30 "bona-fide" single

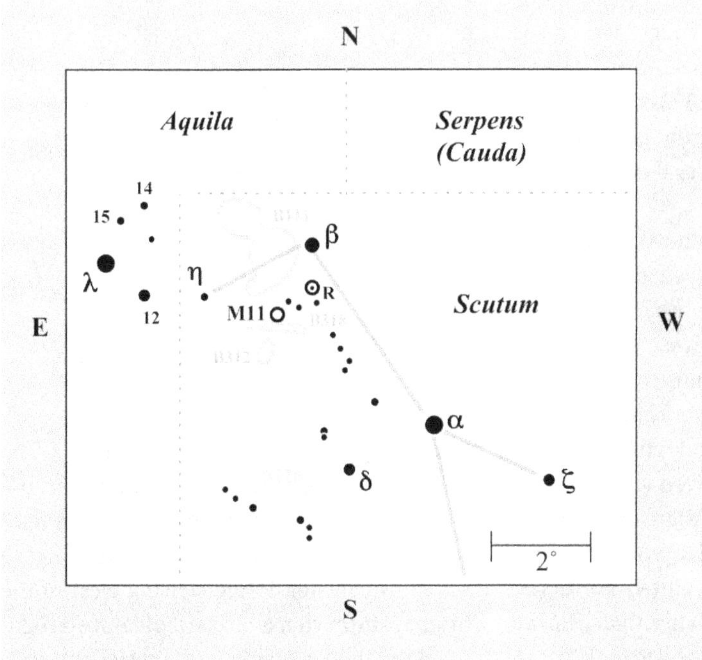

To find M11, use the chart at the back of this book to locate 4th-magnitude Beta (β) Scuti, about 20° (two fists) southwest of brilliant Alpha (α) Aquilae (Altair). Then use the chart here to zero in on it. M11 is not quite 2° southeast of Beta Scuti. The cluster is visible to the naked eye as a fuzzy pellet of light flanked by two 6th-magnitude stars to its northwest. Through binoculars, the view is almost as stunning as it is through a telescope. M11 sits in a notch in the north edge of the Great Scutum Star Cloud, the brightest stellar island outside the Galactic center in Sagittarius. Barnard called this region the Gem of the Milky Way.

At low power, M11 is well resolved and displays a fluid fan of starlight streaming from an 8th-magnitude saffron star near the fan's apex. Although not a true cluster member, this brilliant star is one of many yellow giant stars contained in this cluster, whose age is estimated to be some 500 million years.

When viewing the cluster at 23×, two scenes come to mind: an exploding party favor with countless flecks of confetti fleeing into space, or a volcanic vent blowing out incandescent gas and golden shards of molten rock. Can you see Smyth's "flight of wild ducks" (hence the nickname Wild Duck Cluster) in this V formation of stars? There is a pair of 9th-magnitude stars south of the eastern flock of ducks. Between this double and the saffron star is a wall of faint stars. At

periodic late-type variables. They also determined that the cluster's G-type stars have a mean rotation period of 4.8 days. When these periods were compared with those of the younger stars in the M35 cluster (~150 million years) and with the older stars in the M37 cluster (~550 Myr), they found that G stars rotate more slowly than younger M35 stars and more rapidly than older M37 stars. "The measured variation in the median rotation period," they conclude, "is consistent with the scenario of rotational braking of main-sequence spotted stars as they age." (Main sequence stars lose angular momentum as they age due to a magnetic torque operating on the stellar wind.) "Finally," the researchers continue, "G-type M11 members have a level of photospheric magnetic activity, as measured by light curve amplitude, comparable to what is observed in the younger 125-[million year] Pleiades stars of similar mass and rotation."

moderate power, the cluster's nuclear region appears highly fractured, which was also noticed by the nineteenth-century astronomer Heinrich d'Arrest (1822–1875), who described M11 as "a magnificent pile of innumerable stars. Irregular and as if divided into several agglomerations."

Use high power to examine the cluster's dumbbell-shaped interior (see the accompanying drawing) and the waves of stars flowing away from it to the west. In fact, the outer fringes of the cluster are composed of wild stellar arcs that emanate from the cluster's northern extremes and curve southward. When I look with averted vision, I can see many faint streamers of stars at the limit of vision. With its triangular body, insect-like appendages, and tiny pointed northern "head," M11 looks more like a tick than a flock of wild ducks. (Does the tick's head look fuzzy to you?) If you can view the cluster with north at the top of your field of view, use high power and relax your gaze, or slightly defocus the cluster. I can make out a Jolly Roger (a grinning skull complete with a black eye patch).

If you have a very-wide-field telescope, you must take the time to sweep the Great Scutum Star Cloud, where there is an incredible diversity of bright and dark nebulosity (of various shades of gray). A striking slash of darkness (Barnard 318) lies immediately south of M11. South of B318 lies an eerily dark pond (Barnard 112). By the way, the notch in the northern edge of the Star Cloud (where M11 resides) is created by a massive swatch of darkness (Barnard 11).

One night in August 1995, when the sky was illuminated by the full moon, I turned the world's second-largest refractor – the Lick Observatory 36-inch on Mount Hamilton in California – to M11 and studied it at 588×. The field of view was five times smaller than the cluster, so I had to electrically slew the telescope north and south, east and west, to see the entire cluster. While slewing west of the bright saffron star, a blizzard of faint starlight filled the field. Slewing farther west, the view suddenly began to grow dim. I looked up to see if clouds had moved in, but the Moon and stars were tack sharp against the velvet sky. Returning to the eyepiece, I realized it was dark galactic vapor that had dimmed the starlight. I reversed the slewing direction and started over. This time I slowly "walked" across glistening moonlit fields, wet with starlight, until I gradually sank into that misty moor beyond them.

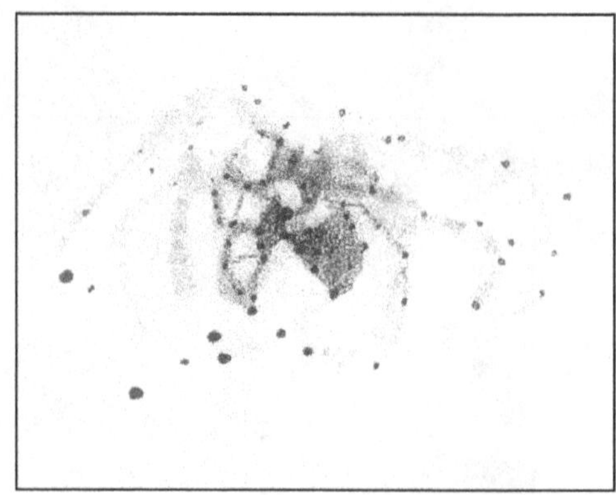

Just 1° northwest of M11 is a famous naked-eye variable, R Scuti, whose semiregular pulsations (magnitude 4.8 to 6.0) repeat about every 143 days. The star occasionally dips below naked-eye visibility.

Here's an interesting note: On June 25, 1995, while viewing M11 at 72×, I wrote in my notebook "very red star nearby." But I didn't indicate exactly where. Can you locate it? Is it a variable?

M12

Gumball Globular
NGC 6218
Type: Globular Cluster
Con: Ophiuchus

RA: $16^h47.2^m$
Dec: -01°57′
Mag: 6.1
Diam: 16′
Dist: ~15,650 light-years
Disc: Charles Messier, 1764

MESSIER: [Observed May 30, 1764] Nebula discovered in Serpens, between the arm and the left side of Ophiuchus. This nebula does not contain any stars, it is circular, and its light is faint. Near the nebula is a ninth-magnitude star. M. Messier plotted it on the second chart of the comet observed in 1769, *Mémoires de l'Académie 1775*, plate IX. Observed again 6 March 1781.

NGC: Very remarkable globular cluster, very bright, very large, irregularly round; gradually much brighter in the middle; well resolved, stars of 10th magnitude and fainter.

THE ENORMOUS SUMMER CONSTELLATION Ophiuchus harbors nearly two dozen globular clusters within range of small telescopes, and seven of them were catalogued by Messier. Small but spectacular, M12 lies only 3 1/2° northwest of M10 (a nearly identical globular) and 2 1/2° east-northeast of 6th-magnitude 12 Ophiuchi. Use binoculars (or your telescope at low power) to view the clusters together for a twin treat. Separated by only 3,700 light-years, these two star-packed clusters are virtually cosmic neighbors. Each would appear as a roughly magnitude 4.5 object as seen by hypothetical inhabitants within the other. By comparison, Omega Centauri, the grandest globular visible from Earth, shines at magnitude 3.9.

M12 is relatively nearby, being only about 15,650 light-years from the Sun and 14,700 light-years from the Galactic center. It lies above the main dusty corridors of the Galactic bulge near the celestial equator. A Class IX object, M12 is not very concentrated, with the brighter stars in the cluster appearing relatively loose. In 1938, Helen Sawyer Hogg of the David Dunlap Observatory noted that both M10 and M12 belong to a group of globular clusters that has a relative scarcity of bright stars, with fewer than 200 stars of absolute magnitude brighter than zero. The

M12

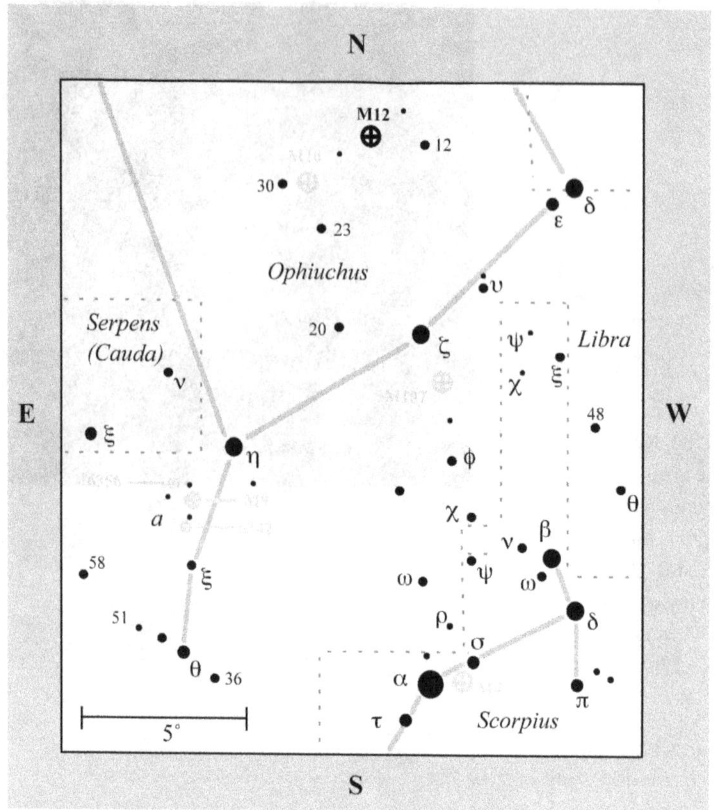

our knowledge, M12 is the first globular where such a signature has been spectroscopically detected," they said.

Again, to find M12, you'll want to locate 6th-magnitude 12 Ophiuchi. Use the chart at the back of this book or the one here to first locate 2nd-magnitude Beta (β) Scorpii, then equally bright Zeta (ζ) Ophiuchi, a little more than 10° (a fist width) to the north-northeast. You'll find 12 Ophiuchi about 8° north of Zeta and about 5 1/2° east-northeast of Beta Ophiuchi. M12 is only 2 1/2° east-northeast of 12 Ophiuchi. First look for the small, diffuse glow with binoculars.

mean visual magnitude of these brightest stars is about 14. Sawyer Hogg found only one variable among its multitude. The cluster spans about 75 light-years in true physical extent, has an integrated spectral type of F8, and is receding from us at 27 miles per second.

In a 2007 paper in *Astronomy & Astrophysics* (vol. 464, p. 939), E. Carretta (INAF – Osservatorio Astronomico di Bologna) and colleagues reported making the first extensive spectroscopic abundance analysis in this cluster. Their results indicated that M12 is very homogeneous as far as heavy elements are concerned. On the other hand, light elements show large variations at all luminosities along the red giant branch. "To

At 23× through the 4-inch, I can see a house- or rocket-shaped asterism just to the north of M12, and the globular looks like a puff of smoke from the rocket's exhaust, with its many bright members strung out like paper streamers in the wind. Admiral Smyth likened these linear features to a "cortège of bright stars," and Lord Rosse saw them as long, straggling tentacles.

Moderate power reveals a very loosely packed nuclear region surrounded by a faint halo of unresolved stars, though high power resolves the halo beautifully. I call M12 the "Gumball Globular" because that's what immediately came to mind when I first saw its wide assortment of bright and colorful stars.

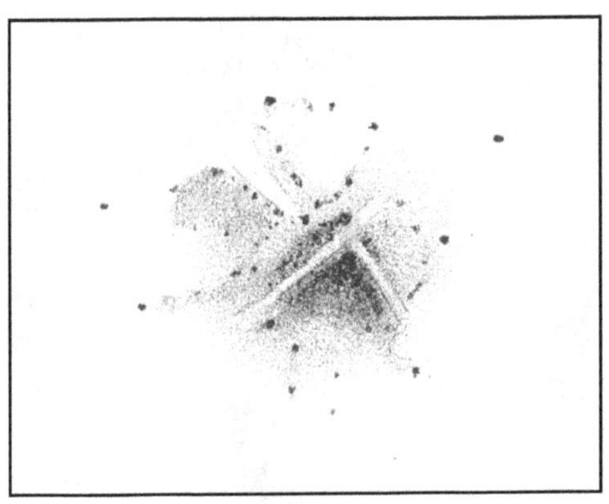

Use 130× to appreciate the central star streams forming a wedge or triangle that fans to the south. A dark "V" borders it to the north. Otherwise, the star patterns favor the south and east. Three of M12's arms enclose dark bays. Can you spot them? Also, M12 should be within naked-eye range. Can you spot it?

M13

Great Hercules Cluster
NGC 6205
Type: Globular Cluster
Con: Hercules

RA: $16^h41.7^m$
Dec: +36°28′
Mag: 5.8; 5.3 (O'Meara)
Diam: 20′
Dist: 23,150 light-years
Disc: Edmond Halley, 1714

MESSIER: [Observed June 1, 1764] Nebula without a star, discovered in the belt of Hercules. It is circular and bright, the center brighter than the edges, and is visible in a one-foot refractor. It is close to two stars, both of eighth magnitude, one above and the other below. The nebula's position has been determined relative to ε Herculis. M. Messier plotted it on the chart for the comet of 1779, which will be included in the Academy volume for that year. Seen by Halley in 1714. Observed again on 5 and 30 January 1781. It is plotted in the English *Atlas Céleste*.

NGC: Very remarkable globular cluster of stars, extremely bright, very rich, very gradually extremely compressed in the middle, stars from 11th magnitude and fainter.

M13 IS GENERALLY CONSIDERED THE finest globular cluster in the northern skies, mainly because it is visible to the naked eye in a well-known grouping of stars that sails high overhead in the summer sky. It is a swollen mass teeming with perhaps 100,000 to a half-million stars spread across 135 light-years or more; a typical globular contains tens of thousands to hundreds of thousands of stars. A relatively close globular (about the same distance as M5), the Great Hercules Cluster is pleasingly bright. I determined its magnitude to be 5.3, which is slightly brighter than that recorded in most references. From dark skies and under good conditions, M13 is easily spotted as a fuzzy "star" with the naked eye, though it can be seen as a perceptible glow even through a light fog.

Edmond Halley discovered the grand cluster in 1714 from the island of St. Helena in the South Atlantic. He went there in 1676 with the sole purpose of cataloguing the southern stars. Halley, who had "poor eyesight," spent 18 months on that remote

at times form a new star, called a 'blue straggler.'"

The cluster lies 23,150 light-years from the Sun, close to its perigalacticon (the point in a star's orbit around our Galaxy that is closest to the Galactic center) of 16,000 light-years, and a halo of extratidal stars with a slight elongation in the direction of motion has been detected. Using Sloan Digital Sky Survey data, Carl Grillmair and S. Mattingly detected leading and trailing tidal tails that appear to emanate from M13 along its orbital path, extending for at least 15° on the sky. They reported in a 2010 paper in the *Bulletin of the American Astronomical Society* (vol. 41, p. 833) that, "The tails agree approximately with the orbit of the cluster, as determined from radial velocity and proper motion measurements. ... On the other hand, initial estimates of the distance to the stream put it at least 5 kpc beyond M 13."

island, plagued by inclement weather, an arrogant governor, and harsh living conditions. In the end, Halley catalogued only 341 stars, but he did discover M13 and the great southern globular star cluster Omega Centauri.

M13 has an integrated spectral type of F6. The cluster is metal-poor, with each of its members containing about 1/34 as much metal per unit hydrogen as the Sun. We see it receding from us at about 150 miles per second. In 2008, the Space Telescope Science Institute released a Hubble Space Telescope image that peered deep into the heart of this teaming metropolis of stars, "where the density of stars is about a hundred times greater than the density in the neighborhood of our Sun. These stars are so crowded that they can,

M13 lies about 2 1/2° south of magnitude 3.5 Eta (η) Herculis, on the western side of the familiar keystone asterism in Hercules. It is an easy target in binoculars and is not difficult to see with the unaided eye from a dark-sky site. Use the accompanying chart to locate it.

At low power, the cluster is bracketed nicely by two 7th-magnitude stars. The western one is of type A2 and the eastern one is type K2, though, when I made my observations, both stars seemed to have a yellowish

M13

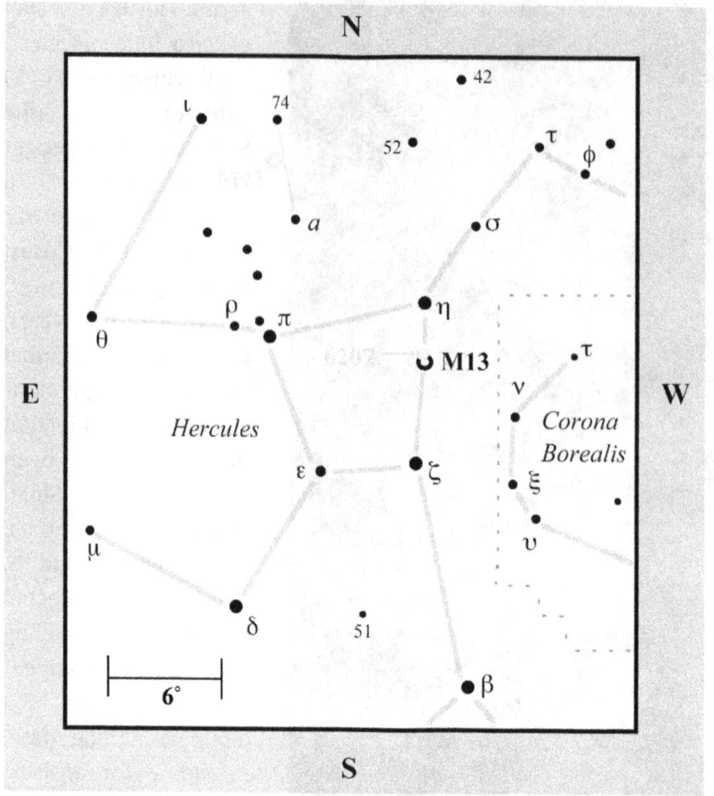

13th-magnitude stars sparkle into view across the cluster's face. The remainder form a barely resolvable haze, like finely crushed sandstone illuminated by a setting sun. Indeed, at times, I swear the center has a blood-like tinge. This is not too difficult to believe, because the cluster's brightest members are red giants. At first, the core appears moderately diffuse, but if you stare long enough, you might see a gradual brightening toward the center.

M13 really packs a punch at high power! With averted vision, the view is almost frightening – a blazing ball of tiny stars. So many more stars, so many more patterns to consider. Arcs of stars to the northeast create an impression of a strong galactic wind blowing from that direction, forming a bow shock on that side of M13. And the forked array of stars to the southwest forms a beautiful cock's-tail wake. Look carefully at the cluster's core; it is fractured, with a definite asymmetry toward the south. Here you can also see many dark patches and rifts. Only once did I recognize Lord Rosse's classic dark Y shape, just southeast of the core, and, surprisingly, that was on a foggy night! I have yet to see it again in the 4-inch, despite having made several attempts to look solely for that feature. Interestingly, I discovered my own Y in the northwest halo, just inside the northern wing, as shown in the drawing. Furthermore, this dark Y shows

tint. Compare the color of the cluster's center with these stars and see if it appears yellow with a slightly greenish halo. John Herschel described the cluster as exhibiting "hairy-looking, curvilinear branches"; Lord Rosse also noted the "singularly fringed appendages ... branching out into the surrounding space." Two of these arms show prominently at 23×. They extend southeast and northwest from the nucleus and look like wings curving to the southwest. A forked "tail" of stars to the southwest completes this bird-like visage. Otherwise, the cluster at low power appears moderately condensed at the center with a gradual spreading out of light.

At moderate power, this 14-billion-year-old cluster shows only a handful of seemingly bright members shining between 11th and 12th magnitude, though about two dozen

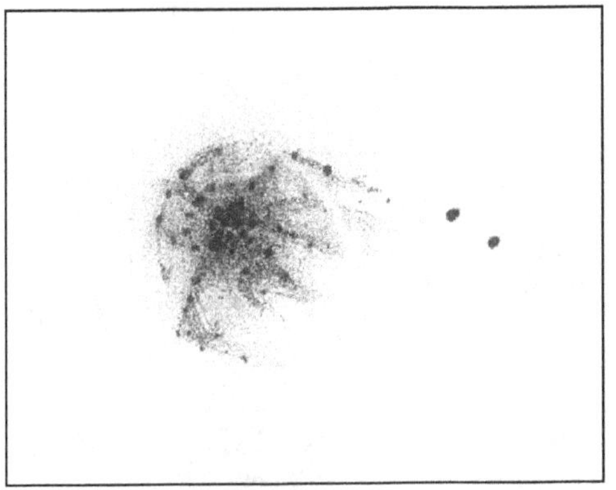

nicely on a photographic plate made with the Lowell Observatory's 13-inch telescope.

M13 is an impressive cluster, but I think its grandeur is slightly overrated for small-telescope users. It certainly is a magnificent cluster in photographs and when seen through large-aperture telescopes. But is it the finest globular cluster in the northern skies for small apertures? In my opinion, through a 4-inch refractor, M13 does not have as strong a visual impact as M5 – even though both clusters have stars of similar brightness (the brightest of which shine around 12th magnitude) and a horizontal-branch magnitude of 15. Furthermore, M5 is, according to my magnitude estimate, only 0.4 magnitude fainter than M13. But globular M5 seems to immediately display an assortment of stellar magnitudes, colors, and star patterns at only 23×, while it takes higher magnifications to fully appreciate M13 – at least through a 4-inch. But everyone will have their own opinion, so try comparing the two clusters some night and see what you think.

Before leaving M13, don't forget to try for the 11th-magnitude spiral galaxy NGC 6207 about 30′ north-northeast of M13's core. The 4-inch at medium power shows it as an obvious elongated haze. These two objects provide a dramatic example of depth of field: NGC 6207 lies at a distance of 46 million light-years – about 2,000 times farther than M13 in the halo of our own galaxy.

M14

NGC 6402
Type: Globular Cluster
Con: Ophiuchus

RA: $17^h37.6^m$
Dec: –3°14′
Mag: 7.6
Diam: 11′
Dist: 33,300 light-years
Disc: Charles Messier, 1764

MESSIER: [Observed June 1, 1764] Nebula without a star, discovered in the drapery that hangs from the right arm of Ophiuchus, and lying on the parallel of ζ Serpentis. This nebula is not large, and its luminosity is feeble; however, it may be seen with a simple three-and-a-half-foot refractor; it is circular. Close to it is a faint star of the ninth magnitude. Its position has been determined relative to γ Ophiuchi, and M. Messier plotted its position on the chart for the comet of 1769, *Mémoires de l'Académie 1775*, plate IX. Observed again 22 March 1781.

NGC: Remarkable globular cluster, bright, very large, round, extremely rich, very gradually much brighter in the middle, well resolved, 15th-magnitude stars and fainter.

FOR THE SAME REASONS THAT M13 is popular, globular cluster M14 is not. It is beyond the normal naked-eye limit and lies in a region of sky belonging to an obscure part of Ophiuchus, in the Serpent Bearer's left arm, largely devoid of bright naked-eye stars. Yet it can be detected through binoculars. Telescopically, M14 is surprisingly detailed and deserves special attention.

In 1783, William Herschel first resolved Charles Messier's "nebula without star" into a cluster of some 300 stars, which appeared in a rich field of stars. "This cluster is considerably behind the scattered stars," he penned, "as some of them are projected upon it."

M14 is a somewhat metal-rich cluster 33,300 light-years from us and only 13,000 light-years from the Galactic center, near the dark folds of the northern extremity of the Galactic bulge. The giant mass of several hundred thousand solar masses spans 106 light-years of space and is receding from us at 41 miles per second.

In one area of globular cluster studies, some astronomers have looked for radial color variations that may result from dynamical processes

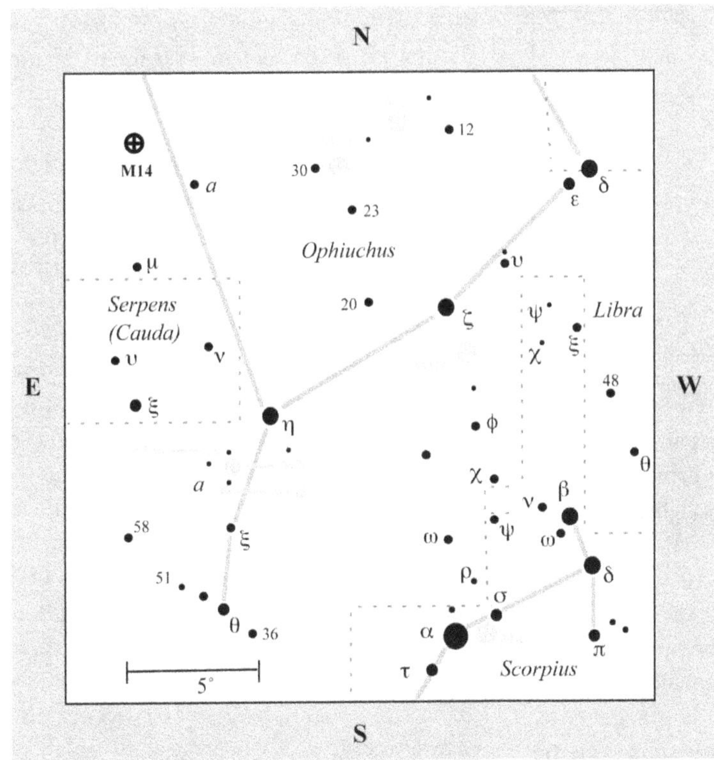

that can "sink" massive stars toward the center. In turn, this nuclear compression can cause close stellar encounters that create massive objects like blue stragglers or binary stars. While M14 does show a color gradient, with bluer stars trending toward its core, no blue stragglers have yet been found in it. The cluster has a well-developed blue horizontal branch and an extended blue tail, and a relatively highly concentrated luminosity distribution.

In a 2012 paper in the *Journal of the Southeastern Association for Research in Astronomy* (vol. 5, p. 34), Kyle E. Conroy and his colleagues confirmed 62 of the 70 previously known catalogued variables in M14 and identified 71 new ones. Of the total number of confirmed variables, they found a total of 112 RR Lyrae stars, 19 variables with periods greater than 2 days, a W Ursae Majoris contact binary, and an SX Phoenicis star.

To find M14, use the chart here and first locate magnitude 2.4 Eta (η) Ophiuchi about 15° northeast of brilliant red Antares in Scorpius. Now look about 6 1/2° east-northeast for magnitude 3.5 Xi (ξ) Serpentis, then nearly the same distance due north for magnitude 4.5 Mu (μ) Serpentis, and train that star in your binoculars. M14 is another 5° hop north; you'll also find a magnitude 4.5 star (Star *a*) in Ophiuchus about 3° to the southwest.

At low power in the 4-inch, this diminutive globular looks as lonely as it does in binoculars. The field is so sparse that the globular appears dim and distant in the void. This is another dusty region of the Milky Way, which contributes to light loss from the globular. But unlike M9, which looks gloomy, M14 shares none of that pallor. In fact, I find it quite luminous, like a comet with a diffuse inner coma surrounded by a diffuse outer coma. Overall, the cluster has a pale straw color, which I find interesting in that the nineteenth-century English astronomer Admiral William Henry Smyth referred to it as "lucid white."

Although there was not even a hint of resolution at low power, I noticed that what at first appeared to be a circular glow actually had a curved section of hazy light cutting north to south through the tiny, faint nuclear region. It also has a quite extensive halo, which was

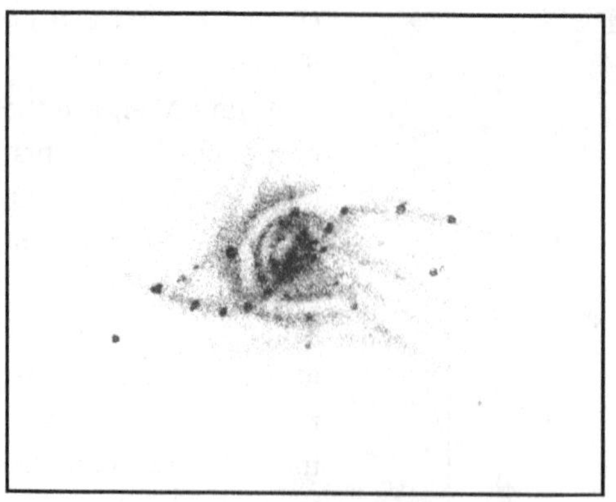

diminished greatly one night under high, thin clouds, so I wondered how city lights must affect this delicate object. Indeed, though M14 is catalogued as having a diameter of 11′, some popular references list a diameter as small as 3′.

The cluster starts to reveal itself at moderate power, when at first glance the center appears as a bulging mass, a cracking shell of starlight. Even the outer halo, which is bracketed by two roughly 12th-magnitude stars, can be partially resolved. The cluster now bursts with faint starlight; the brightest stars in the cluster shine at magnitude 12.8, and decent resolution occurs around magnitude 14. That hazy north–south curve reveals patches of nicely resolved stars, like star-forming regions in the arms of a spiral galaxy. About seven distinct arms extend in various directions.

The cluster's center is loose enough to start resolving at moderate power, and how nice it is to see colorfully bright stars against the core. Higher power immediately produces a most interesting sight: a tiny stellar core emerges, tangerine in color. Perhaps it is the "faint sparkle" noted by Luginbuhl and Skiff in their *Observing Handbook and Catalogue of Deep-Sky Objects*. The rest of the nucleus is a jumble of bright chunks of starlight whose north and south edges are slightly swollen, giving that region a slight dumbbell shape. Perhaps the spaces between these chunks are what astrophotographer Isaac Roberts was alluding to when he saw "vacancies in the centre" of the object on his glass plates.

The core is also surrounded by a ring of stars – a garland or rosette. Look closely and you might see ripples of stellar halos radiating from the nucleus to the southwest, as if a pebble were being tossed at a sharp angle into that pond of stars. A dark lagoon to the southeast of the outermost halo is enclosed by a shoal of faint stars that connects to the southern arm. Finally, avert your gaze way off to one side and then relax. Do you see the nucleus burning as a lens-shaped mass of equally bright stars?

By the way, only one nova and one nova candidate are known to have occurred in globular star clusters in our Galaxy. The first was a classical nova eruption of T Scorpii in M80 discovered visually in 1860. The second was a 16th-magnitude classical nova candidate that appeared in 1938 but was not discovered until 1963, when Amelia Wehlau of the University of Western Ontario found the variable on photographic plates taken by Helen Sawyer Hogg in 1938 with the 72-inch telescope at the David Dunlap Observatory.

In 1986, Michael Shara of the Space Telescope Science Institute and his colleagues proposed an optical candidate of the nova in M14 based on ground-based CCD camera observations. In 1990, the Space Telescope Science Institute announced that Bruce Margon of the University of Washington and his colleagues had used the Hubble Space Telescope Faint Object Camera to survey a tiny field containing the area of Shara's candidate star and showed it to be multiple, with at least six stars – none of which were particularly blue (which the nova should have been). As this nova candidate was discovered long after maximum light, Shara notes that there is a "modest probability that it was a background supernova."

M15

Great Pegasus Cluster
NGC 7078
Type: Globular Cluster
Con: Pegasus

RA: $21^h30.0^m$
Dec: +12°10′
Mag: 6.3; 6.0 (O'Meara)
Diam: 18′
Dist: ~33,900 light-years
Disc: Jean-Dominique Maraldi II, 1746

MESSIER: [Observed June 3, 1764] Nebula without a star between the head of Pegasus and that of Equuleus. It is circular and the center is bright. Its position was determined relative to δ Equulei. M. Maraldi mentions this nebula in the *Mémoires de l'Académie 1746*. "I noted," he said, "between the stars ε Pegasi and β Equulei, a fairly bright, nebulous star, which consists of several stars. Its right ascension is 319°27′6″, and its declination +11°2′22″."

NGC: Remarkable, globular cluster, very bright, very large, irregularly round, very suddenly much brighter in the middle, well resolved into very [faint] stars.

NEARLY A TWIN OF M2 IN AQUARIUS, this glittering gem in the winged horse Pegasus is one of six beautiful globular star clusters brighter than 7th magnitude that grace the northern sky (the others being M2, M3, M5, M13, and M92). The Great Pegasus Cluster, M15, can be spotted without difficulty as a "fuzzy star" with the unaided eye, lying just 4° northwest of the topaz (type K2 I) 2nd-magnitude star Epsilon (ε) Pegasi (Enif). M15 is ~33,900 light-years distant (10,750 light-years farther away than M13) from both the Sun and the Galactic center. It measures ~180 light-years in diameter. Like M13, it contains many red giant stars. But because of its greater distance, M15 appears fainter and more compact than M13. Actually, M15 is one of the densest globular star clusters known in our Galaxy, with a collapsed core measuring only ~1.4 light-years in true physical extent.

Jean-Dominique Maraldi II discovered this stellar metropolis on September 7, 1746, while he was looking for Chéseaux's comet, which also led him to discover M2; he described it as "a fairly bright, nebulous star, which consists of several stars."

The cluster is easy to see through binoculars, appearing as a small but brilliant,

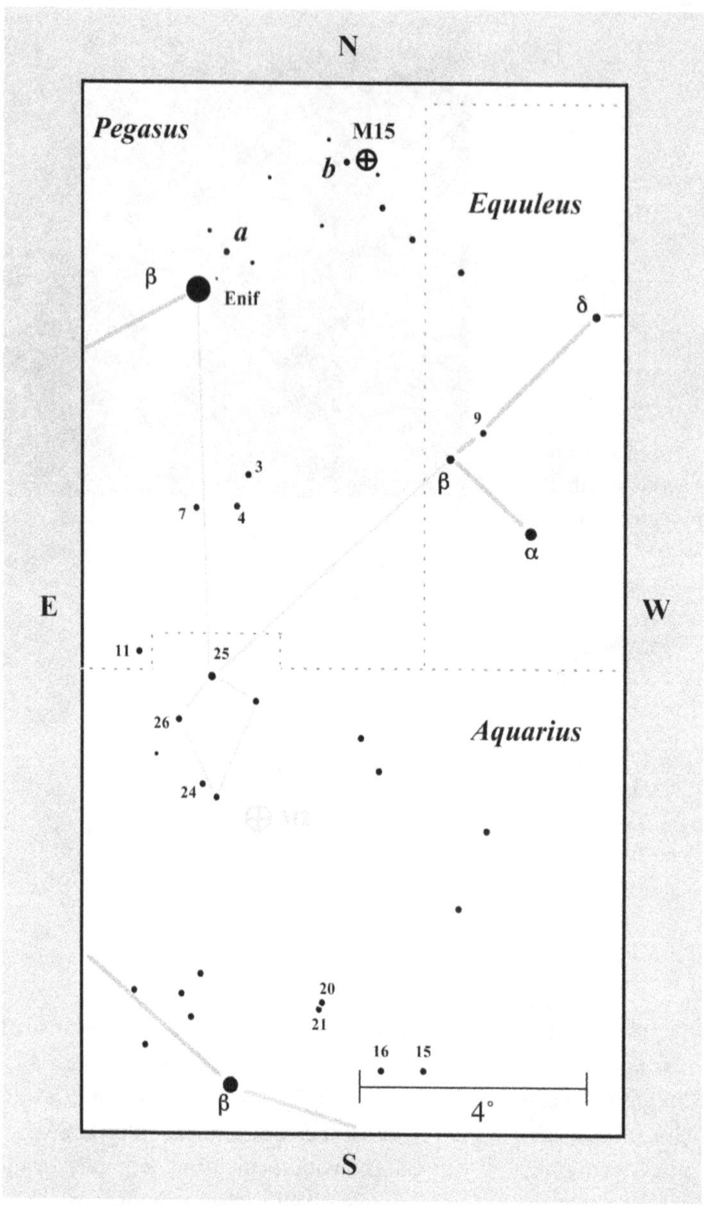

8th-magnitude stars. Hazy, spider-like arms are already apparent, and the cluster brightens rapidly toward the center. Otherwise, like M13, most of M15's stars evade direct gaze and require averted vision. While the cluster's brightest stars shine at magnitude 12.6, you have to be able to see to magnitude 15.9 to resolve it well. William Herschel rated this a good test object for resolution. But subtle details do show through. For example, it has a definite asymmetry.

The late Harvard University astronomer Harlow Shapley first confirmed this symmetry by noting the oblateness at the cluster's central bulge, which is surrounded by a spherical shell of stars. M15, which is currently close to its apogalacticon (the point in a star's orbit around our Galaxy that is farthest from the Galactic center), shows some tidal distortions at its tidal radius, and Katrin Jordi and Eva K. Grebel (Astronomisches Rechen-Institut, Heidelberg, Germany) found an extratidal extension in M15 in the southwest direction – though these stars have yet to leave the cluster's gravitational potential.

6th-magnitude orb of diffuse light, with a highly condensed center. To starhop to it with a telescope, start at Enif and move nearly 1° north-northeast to 6th-magnitude Star *a*; M15 is about 3° west-northwest, next to magnitude 6.5 Star *b*. At 23× through the 4-inch, the cluster hides inside a triangle of three 7th- to

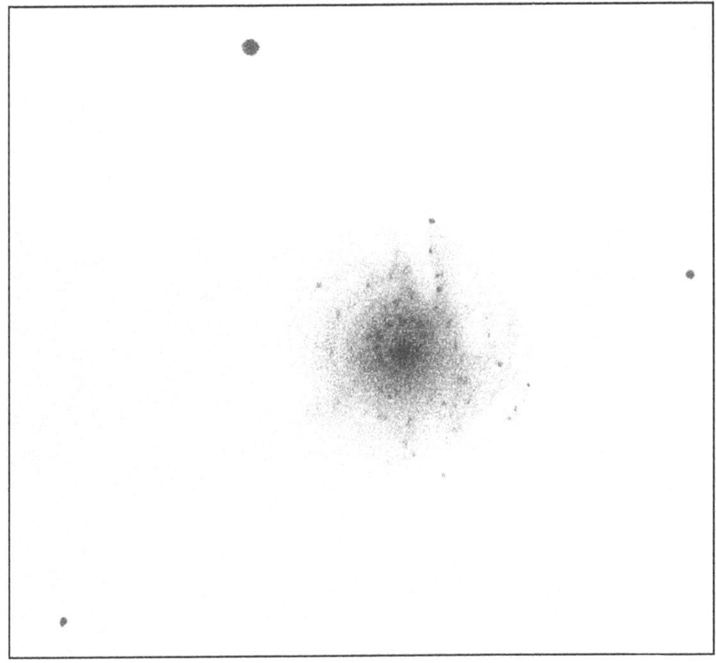

to detect in backyard telescopes. I did glimpse it, though, with Brent Archinal through the 31-inch Warren Rupp reflector in Mansfield, Ohio (or at least its central star), which shines at magnitude 15.0. Both the nebula and the star eluded my gaze in the Genesis. The cluster is also host to a wealth of variable stars (more than 100 are known).

M15 has long been thought to possess an intermediate-mass black hole at its highly collapsed core. In addition, the stars at the cluster's center display high central velocity dispersion. In the early 2000s, the Hubble Space Telescope provided mounting evidence in support of this claim, with estimates of the black hole's mass ranging from between 500 and 3,200 solar masses. But, in 2012, Brian Murphey and his colleagues developed a set of dynamically evolving models for M15 to directly address the issue of whether a central black hole is required to fit Hubble Space Telescope (HST) observations of the cluster's stellar spatial distribution and kinematics (*The Astrophysical Journal*, vol. 732, p. 67, 2011.) The researchers found a central black hole is not needed. Their models contain a substantial population of neutron stars and massive white dwarfs that range in mass up to 1.2 solar masses. "The combined contribution by the massive white dwarfs and neutron stars provides the gravitational potential needed to reproduce HST measurements of the central velocity dispersion profile," they say. So the jury is still out on this lingering subject.

M15 also displays dark patches. One obvious dark feature appears next to a detached string of stars on the northeast edge of the cluster's inner shell. Another tighter arc of stars to the east of the nucleus makes the entire central region appear warped in that direction, something noticed by d'Arrest in the nineteenth century. According to the Rev. Thomas W. Webb (1807–1885), "Buffham, with a 9-inch [mirror] finds a dark patch near the middle with two faint, dark lines or rifts like those in M13." I did not notice these. My view is more like the one Isaac Roberts described at the turn of the twentieth century, in which stars are arranged "in curves, lines and patterns."

The cluster has an unusual resident, a 14th-magnitude planetary nebula, Pease 1, on its northeast side. In fact, M15 is one of two globular clusters known to contain a planetary; the other is a 10″ × 7″ object, GJJC-1, in M22. F. G. Pease discovered the M15 planetary in 1928. But measuring only 1′ in diameter and awash in a sea of stars, Pease 1 is very difficult

M16

The Ghost, Eagle Nebula, **or** *Star Queen Nebula*
NGC 6611 (cluster)
IC 4703 (nebula)
Type: Open Cluster and Emission Nebula
Con: Serpens (Cauda)

RA: $18^h18.75^m$ (cluster)
Dec: –13°48′ (cluster)
Mag: 6.0 (cluster)
Dim: 35′ × 28′ (nebula)
Diam: 8′ (cluster)
Dist: ~5,700 light-years
Disc: Philippe Loys de Chéseaux, 1745 or 1746 (cluster); Charles Messier, 1764 (nebula)

MESSIER: [Observed June 3, 1764] Cluster of faint stars, mingled with faint luminosity, close to the tail of Serpens, not far from the parallel of ζ in that constellation. With a small telescope this cluster appears to be a nebula.

NGC: Cluster, at least 100 bright and faint stars.

IC: Bright, extremely large, cluster in the middle with 16 stars involved.

LOCATED ABOUT 2.5° WEST-NORTHWEST of magnitude 4.7 Gamma (γ) Scuti, in the spine of the Milky Way, M16 is a most tantalizing sight – a fan of nebulosity (about 60 light-years in true physical extent) centered on a loosely scattered cluster of stars that spans 13 light-years.

Swiss observer Philippe Loys de Chéseaux discovered M16 in 1745 or 1746, describing it simply as a "cluster of stars" through his simple telescopes. Messier independently swept up the cluster in 1764 and found it "mingled with faint luminosity." While it's clear that Messier saw the nebulosity, his terminology for it is interesting. Note that he goes on to say that through a "small telescope this cluster appears to be a nebula." In other words, Messier believed that nebulae were unresolved clusters of stars. Since his larger telescope resolved the "nebula" into a cluster, there remained a certain unexplained "luminosity"; at least that's one interpretation.

Note, too, that the nebula does not have an NGC designation. For some reason, the nebula escaped the gaze of William and John Herschel, so it appears only as a star cluster in the 1888 *New General Catalogue*, with the designation NGC 6611. Only after Welsh amateur astronomer Isaac Roberts (1829–1904) imaged it nearly a decade later did the nebula finally appear in the 1908 2nd *Index Catalogue* as IC 4703.

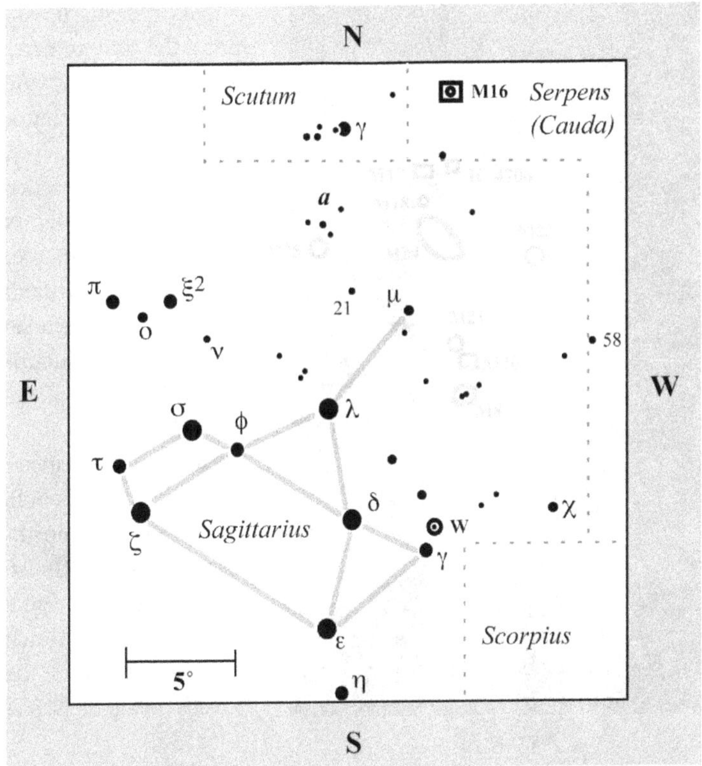

To find this amazing complex, start by using the chart at the back of the book to locate 4th-magnitude Mu (μ) Sagittarii, then look about 2 1/2° north for the Small Sagittarius Star Cloud (M24). If you have binoculars, place M24 at the bottom of the field of view and you'll see two small, diffuse glows: M17 (the Swan Nebula), about 2° north-northeast from the center of M24; and M16, about 2° north-northwest of M17. And, as mentioned, M16 also lies about 3° west-northwest of magnitude 4.7 Gamma (γ) Scuti.

As photographic emulsions became increasingly sensitive to light, the nebula began to reveal a cornucopia of elaborate detail. In his *Celestial Handbook*, Robert Burnham Jr. calls its photographic appearance "one of the great masterworks of the heavens," and his description of the nebulous opulence bears repeating:

> Thrusting boldly into the heart of the cloud rises a huge pinnacle like a cosmic mountain, the celestial throne of the Star Queen herself, wonderfully outlined in silhouette against the glowing fire-mist. ... In the vast reaches of the Universe, modern telescopes reveal many vistas of unearthly beauty and wonder, but none, perhaps, which so perfectly evokes the very essence of celestial vastness and splendor, indefinable strangeness and mystery, the instinctive recognition of a vast cosmic drama being enacted, of a supreme masterwork of art being shown.

Visible as a hazy patch with the naked eye, the M16 complex occupies the northern end of a large S-shaped asterism of stars that, if viewed together at 23× through the 4-inch with north up, looks very much like a seahorse. In fact, the emission nebula itself, with north up, resembles a hand puppet, or a cartoon-like ghost flying through the night with outstretched arms and bulging eyes (thus my nickname the Ghost for the nebulosity).

At the center of the nebula, a stalagmite of blackness (20 trillion miles long) to the northwest and a wedge of darkness to the north together create one of the most mystifying sights visible in galactic nebulae – tidal waves of dark matter that appear to be scrubbing away the bright coast of gas with their heavy ebb and flow. Earlier this century, astronomer Fred Hoyle believed that we were witnessing the expansion of hot gas into cooler gas, with

is overexposed. If you spend the time to carefully scrutinize the southeastern extremities of the nebula, you might notice an absence of gas there – Burnham's "cosmic mountain." Can you trace out the faint wisps of dark nebulae between the mountain and the top of the throne?

The young, hot cluster illuminating the nebula dominates the northwest part of the Ghost. There is a challenging double for the 4-inch near the center of the Ghost's head. It is just

the hot gas erupting like an exploding bomb. And that is largely what recent images from the Hubble Space Telescope, such as the one shown here, have revealed: ebony pillars trillions of miles long dramatically being boiled away by the ultraviolet radiation of nearby stars. Shooting off the column tips are tapered nodules of gas (EGGs), each as wide as the Solar System, where star formation is occurring.

The Ghost and its associated dark clouds are an extreme visual challenge in small telescopes, unlike the dark channels and swirls so clearly visible in M8. Do take up the challenge, however. From Hawaii, the 4-inch can easily pick out the location of the dark northern wedge and reveal a definite V-shaped hole in the heart of the Ghost (the top of the Queen and her throne). A curious bright patch of nebulosity or unresolved cluster of stars lies just to the south of that V (see the drawing). In most photographs I have seen, this region

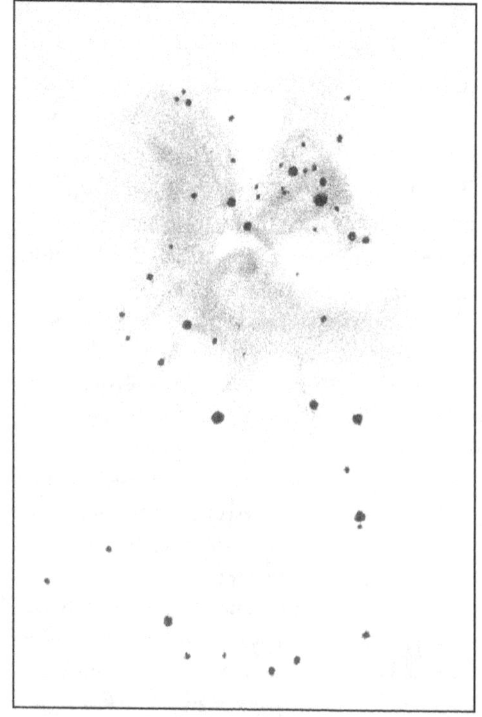

below its left "eye," in a faint stream of nebulosity. In recent papers, the age of the M16 cluster has been given as between 2 million and 6 million years. The cluster harbors 93 young and hot OB stars, together with a population of low-mass stars. Eighty-five percent of the O stars of NGC 6611 have been detected in x-rays. In a 2012 publication of the American Astronomical Society (AAS Meeting No. 219), Guido De Marchi (ESA, Netherlands) and his colleagues published their discovery of pre-main-sequence stars with sizable active accretion disks with ages ranging from 8 million to 30 million years, with a median value of 15 million years. "This is the largest homogeneous sample to date of Galactic [pre-main-sequence] stars considerably older than 10 [million years] that are still actively accreting from a circumstellar disc. ... These values imply a characteristic exponential lifetime of 5 [million years] for disc dissipation," the authors say.

M17

Omega, Horseshoe, **or** *Swan Nebula*
NGC 6618
Type: Emission Nebula and Open Cluster
Con: Sagittarius

RA: $18^h20.8^m$
Dec: –16°11′
Mag: 6.0 (cluster)
Dim: 20′ × 15′ (nebula)
Diam: 27′ (cluster)
Dist: ~5,500 light-years
Disc: Philippe Loys de Chéseaux, 1745 or 1746 (nebula); William Herschel, 1785 (cluster)?

MESSIER: [Observed June 3, 1764] Streak of light without stars, five to six minutes long, spindle-shaped, and rather similar to that in the belt of Andromeda, but very faint. There are two telescopic stars nearby, lying parallel to the equator. Under a good sky, this nebula can be seen very clearly with a simple three-and-a-half-foot refractor. Observed again 22 March 1781.

NGC: Magnificent, bright, extremely large, extremely irregular in shape, hooked like a "2."

MORE MESSIER OBJECTS (15) ARE located in Sagittarius than in any other constellation. And for good reason. The mythical Archer stands vigil in the direction of the center of our Galaxy, the area most crowded with stars, dust, and gas. No wonder, then, that this parcel of sky yields the greatest variety and concentration of galactic star clusters and nebulae, including M17, which is a combination of both. With the exception of the Orion Nebula (M42), M17 is the brightest galactic nebula visible to observers at mid-northern latitudes.

Philippe Loys de Chéseaux discovered it in 1745 or 1746, calling it a "nebula" in the "perfect form of a ray or the tail of a comet." William Herschel first detected stars within the nebulosity, but he suspected they were foreground objects. But M17 does have a loose cluster of stars associated with it.

The Swan, as this emission nebula is often called, can be seen with the naked eye as a 6th-magnitude patch of light 2 1/2° south of M16 and 2 1/2° southwest of Gamma (γ) Scuti. It is one of the most massive molecular clouds and most luminous H II regions in the Galaxy, spanning some 30 light-years of space in its longest direction. The H II region is in fact a "blister" erupting from the side of a giant molecular cloud.

some material to stream away from the surface, creating the glowing veil of even hotter gas that masks background structures in the accompanying image.

In infrared light, the NASA Spitzer Telescope has revealed a dragon-shaped cloud of dust (called M17 SWex) forming stars at a furious rate. This youngest episode of star formation is playing out as the dusty dragon passes through the Sagittarius spiral arm of the Milky Way, "touching off a galactic 'domino effect,'" as the NASA image release put it. The release noted that, "Over time, this area will flare up like the bright M17 nebula, glowing in the light of young, massive stars." Fifteen very young OB stars have since been identified in the nebula; some still have circumstellar disks, while others appear to have cleared out their surroundings.

The Hubble Space Telescope imaged a swath of the nebula's interior 3 light-years wide, revealing it to be a turbulent sea of nebulous splendor with a crib of newborn stars swaddled in nascent blankets of glowing gas. Intense ultraviolet radiation fleeing from these hot, young stars not only causes the nebula to fluoresce but also eats away at the dark waves of cold gas from which these stars formed. The intense heat and pressure cause

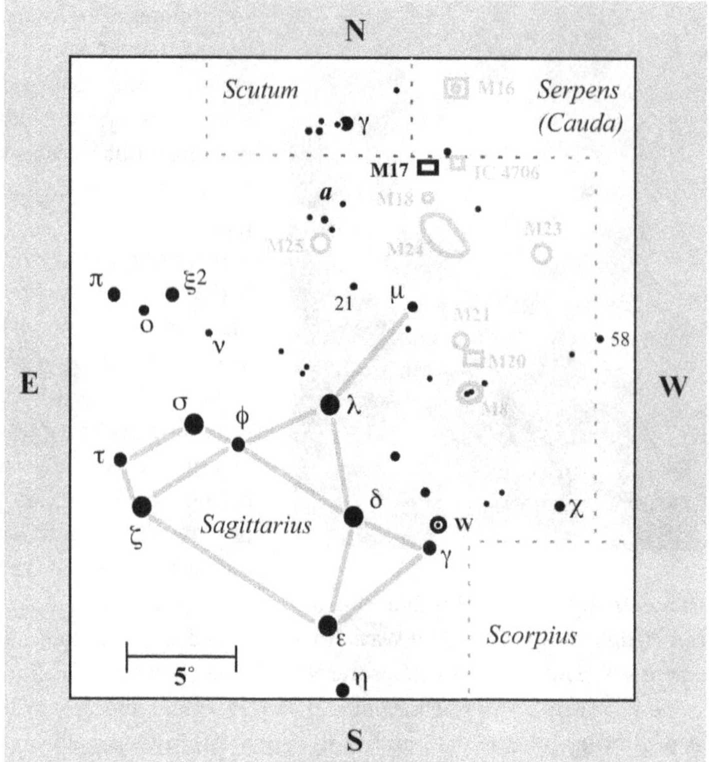

southwestern end by a faint hook (the swan's curved neck). But with a glance at 23× through the 4-inch, M17 first appears as a long blaze of gas and starlight slashed by dark lanes of obscuring matter. Consider that this bar spans 12 light-years, or 72 trillion miles, of space. Camille Flammarion likened this lengthy feature to a "smoke-drift, fantastically wreathed by the wind," a wonderfully believable impression.

The faint hook of the swan's neck should materialize soon after you survey the bar. Stay with low power and let your eye drift across the field in all directions. The swan appears to be swimming in a faint mist rising from a black pool. With medium power, concentrate on the southern half of the swan and you might see long vapors rising off its back and neck. A prominent "check mark" of dark nebulosity forms the crook in the swan's neck. (This is not to be confused with the bright nebula, which also has the appearance of a check mark.) Now use high power to look at the star marking the western end of the bright bar. Immediately encompassing it are four bright knots of gas forming a Celtic cross.

The ghostly hook has given rise to the nebula's other nicknames – the Omega Nebula, because of its resemblance to the Greek letter "omega" (V), or the Horseshoe Nebula – names introduced by Smyth in the nineteenth century. Others have commented on the

To find the ghostly Swan, start by using the chart at the back of the book to locate 4th-magnitude Mu (μ) Sagittarii. Then look about 2 1/2° north for the Small Sagittarius Star Cloud (M24). If you have binoculars, place M24 at the bottom of the field of view and you'll see M17 about 2° north-northeast from M24's center.

Through the 3-inch at low power, the Swan Nebula is a stunning sight: a little bar of light with a loop and dark hole to the west. Fainter material washes to the south, blending in with a patchwork of starlight. At 72×, the object stands out boldly. In his 1889 *Celestial Handbook*, George F. Chambers was the first to compare this peculiar-shaped nebulosity to a swan floating on water. He was alluding to M17's brightest features, namely a long bar of gas (the swan's body) topped on the

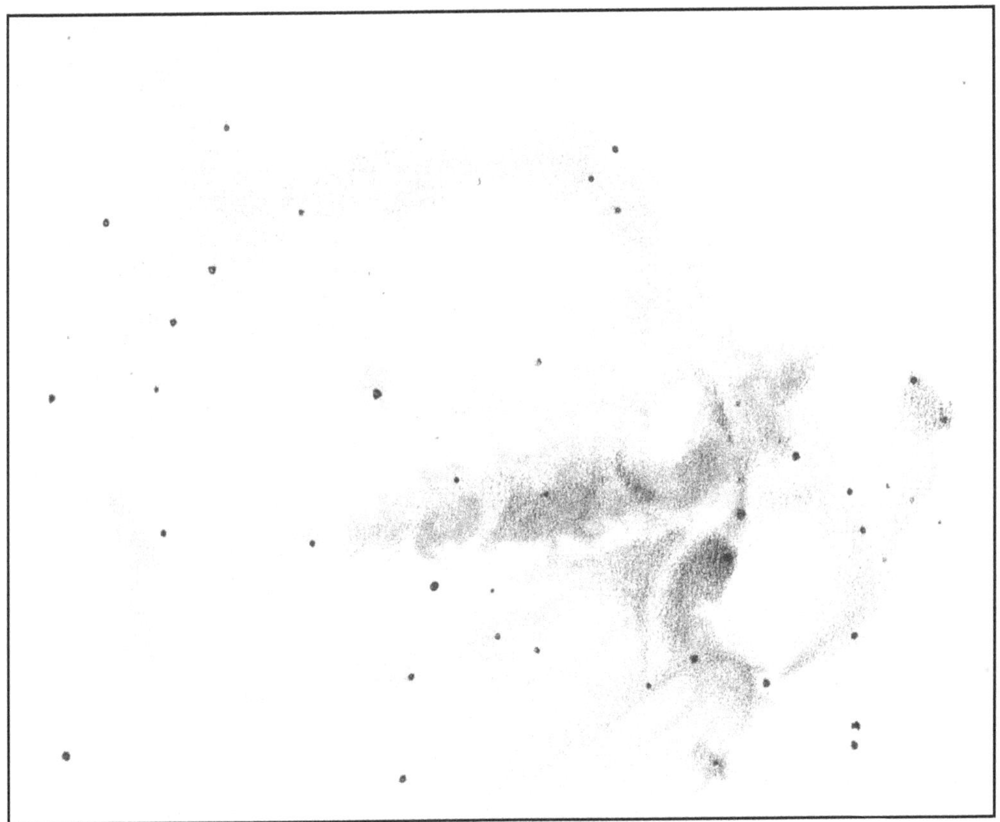

nebula's resemblance to the number 2. But, as my drawing shows, the 2, the Horseshoe, or the Swan is but a part of a vast and elaborate network of gas and dust. It takes a discerning eye and a combination of moderate and high powers to bring out the finest details within the nebulous regions.

For example, notice how the hook actually forms a complete loop, the northernmost portion being the most difficult to make out, requiring averted vision and patience. Note also the apparent absence of starlight within the loop. This is probably caused by a cloud of obscuring matter. Certainly this is the darkest region in the entire nebula; it looks like an ink stain in very-long-exposure photographs. If you return to low power and really study the faint envelope of nebulosity surrounding the swan, which measures 20′ × 15′, you will discover that it is not symmetric. It ends abruptly to the west of the swan's head and the "black hole." It's as if the swan has sailed westward across a horseshoe-shaped pond to a shore of black sand. Curiously, about one swan diameter to the northwest is the tiny glow of emission nebula IC 4706. Could these glows be related, being separated only visually by a swath of foreground dark nebulosity? If so, you can imagine the swan looking across this dark gulf at its isolated cygnet.

Unlike the obvious star clusters found within M8 and M16, the one inside M17 appears (visually through small telescopes) to be nothing more than a weak cast of 35 stars of 9th magnitude or fainter. But a recent study of the M17 H II region with the Chandra X-ray Observatory has revealed 886 x-ray point sources centrally concentrated in the field, implying a total population of 8,000–10,000 stars in NGC 6618, the M17 ionizing cluster. And wider-field near-infrared imaging has provided evidence of ongoing star formation in the molecular cloud. It's estimated that at least 1,250 stars are currently forming in the M17 giant molecular cloud. According to Matthew Povich (University of Wisconsin, Madison) and his colleagues, the cloud continues to form stars at greater than 25 percent of the star-formation rate that produced the massive million-year-young NGC 6618 cluster (*The Astrophysical Journal Letters*, Vol. 714, pp. L285–L289, 2010).

M18

Black Swan
NGC 6613
Type: Open Cluster
Con: Sagittarius

RA: $18^h20.0^m$
Dec: -17°06′
Mag: 6.9
Diam: 7′
Dist: ~3,900 light-years
Disc: Charles Messier, 1764

MESSIER: [Observed June 3, 1764] Cluster of faint stars, slightly below the previous one, number 17, surrounded by faint nebulosity. This cluster is less obvious than the penultimate one, number 16. With a simple three-and-a-half-foot refractor, this cluster looks like a nebula, but with a good telescope only stars are visible.

NGC: Cluster, poor, very little compressed.

M18 IS A SPARSELY POPULATED OPEN star cluster only 1° south of M17, near the extreme northern edge of the Small Sagittarius Star Cloud (M24). Absorption along our line of sight is quite appreciable, being 1.4 magnitudes in front of the cluster. M18 lies in the Sagittarius spiral arm, about 70 light-years from the Galactic plane, and has an age of ~30 million years. Its brightest main-sequence member is of spectral type B.

In his *Celestial Handbook*, the late Robert Burnham Jr. calls this loose, 7′-wide gathering of stars one of the most neglected Messier objects, which is sometimes omitted from lists of galactic star clusters. Although credited with having only 40 members in the telescopic field, faint background stars surround the cluster and add to the visual pleasure. It has a Trumpler class of II3pn – meaning a poorly populated detached cluster of bright and faint stars with little central concentration; Trumpler also suspected nebulosity associated with it. In true physical extent, M18 spans only 8 light-years of space.

At 23× through the 4-inch, M18 shares the field with M17, and the two are separated by a large but faint S-shaped string of similarly bright stars. I call M18 the Black Swan because the main body of bright stars forms a pattern reminiscent of the bright nebulosity in M17, but unlike the rather sleepy-looking M17 swan, the M18 swan is raising one of its large wings – a black wing (a region devoid of bright stars) outlined by five roughly 10th-magnitude stars. Black Swan is a double entendre because it also refers to the fact that this cluster is often ignored or neglected.

Contrary to what is sometimes stated – that M18 is best viewed at low power, when

M18

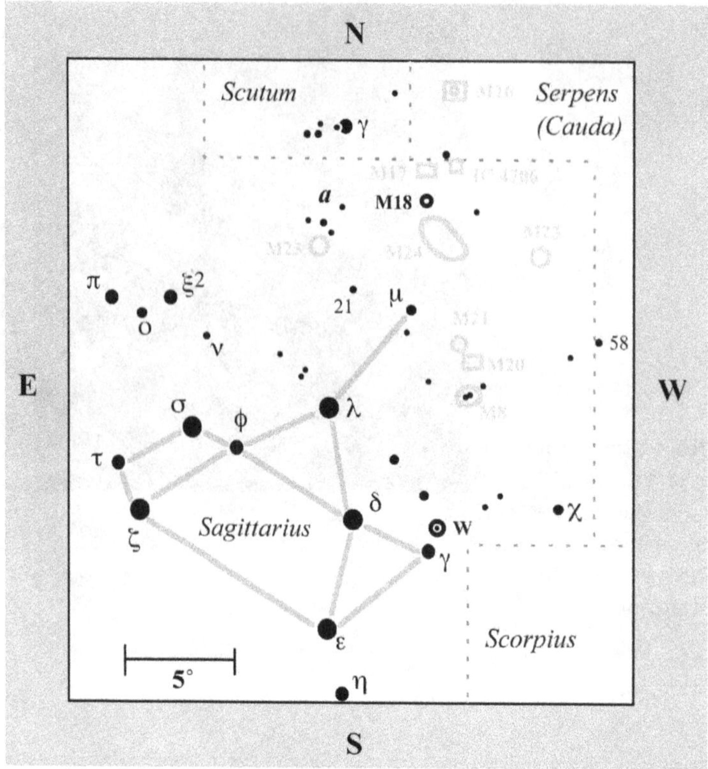

it appears small and concentrated – I find the cluster comes to life at 130×, when the faint background stars of this rich Milky Way region enhance the view of the otherwise subtle grouping of 10th- to 11th-magnitude stars. A nice double star within it also adds a bit of sparkle. Admittedly, low power places M18 in a very favorable light, with the Big Swan (M17) to the north, a rich swath of nebulosity (IC 4706) to the northwest, and the Small Sagittarius Star Cloud (M24) to the south. This tiny, seemingly insignificant cloud of meager starlight is surrounded by dazzling cosmic giants.

Although I did not notice any of Trumpler's nebulosity visually, photographic plates made with the 48-inch Schmidt telescope on Palomar Mountain in California reveal that the cluster is bathed in a faint nebulous glow. I wonder what size telescope is required to see this. Meanwhile, can you make out the wishbone pattern of stars to the southwest (the swan's head) and the dim stream of 12th- to 13th-magnitude stars outlining the southern tip of the upraised wing?

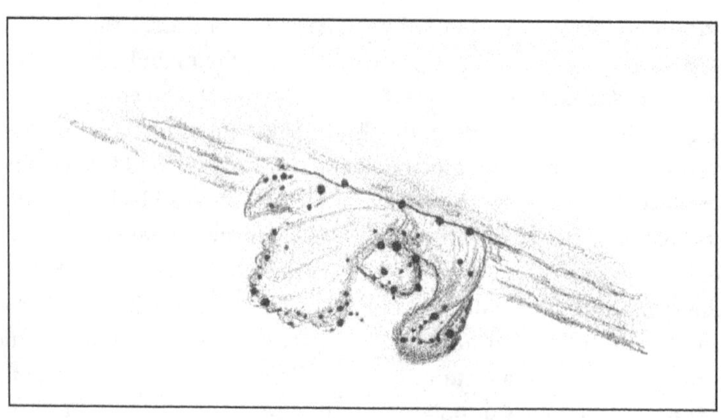

M19

NGC 6273
Type: Globular Cluster
Con: Ophiuchus

RA: $17^h02.6^m$
Dec: –26°16′
Mag: 6.8
Diam: 17′
Dist: ~28,700 light-years
Disc: Charles Messier, 1764

MESSIER: [Observed June 5, 1764] Nebula without stars, on the same parallel as Antares, between Scorpius and the right foot of Ophiuchus. This nebula is circular; it is clearly visible with a simple three-and-a-half-foot refractor. The known star closest to this nebula is sixth-magnitude Flamsteed 28 Ophiuchi. Observed again 22 March 1781.

NGC: Globular, very bright, large, round, very compressed in the middle, well resolved. It consists of stars of 16th magnitude and fainter.

DESPITE WHAT THE NGC'S description says, M19 in Ophiuchus is a challenging object to resolve. Although the cluster shines with a total magnitude of 6.8, the average brightness of its most luminous stars is about magnitude 14. From dark skies, keen-eyed observers can spot M19 with the naked eye nearly 3° west of 4th-magnitude 36 Ophiuchi and 1 1/4° south-southeast of the 6th-magnitude pairing of stars 26 Ophiuchi.

M19 is the most elongated globular cluster known. Harvard astronomer Harlow Shapley (1885–1972) estimated that the cluster had a degree of ellipticity of 6 on a scale of 10 and contained twice as many stars on its major axis as on its minor axis. The reason for this great degree of ellipticity appears to be the cluster's extreme proximity to the Galactic center (only ~5,500 light-years from it) and only 9° from the Galactic plane. The cluster has 1/54 as much metal per unit hydrogen as the Sun and has an integrated spectral type of F7. We see it approaching us at about 85 miles per second.

To find it, first locate brilliant Alpha (α) Scorpii (Antares) in the heart of the Scorpion and then look a little more than 10° east-northeast for magnitude 3.5 Theta (θ) Ophiuchi and 36 Ophiuchi about 2° to its southwest. Again, M19 is just a little less than 3° west of 36 Ophiuchi. Use binoculars first to spot the small glow.

Even a glimpse through the 4-inch telescope at low power reveals the cluster's ellipticity,

M19

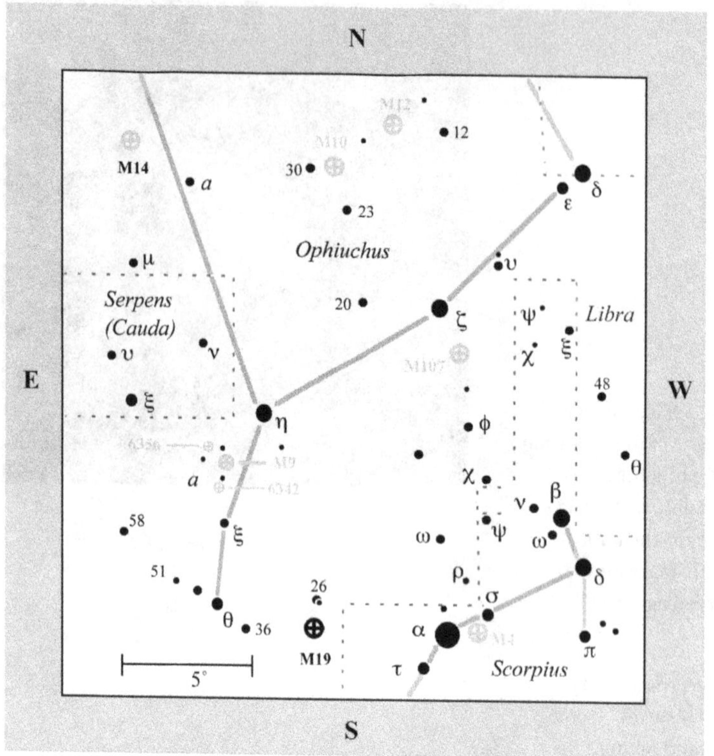

though the telescope probably only reveals half of the ~140-light-year-wide orb. At 72×, there is not much difference: an unresolved haze gradually diffuses out from a bright stellar nucleus. High power shows about a half dozen stars, most of which hug the outer fringe of the cluster's halo at the main cardinal directions. The cluster has a horizontal-branch magnitude of 17, so large telescopes are required to resolve it well. M19 is oblate north to south, though the multitude of unresolved stars surrounding the cluster's core favor the west. Noteworthy are some spiral-like arms of stars, which in the south appear to curve counterclockwise, whereas those in the north curve clockwise. The core is highly concentrated, having a diameter of only 1′.

The globular is also colorful: I see a topaz core surrounded by swirls of blue smoke. Admiral William Henry Smyth found the stars to be of creamy white tinge, and slightly lustrous in the cluster's center. Do you see the numerous dark patches that stain the cluster, making it appear mud-splattered? I find these patches particularly intriguing because the cluster resides in a bright gulf of starlight surrounded by the vast naked-eye rivers of darkness that cut through the Scorpius Milky Way region. Could these stains be tiny black clouds silhouetted against the more distant globular?

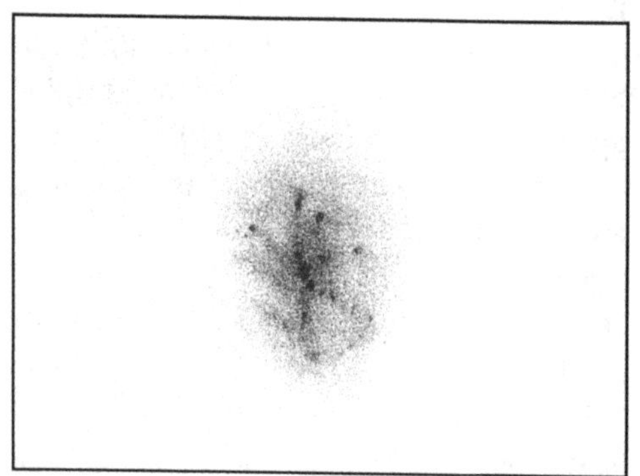

M20

Trifid Nebula **or** *The Clover*
NGC 6514
Type: Nebula and Cluster
Con: Sagittarius

RA: $18^h02.5^m$
Dec: –23°02′
Mag: 6.3 (cluster)
Dim: 20′ × 20′ (nebula)
Diam: 28′ (cluster)
Dist: ~5,400 light-years
Disc: Charles Messier, 1764; and possibly Guillaume Le Gentil around 1747

MESSIER: [Observed June 5, 1764] Star cluster, slightly above the ecliptic, between the bow of Sagittarius and the right foot of Ophiuchus. Observed again 22 March 1781.

NGC: A magnificent object, very large and bright, trifid, a double star involved.

M21

NGC 6531
Type: Open Cluster
Con: Sagittarius

RA: $18^h04.2^m$
Dec: –22°30′
Mag: 5.9
Diam: 16′
Dist:~ 4,200 light-years
Disc: Charles Messier, 1764

MESSIER: [Observed June 5, 1761] Star cluster close to the previous one [M20]. The known star closest to these two clusters is seventh-magnitude Flamsteed 11 Sagittarii. The stars in these two clusters are of eighth and ninth magnitude, and are surrounded by nebulosity.

NGC: Cluster, pretty rich, little compressed, stars from magnitude 9 to 12.

M21

The 2°-wide expanse of Milky Way encompassing M8, M20, and M21 is the most dramatic Messier field in the entire sky. At 23×, these objects and several other clusters and nebulous patches (both bright and dark) fill the field. Such a tight gathering of nebulous splendor might be nothing more than a chance alignment. M8 and M20 are possibly part of the same complex; certainly their distances (5,200 and 5,000 light-years, respectively) suggest they could be. And even though M20 and M21 are separated by about a thousand light-years, I cannot discuss one without also discussing the other, because they look like they belong to the same celestial microcosm.

If you look at this region – which is about 5° (roughly two finger widths) west-northwest of 3rd-magnitude Lambda (λ) Sagittarii – with the naked eye, you should immediately see two hazes (M8 and M20) making an arc with the 5th-magnitude star 4 Sagittarii 1/2° to the west. Use binoculars or a wide field of view to see the "fishhook" of Milky Way (NGC 6526) between M20 and M8. Now concentrate on the interior of the fishhook. Do you see a "black hole"?

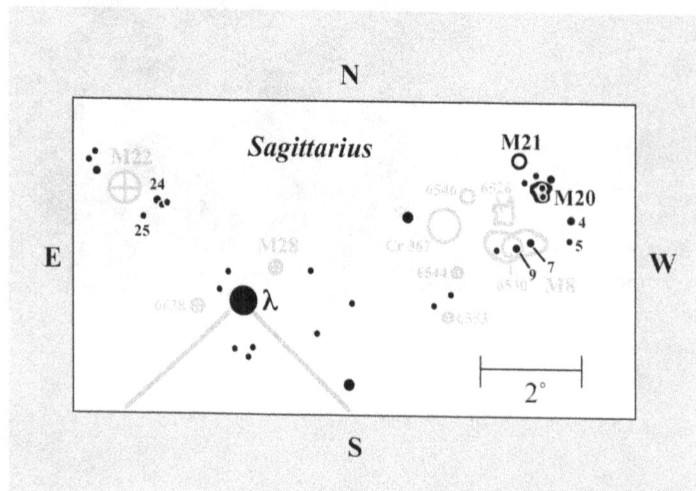

My guess is that Messier and his contemporaries must have first noticed these hazes with their naked eyes (as they probably did with M24, M42, M44, M45, etc.). But when Messier turned his telescopes on M20, he noticed not a nebula but a cluster of stars. Only in his description of M21 does he say that both M21 and M20 are enveloped in nebulosity. Messier never resolved the tiny glow of the Trifid in his small telescope, and the cluster he referred to is in the southern portion of the "cruciform group" cited by Webb.

Webb's Cross is a fine collection of roughly a half dozen 6th- and 7th-magnitude stars (shaped like a cross with warped arms) that appears to the naked eye as a single nebulous patch; M21 marks the northern tip of the cross and M20 the base. So I can understand Messier's confusion: to the naked eye, these two clusters appear to be immersed in a cocoon of galactic gas, which vanished when he employed his optically inferior telescope. By the way, I prefer to recognize Webb's Cross as an asterism, like the crossbow of stars in M8, though I wonder if it is really a cluster.

There is a minor mystery here. Messier notes that the brightest star in M21 is 11 Sagittarii, but that star lies 2° to the southeast of the presently recognized cluster! He obviously misidentified the star, but someone who likes to model history from shreds of evidence might have fun trying to decipher exactly what happened here. I precessed Messier's 1764 positions for M20 and M21, and everything seems to be in order.

John Herschel is credited with being the first

to call M20 the Trifid Nebula. A small but noble cloud of glowing gas, M20 looks more like a four-leaf clover glazed by frost than a tri-lobed nebula. The name "trifid" is very deceiving (especially to anyone who first sees it in a photograph); it refers only to the nebula's three brightest southern portions. Do you find it strange that John Herschel would have called this nebula a trifid when his father, William, had already catalogued it in four portions? Anyway, I prefer the clover metaphor for two reasons. First, M20 has the shape of one, and second, because the fourth lobe is faint in small telescopes, you should feel lucky if you glimpse it!

M20 is an isolated H II region comprised of two parts: (1) a remarkable reddish nebula whose gas is ionized by an O7 star (HD 164492; HN 40) and trisected by obscuring dust lanes, and (2) a blue reflection nebula in the north. The region lies about 5,400 light-years distant and spans about 30 light-years. Different stages of star formation have been detected at various wavelengths, as well as optical jets, mid- and far-infrared protostars, and near-infrared young stellar objects.

The Trifid is a wonderful example of a massive star-forming region in a turbulent, filamentary molecular cloud. As B. Lefloch (Laboratoire d'Astrophysique de l'Observatoire de Grenoble) and colleagues explain in a 2008 paper in *Astronomy and Astrophysics* (vol. 489, p. 157), the cloud material is distributed in fragmented dense gas filaments with sizes ranging from about 3 to 30 light-years. One massive filament connects M20 to a nearby supernova remnant known as W28. "These filaments pre-exist the formation of the Trifid and were originally self-gravitating," the authors say. "The fragments produced are very massive (100 M_{sun} or more) and are the progenitors of the cometary globules observed at the border of the HII region." They identified 33 cores, 16 of which are currently forming stars. They go on to propose that W28 could have triggered the formation of protostellar clusters in nearby dense cores of the Trifid.

Telescopically, the leaves of the Trifid's clover are clearly separated by dark lanes of obscuring matter emanating from a central well of darkness. Use high magnification to peer into the well and, given stable atmospheric conditions, you might see HN 40 – a discrete triple star. The two brightest members (magnitude 7.6 and 8.7) display a striking color difference. I see the brighter one, the O7 star, for some reason with a mustard hue, while the other is a dying ember, charcoal-colored with a spark of red. The third star shines at magnitude 10.7 and looks colorless in the 4-inch.

While HN 40 appears to be M20's main source of illumination, other hot stars,

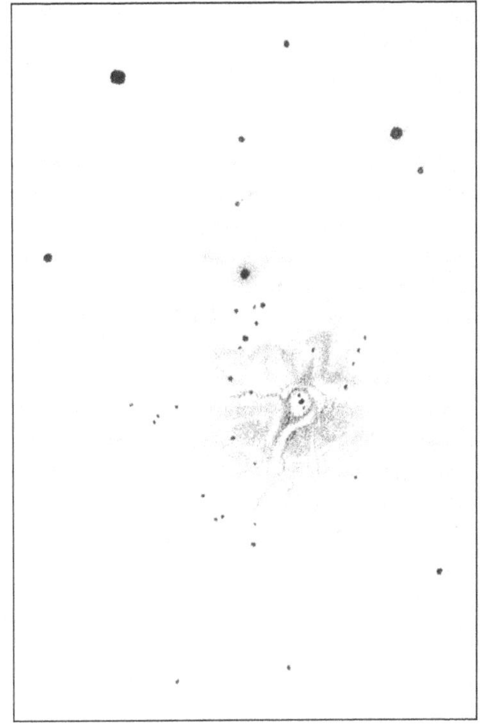

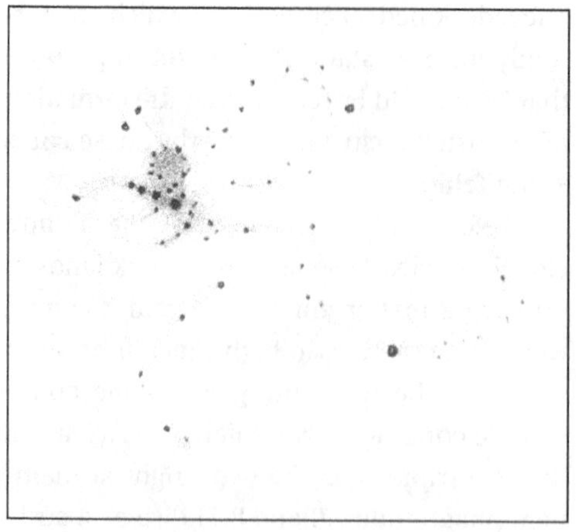

cloaked by dark matter, could contribute energy as well. As K. Torii (Nagoya University, Japan) and colleagues report in a 2011 paper in *Astrophysical Journal* (vol. 738, p. 46), their large-scale study of the molecular clouds toward the Trifid Nebula revealed two molecular components at separate velocities toward the center. They identify the two clouds as the parent clouds of the first-generation stars in M20, saying: "We argue that the formation of the first generation stars, including the main ionizing O7.5 star, was triggered by the collision between the two clouds in a short timescale of ~1 million years, a second example alongside Westerlund 2, where a super-star cluster may have been formed due to cloud–cloud collision triggering."

When I use 23× in the 4-inch and let my imagination fly, I do see a trifid nebula, though it is not the one Herschel refers to. The Trifid itself is one leaf of that; a ball of gas surrounds the magnitude 7.4 star just 10′ to the north (the second leaf). Now, use averted vision to follow the streamers of celestial fog that waft to the west. Do you see where more gas is illuminated by the 6th-magnitude star marking the western arm of Webb's Cross? All the nebulosity associated with the Trifid spans a total apparent size two-thirds that of the full moon.

Less than a degree northeast of the Clover glimmers M21, a bright spread of young stars, about 50 of which are easily visible in small telescopes. The cluster is located in the Galactic disk near the Sagittarius star-forming region and the Trifid Nebula, though it is not associated with any nebulosity. The cluster lies about 4,200 light-years distant and has many early B-type stars. In 2001, Byeong-Gon Park and colleagues did a photometric study of 56 main-sequence (MS) members of M21 and derived the MS turnoff age of the cluster as 7.5 million years, the pre-main-sequence age of the cluster as between 3 million and 3.5 million years, and the age spread of the cluster as about 4 million years (*Journal of the Korean Astronomical Society*, Vol. 34, pp. 149–155, 2001).

Although it is listed as covering an area of 16′ (~20 light-years), M21's boundaries are poorly defined. Use high power to pierce the hazy heart of M21, the haze being an illusion created by numerous unresolved stars in its strong central concentration. Beyond that, do you see the cluster's "spiral" structure? I can follow arms of stars flowing away from the central triangle of stars. One of these stars looks blue with a hint of yellow, while another looks blue with a hint of red. Overall, the cluster's stars appear white with just tinges of color. Use low power to appreciate the linear alignment of many of its fainter members. With a little flight of fancy, I see the hazy heart with its flashes of star color as a firefly, and the swirls of stars as its whimsical path across the heavens.

Here is a naked-eye challenge for you. Can you resolve Webb's Cross and separate M21 from M20?

M22

Great Sagittarius Cluster **or** *Crackerjack Cluster*
NGC 6656
Type: Globular Cluster
Con: Sagittarius

RA: $18^h36.4^m$
Dec: –23°54′
Mag: 5.2
Diam: 32′
Dist: ~10,400 light-years
Disc: John Hevelius, apparently noticed it prior to 1665

MESSIER: [Observed June 5, 1764] Nebula, below the ecliptic, between the head and bow of Sagittarius, close to the seventh-magnitude star Flamsteed 25 Sagittarii. This nebula is circular, does not contain any stars, and is clearly visible in a simple three-and-a-half-foot refractor. The star λ Sagittarii was used to determine its position. Abraham Ihle, the German, discovered it in 1665 when observing Saturn. M. le Gentil observed it in 1747, and published a drawing, *Mémoires de l'Académie 1759*, page 470. Observed again 22 March 1781. It is plotted in the English *Atlas Celéste.*

NGC: Very remarkable, globular cluster, very bright, very large, round, very rich, very much compressed, stars from 11th to 15th magnitude.

WITHOUT QUESTION, M22 SHOULD BE called the "Great Sagittarius Cluster." It is a bonfire of a half-million stars that blazes at magnitude 5.2 and measures 32′ across – about the apparent diameter of the full moon. Among globulars, it ranks third only to Omega Centauri (Southern Gem 62) and 47 Tucanae (Southern Gem 2) in brightness and apparent size. The late Robert Burnham Jr. quipped that J. R. R. Tolkien penned an exquisite description of M22 in *The Hobbit* when he spoke of the fabulous jewel called the Arkenstone of Thrain: "It was as if a globe had been filled with moonlight and hung before them in a net woven of the glint of frosty stars."

M22 lies only 2 1/2° northeast of the K-type star Lambda (λ) Sagittarii, which marks the tip of the celestial teapot. It has the distinction of being the first globular cluster to be discovered. Edmond Halley (1656–1742) ascribed its discovery in 1665 to the obscure German astronomer Abraham Ihle, who, Halley said, found it "whilst he attended the Motion of *Saturn* then near its *Aphelion.*" But, according to the late Kenneth Glyn Jones, "Cassini II reported that, according to [Gottfried] Kirch [(1639–1710)] Hevelius had seen M22 'a long time before.'" In August 1747, Guillaume Le Gentil of France made a drawing of the object as seen through his telescope of 18-foot focal length, seeing it as "very irregular, long-haired and spreading some kind of rays of light all around its diameter."

Of these rays, Admiral Smyth commented that "I cannot say that I clearly understand how or why his telescope exhibited these [rays of light]." But I believe the rays might have been long streamers of unresolved stars radiating from the cluster – the nature of which was not yet known to these early observers. Then again, Charles Messier made no mention of them when he reviewed the object on June 5, 1764. It took one of William Herschel's powerful reflectors to resolve the object into stars, though Herschel, too, made no mention of the "rays," referring to the object's form as merely a round cluster.

At a distance of only 10,400 light-years, M22 is among the closest globular clusters – closer than any visible in the northern sky. Its location 16,000 light-years from the Galactic center and 9° south of the Galactic plane, however, diminishes its true radiance, because of the intervening dust. In true physical extent, the cluster spans about 70 light-years of space.

In a 2011 paper in *Astronomy and Astrophysics* (vol. 532, p. 8), A. F. Marino (Max-Planck-Institut für Astrophysik) and colleagues present a detailed chemical composition analysis of 35 red giant stars in M22. High-resolution spectra for this study were obtained at five observatories and analyzed in a uniform manner. One of the principal results of this study is the finding of a substantial star-to-star metallicity scatter (1/100 to 1/40 as much metal on average as the Sun), which provides evidence that many, and maybe all, globular clusters host multiple stellar populations.

To the naked eye, M22 is a tight bead of light lying about 2 1/2° northeast of Lambda Sagittarii. Through 10 × 50 binoculars, it is a diffuse cometary glow with a bright core and fluffy halo. At 23× through the 4-inch, M22 appears more oval than round, with the major axis running slightly east of north. Let your eye relax, and with time you will notice that the oval glow is surrounded by a faint halo of unresolved stars that has prominent extensions to the north and south. A powerful orange star punctuates the northern arm.

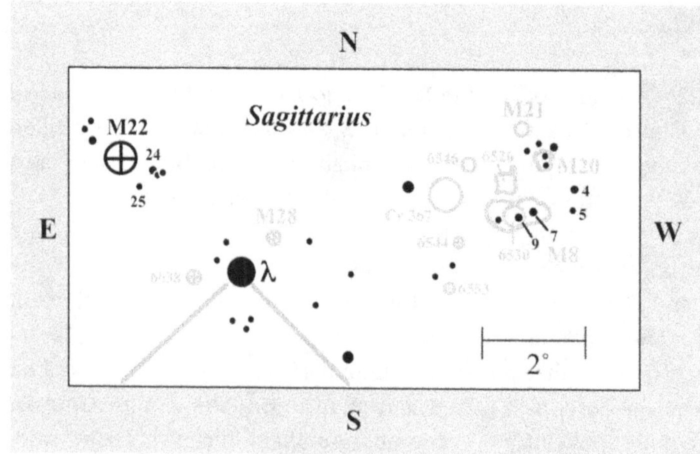

My pencil drawing shows the view mainly with high power, so it does not capture the cluster's many hazy extensions and its enormous halo. I call M22 the Crackerjack

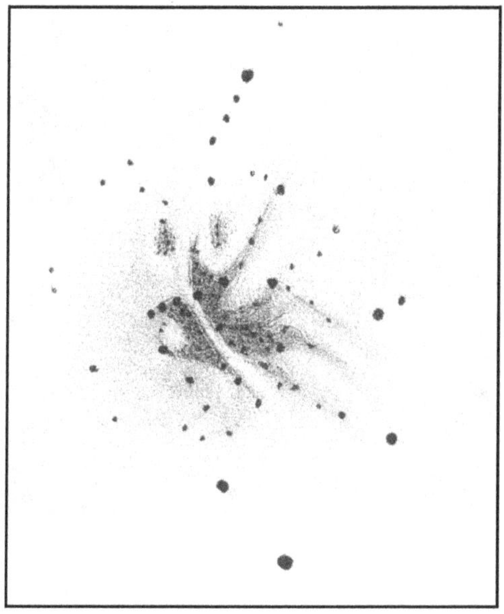

Cluster (after the sweet popcorn treat with the prize in every box) because, at high power, after you penetrate the outer swarms of stars, many surprises await your gaze.

The 3′ core of the cluster appears to sit in a hook of 10th- and 11th-magnitude stars on the south and west sides. It looks as if the cluster has been snagged by this hook of stars while drifting through space, like flotsam in a river getting snared by a hanging branch. And, as Brian Skiff logs, a prominent clump of five stars – a fuzzy cluster within the cluster – can be found in the northeast quadrant of the core. Most striking, however, is the dark gash running southwest to northeast across the core. It is the most prominent rift in any globular I have seen, and it cleanly splits the nuclear region in half! Furthermore, dark bays run into the southern part of the nucleus from the southeast, and two stellar deltas seem to have formed in the mouth of the dark river to the south.

To view the inner details most effectively, I observe the cluster in twilight, when the outer distraction of stars fades into the emerging dawn. This visual trick works for other bright globulars as well.

Like the globular cluster M15, M22 contains a planetary nebula, PK 009-07.1 (also known as GJJC-1, which has been identified as the infrared source IRAS 18333-2357). Its tiny disk measures only about 3″, and its central star shines at around magnitude 14.5. Proper-motion studies have shown the planetary to be part of the cluster. Its precise epoch 2000.0 position is $18^h36^m22.82^s$, –23°55′18.3″.

M23

NGC 6494
Type: Open Cluster
Con: Sagittarius

RA: $17^h56.9^m$
Dec: -19°01′
Mag: 5.5
Diam: 25′
Dist: ~2,000 light-years
Disc: Charles Messier, 1764

MESSIER: [Observed June 20, 1764] Star cluster between the tip of the bow of Sagittarius and the right foot of Ophiuchus, very close to the star Flamsteed 65 Ophiuchi. The stars in this cluster are very close to one another. Its position was determined relative to μ Sagittarii.

NGC: Cluster, bright, very large, pretty rich, little compressed, stars of 10th magnitude and fainter.

THE SAGITTARIUS MILKY WAY CONTAINS so many notable Messier objects – the Lagoon Nebula (M8), the Trifid Nebula (M20), the Swan Nebula (M17), the Great Sagittarius Cluster (M22) – that it is easy to overlook some of the smaller treats. M23, for example, all but hides in the southwest corner of a dark and inconspicuous valley that slices through the hub of our Galaxy. Yet this open cluster of 150 or so stars spread across 25′ (15 light-years) is deceivingly dynamic.

Train your binoculars on this region, about 5° due west of the Small Sagittarius Star Cloud (M24) or 4 1/2° northwest of 4th-magnitude Mu (μ) Sagittarii, and see if the cluster reminds you of a spider lying in wait for its prey. That M23 lies at the center of an apparent well in the Milky Way might also lead one to think that it has devoured the stars around it.

The loneliness of M24 and its dark surroundings is somewhat befitting. The cluster is intermediate in age, being around 300 million years old, and has spent about half of its life in a gas-poor interarm region of the Milky Way. Nevertheless, W. L. Sanders (*Astronomy and Astrophysics Supplement Series,* vol. 84, pp. 615–618, 1990) determined that this environment has not influenced the cluster's metallicity, which is on the metal-rich end of the spectrum. Sanders cautions, however, that "this finding does not rule out the possibility that the cluster has passed through a very metal rich molecular cloud, thereby experiencing significant environmental metal enrichment."

Jeffrey Wilkerson of Luther College and his colleagues report in a 2007 paper in the *Bulletin of the American Astronomical Society* (vol. 39, p. 111) how they surveyed the half-degree-square field containing open cluster M23 and found 30 variable stars (28 of them new) ranging in magnitude from about 10 to

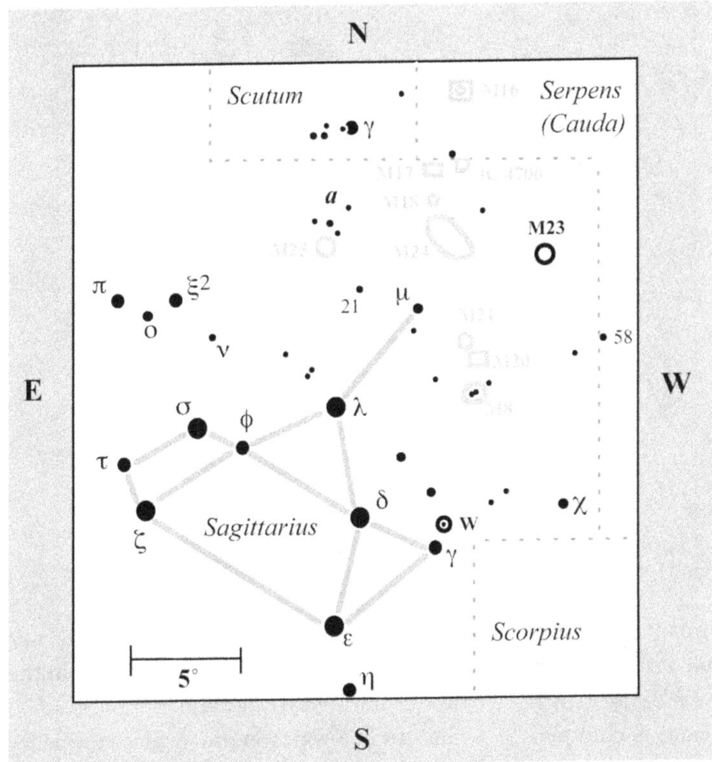

of stars to form the bat's tail and body. Its wings arc to the southwest and northeast. Opposite-curving arcs southeast of the stellar ellipse seem to flow away from the head of the bat. With a stretch of the imagination, I see these waves of starlight as the bat's ultrasonic screams. A tight triangle of stars in front of the bat could be the insect it has echolocated. Can you see the dark stream of nebulosity between the bat's head and the first ultrasonic wave?

Creative perception is an important part of astronomy's heritage. What are the constellations but figments of some creative individuals' minds? Some familiar star patterns, like Orion's belt and the Big Dipper, are even household names. Looking for shapes, likenesses, and fanciful imagery in patterns of stars and nebulosity makes stargazing more fun and accessible, and easier to share with others. It can also help you remember subtle details of an object or region of sky. Memorizing star patterns through the telescope can be the first step to making discoveries. If, for example, a new star (a nova) were to appear near M23 – disrupting the square, or the boat, or the bat motif you had become familiar with – you would recognize that something was different; the shape you created would appear altered. This is how nova hunters George Alcock and Peter Collins have each discovered four new stars – using binoculars!

17. Seven of the stars are eclipsing binaries, including two apparent W UMa contact binaries and one additional eclipsing binary. The remaining 23 variables are likely semiregular variables.

The arcing patterns of the cluster's 75 brightest stars entice the telescopic viewer to form a creative mental image of some kind. Robert Burnham Jr. suggests an outline of a Chinese temple, or a bit of calligraphy. John Mallas saw a bat in flight, and I couldn't agree more. Admittedly, I am partial to bats, so I need no further convincing.

You really need low power and a north-up orientation of the cluster to fully appreciate the bat motif (see the drawing). A faint string of stars flowing southeast of an 8th-magnitude star meet with a decidedly elliptical pattern

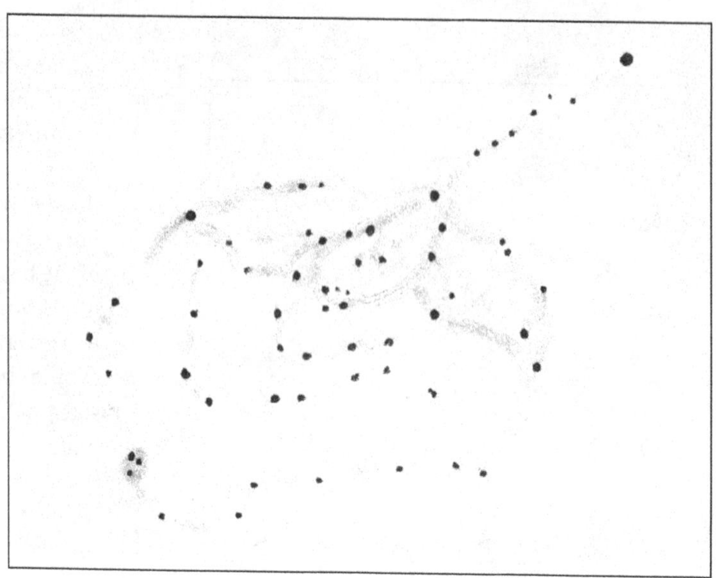

At 72×, the bat loses its impact. Try, however, to locate what looks like a little spoon holding a peanut (again, view with north up). At high power, the entire cluster is clumped in little isolated groups of stars separated by meandering rivers of dark nebulosity, which seem to empty into two pools of darkness (on either side of the bat's tail). Using 130×, see if you can locate a trio of double stars in the cluster's northwest quadrant.

Return to low power and explore the Milky Way around M23, especially to the west, where you'll find a great crack in the stellar expanse – the cosmic cave from which the bat emerged.

M24

Small Sagittarius Star Cloud
Type: Star Cloud (M24)
Con: Sagittarius

RA: $18^h17.4^m$
Dec: −18°36′
Mag: 4.6; 2.5 (O'Meara)
Dim: 1° × 2°
Dist: ~10,000 light-years
Disc: Charles Messier, 1764

MESSIER: [Observed June 20, 1764] Cluster on the same parallel as the previous one [M23], and above the tip of the bow of Sagittarius, in the Milky Way. A large nebula, within which there are several stars of different magnitudes. The luminosity that is spread throughout the cluster is concentrated in several regions. The position given is that of the center of the cluster.

NGC: None

BETWEEN THE LAGOON NEBULA (M8) and the Swan Nebula (M17) lies one of the most impressive stellar cities visible in small telescopes. M24 is not a true galactic cluster but a rectangular-shaped star cloud measuring 2° × 1°. Of all the Messier objects, M24 is second only to the Andromeda Galaxy, M31 (3° × 1°), in apparent size. Commonly called the Small Sagittarius Star Cloud, M24 is a virtual carpet of stellar jewels, laid out across 350 light-years of space. To the eye, it is so big that estimating its brightness is tricky: you have to defocus comparison stars until they've ballooned to four moon diameters, so that the comparison star's light is spread over an area of sky equal to that of M24. Most references list the star cloud's magnitude as 4.5, but this may be too faint; I place it a full two magnitudes brighter, at 2.5. Because its light is spread out over such a large area, its surface brightness is low, making the cloud appear dimmer than it really is. Light pollution has robbed many of us of the privilege of enjoying this summertime wonder. If you are a city dweller, plan to spend some time admiring this galactic treasure under a truly dark-sky site.

M24 is part of the naked-eye fabric of the Milky Way. Messier, however, appears to have been the first to resolve its light into a cluster-like patch of stars divided into several regions. Many early references credit Messier with discovering not the Small Star Cloud but the 4′-wide, 11th-magnitude open star cluster NGC 6603 within this rich segment of the Milky Way's Sagittarius-Carina spiral arm, but John Herschel first discovered that on July 15, 1830. The NGC description for the cluster is: "Remarkable cluster, very rich and

of M24. Sweep these dusty cobwebs away and you'd see the Large and Small Sagittarius Star Clouds blend and bloom into a garden of delight.

Through the Genesis at 23×, no sight in the visible universe shares M24's mystical qualities. I was amazed to find this seemingly three-dimensional patch of Milky Way beaming with a distinctive pale green sheen, suggesting a composition more of gelatin than of starlight, a giant euglena wrapped in stellar filigree.

The entire star cloud is blotted with pools of dark nebulosity. One prominent pool (Barnard 92) resides on the cluster's northwest side; it is surrounded by a succession of moderately bright stars. Use high power to concentrate on the center of the pool. Do you see your "reflection" – an illusion created by the light of a solitary 12th-magnitude star shimmering in the center? A wide canal of darkness to the north of M24 creates a sharp northern border, but the cloud seems to continue beyond the canal in a tapering spit toward the Swan Nebula.

Equally remarkable are the vast tracts of dark nebulosity to the west and east. These rival the star cloud in splendor. When I place M24 in the center of the field of view at 23× and then move one field to the northwest, I feel as if I've stepped off a steep cliff of starlight and am free-falling into a dizzying

very much compressed, round, stars of [12th] magnitude and fainter, in the Milky Way." The equinox 2000.0 coordinates for NGC 6603 are RA: $18^h18.5^m$; Dec: −18°24′. It lies about 9,400 light-years distant.

M24 is one of my seven wonders of the naked-eye Milky Way. It is a vast citadel of starlight about 10,000 light-years distant in the plane of the Milky Way and helps to define one of its spiral arms. But unlike an open star cluster, which also lies in the Galactic plane, M24 is not an isolated "patch" of starlight but the effect of an optical illusion caused by the chance framing of dark dust in the line of sight. The dust forms the thick walls of a three-dimensional tunnel through which we happen to glimpse a section of one of the Milky Way's inner spiral arms at the position

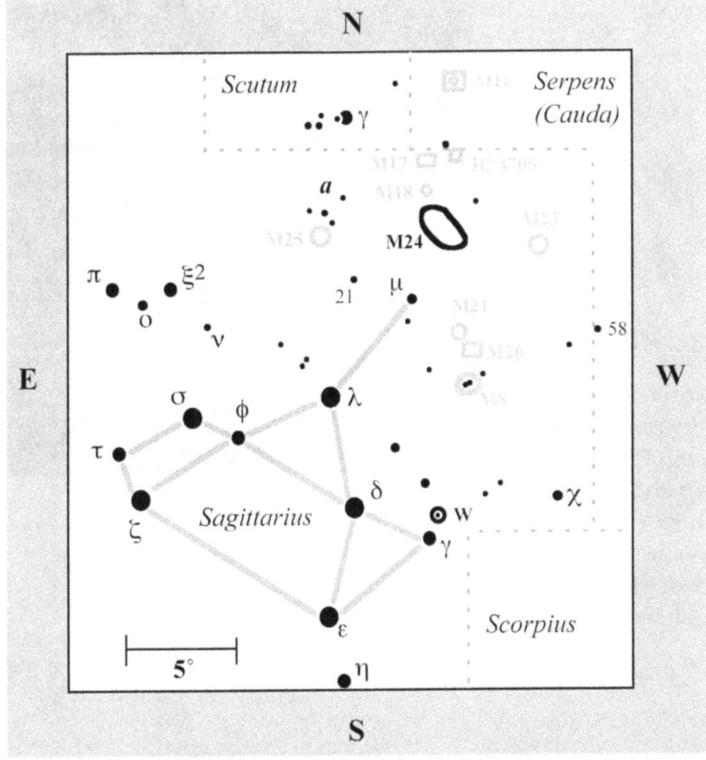

northern portion of M24, just 15′ north-northeast of an orange K5 star of magnitude 6.4. Messier probably would not have seen this tiny, compact glow, which shines feebly with the light of an out-of-focus 11th-magnitude star. Even from the dark skies of Hawaii, I nearly overlooked that minuscule fuzzy knot, which at low power affords only a hint of resolution. High power reveals a semicircular dark patch just south of it. Whereas for most objects I recommend a long, steady gaze to distinguish a particular feature, to see this dark loop you must give a quick, brief glance. That's because, for dark nebulae, the longer you look in the eyepiece, the more dark nebulosity you see, and the dark feature you're interested in soon gets lost in a complex web of dark streamers and spots that emerge from the background, especially with averted vision. In fact, if you spend a few minutes staring at M24, the star cloud will suddenly be swarming with dark veins.

Finally, compare the color of M24 with that of a smaller star cloud south of Mu (μ) Sagittarii. The Mu Sagittarii cloud looks gray and diffuse; I once thought I had breathed on the lens. It contrasts with the greenish glow of M24!

celestial abyss. If you relax your gaze, you might see the ever-widening wedge-shaped array of dark veins running off to the southeast. Wide-field photographs of this region reveal a dark corridor about the size of M24 to the southeast. It looks as if someone has lifted M24 like a log from that spot and dropped it in its present location, leaving behind its barren imprint in the celestial "soil."

Notice the description of M24 in the NGC, which obviously could not have been referring to the same object as Messier's "large nebulosity" with a diameter of 2°. M24 was formerly misidentified as the magnitude 11.4 star cluster NGC 6603, which resides in the

M25

IC 4725
Type: Open Cluster
Con: Sagittarius

RA: $18^h31.75^m$
Dec: –19°07′
Mag: 4.6
Diam: 26′
Dist: ~1,950 light-years
Disc: Philippe Loys de Chéseaux, 1745 or 1746

MESSIER: [Observed June 20, 1764] Cluster of faint stars near the two preceding ones [M23 and M24], between the head and the tip of the bow of Sagittarius. The known star closest to this cluster is sixth-magnitude Flamsteed 21. The stars in this cluster are difficult to see with a simple three-foot refractor. No nebulosity is visible. Its position has been determined relative to μ Sagittarii.

IC: Cluster, pretty compressed.

ADMIRAL SMYTH, IN HIS *CYCLE OF CELESTIAL Objects*, captured the immediate impression of M25: "a loose cluster of large and small stars. … The gathering portion of the group assumes an arched form, and is thickly strewn in the south, or upper part, where a pretty knot of minute glimmers occupies the centre, with much star-dust around." This bright open cluster in Sagittarius makes an easy naked-eye target and is finely resolved in small telescopes. It lies about 3 1/2° east-southeast of the Small Sagittarius Star Cloud (M24) and just south of an inverted spoon of binocular stars. Its estimated 600 stars stretch across 15 light-years of space, and the 30 or so bright ones that we can see cover a 1/2° area of sky.

M25 does not have an NGC designation, obviously an oversight by John Herschel, as his father observed it in 1783 after Messier catalogued it in 1764, though Philippe Loys de Chéseaux discovered it in 1745 or 1746. It made it into the *Index Catalogue* after Harvard astronomer Solon Bailey (1854–1931) determined its position on photographic plates, as reported in 1908.

The cluster is young, with an age of about 95 million years, and is located in the direction of the Galactic center, about 170 light-years below the Galactic plane, where its light is reddened by 0.5 magnitude. A. L. Tadross and colleagues (*Astrophysics and Space Science*, Vol. 282, pp. 607–623, 2002) determined the cluster's total mass to be 1,937 solar masses, noting that about 24 percent of the material mass of the cluster has remained as interstellar matter after the processes of formation.

Although the cluster contains thousands of stars, the main body of bright telescopic members forms a north-south-oriented figure eight with legs extending from all sides; in this way the cluster looks very beetle-like.

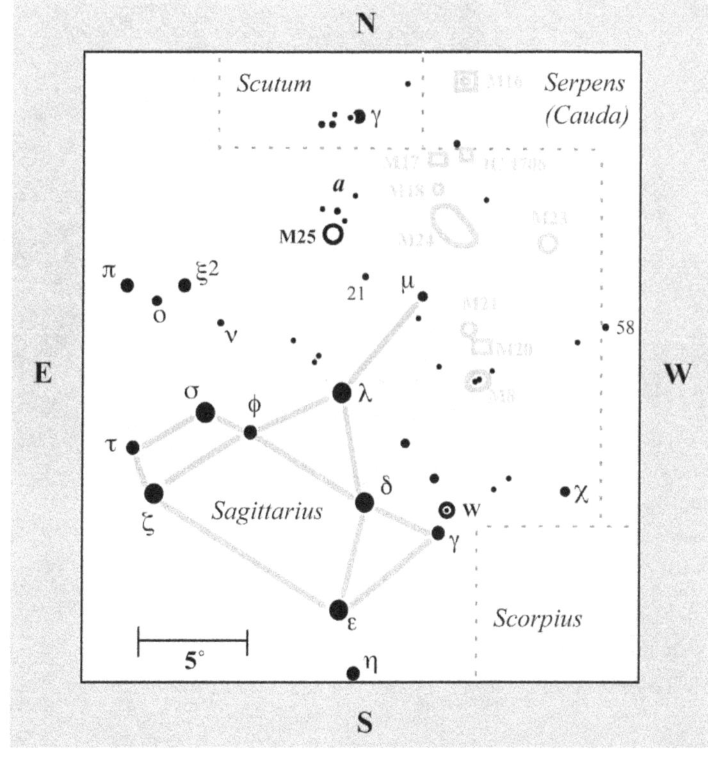

The western half of the southern section of the "8" is filled with stars; this curious-looking arrangement is Smyth's pretty knot of minute glimmers. Just east of the knot blazes the Cepheid variable star U Sagittarii, a member of M25, whose magnitude waxes and wanes between 6.3 and 7.1 about every 6 days, 18 hours; near minimum, the star's light becomes yellowed. With a little effort, you can chart its subtle variation in intensity and derive its period. Can you see this star with the naked eye when it is at maximum?

At medium power, several strong arcs of starlight spread across the field of view like rippled dunes glistening in a midday sun. Shadows fill the troughs between each crest and fade into the misty light of the stellar background. (On the off chance that you have had the opportunity to study an alligator's foot up close, the tapered digits of these star streams will look surprisingly familiar.)

M25's stellar arcs are so visually stunning that the cluster is beautiful for that reason alone. Once again, we have the opportunity to focus attention not on the bright stars but on the wisps of dark nebulosity – it's a skill worth developing, because it will add a new dimension to your observing. First use low power and let your eye drift until it finds a comfortable position for averted vision. Then relax your gaze. M25's dark central stream will become inky black and seem to lengthen well beyond the eastern and western edges of the cluster. The rest of the cluster looks like a child has taken a felt-tip pen to it and begun inking out the stars row by row.

After studying M25, look immediately to its east. There you will see a patch of Milky Way about the same angular size as M25 that looks like an afterimage of the cluster.

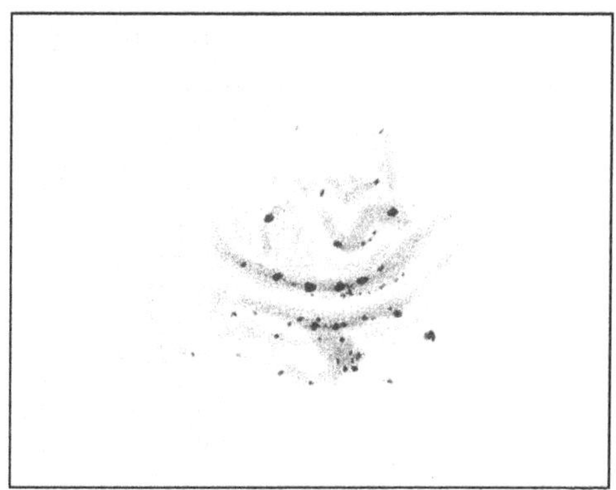

M26

NGC 6694
Type: Open Cluster
Con: Scutum

RA: $18^h45.25^m$
Dec: -09°23′
Mag: 8.0
Diam: 10′
Dist: ~5,200 light-years
Disc: Charles Messier, 1764, though Guillaume Le Gentil may have spied it first in 1749

MESSIER: [Observed June 20, 1764] Cluster of stars close to the stars *n* and *o* of Antonius [now ε and δ Scuti], and among which there is one that is brighter. With a three-foot refractor they cannot be detected; a good instrument must be used. This cluster contains no nebulosity.

NGC: Cluster, quite large, pretty rich, pretty compressed, stars from 12th to 15th magnitude.

I FIRST ENCOUNTERED M26 DURING A comet sweep in the rich Milky Way region in and around Scutum – perhaps just as Messier had done more than 200 years earlier. Using 23×, I noticed the cluster's concentrated glow enter the eyepiece, but it looked so small that I paid it little attention, figuring it was just an asterism; more intriguing was the 8th-magnitude globular cluster NGC 6712 about 2° to the northeast. After admiring the globular, a twinge of curiosity caused me to return to that tiny asterism, in which an ice-blue gem sparkled conspicuously amid a sprinkling of diamond dust. Suddenly I realized that this speckled glow was M26.

From brighter skies, this object might well be overlooked because its tantalizing gleam of background stars would be all but washed out. M26 is an innocuous knot of stars that spans an area of about 15 light-years and contains 120 telescopic stars in a disk 10′ in apparent diameter, though only about two dozen of the stars are readily seen in the 4-inch. That's pretty condensed compared with other Messier open clusters. Note that M25 is about 2 1/2 times larger in apparent diameter but also measures 15 light-years in true physical extent. That's because M25 is 2 1/2 times closer to us than M26. The cluster is young, having an age of about 85 million years.

While Charles Messier clearly discovered M26 in 1764, French astronomer G. Bigourdan (1851–1932) suggests the possibility that Guillaume Le Gentil may have spied it first in 1749. Le Gentil notes the discovery of the object on October 10, describing it as "the nebula which precedes the right foot of Antonius [now Scutum]." Le Gentil also goes on to discuss Kirch's discovery of the cluster

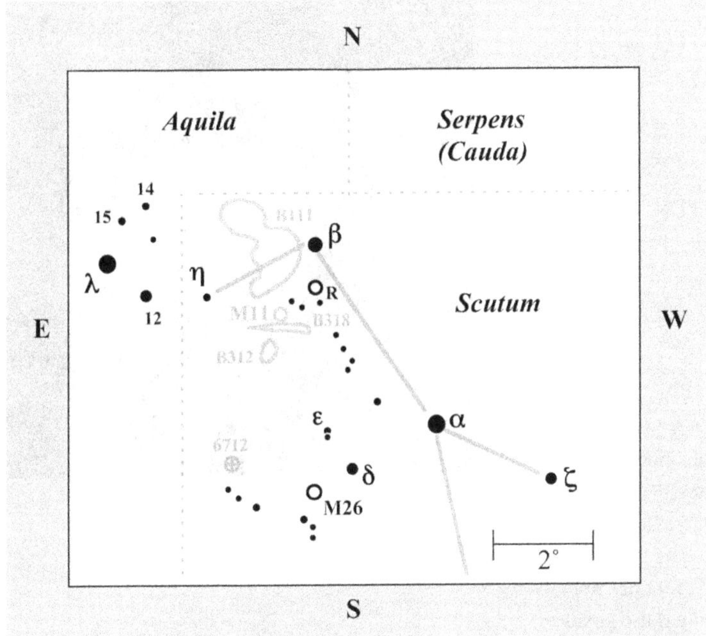

of stars now known as M11, which precedes the *left* foot of Antonius. This shows that Le Gentil is referring to a new object, which he thought "a true nebula." Bigourdan argues that the only object preceding Antonius's left foot is M26, though it is clearly not a nebula.

You'll find the M26 star cluster about 3 1/2° south-southwest of M11 in the stunning Scutum Star Cloud, less than 1° east-southeast of 5th-magnitude Delta (δ) Scuti. The brighter of M26's stars shine at only magnitude 10.3, for a combined magnitude of 8.

The entire region surrounding M26 appears mired in dark nebulosity, and the low-power field is haunted by fleeting hazes of unresolved stars. Moderate power shows two rows of stars extending to the north from the cluster's central diamond of stars. These pincers are clearly divided by a lane of obscuring matter, though other observers have documented it as a "hole," bereft of stars. James Cuffey of Indiana University first noticed the black heart of M26 in 1940, estimating that the star density at the cluster's core was 13 percent less than in the regions immediately surrounding this 3′-diameter zone of darkness. That's because M26 is seen through one of the densest parts of obscuring matter in the Scutum Star Cloud, where the stars are reddened by 0.6 magnitude.

The diamond is further caged to the southwest and northeast by two thicker bars of blackness. With averted vision and a relaxed gaze, the entire cluster seems to smolder with smoke. Try defocusing the telescope slightly and concentrating on the dark areas, and a most remarkable vision materializes: the shadow of a crucifix with light and dark rays radiating from its center. Look roughly one cluster diameter away from M26 in all directions and see if you can pick out a "sacred circle" of similarly bright stars enclosing the cluster and cross.

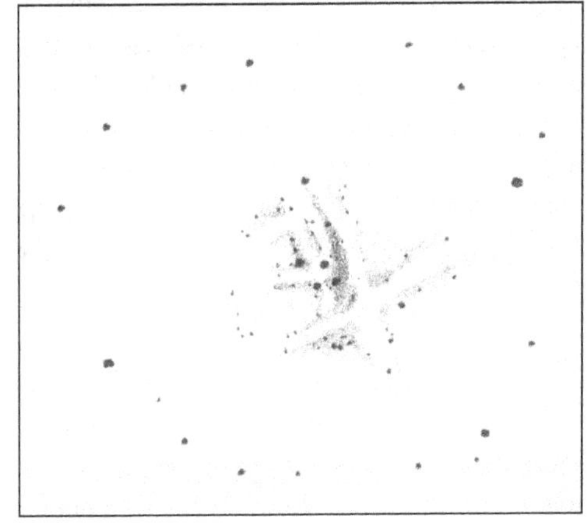

M27

Dumbbell Nebula
NGC 6853
Type: Planetary Nebula
Con: Vulpecula

RA: $19^h59.6^m$
Dec: +22°43′
Mag: 7.3
Dim: 480″ × 340″
Dist: ~1,200 light-years
Disc: Charles Messier, 1764

MESSIER: [Observed July 12, 1764] Nebula without a star, discovered in Vulpecula, between the two forepaws, and very close to the fifth-magnitude star Flamsteed 14 in that constellation. It can be seen clearly in a simple three-and-a-half-foot refractor. It appears oval-shaped and does not contain any stars. M. Messier plotted the position on the chart of the comet of 1779, which will be published in the Academy volume for that year. Observed again 31 January 1781.

NGC: Magnificent object, very bright, very large, binuclear, irregularly extended (Dumbbell).

ALTHOUGH NO STAR IN VULPECULA shines brighter than magnitude 4.4, this summer constellation does boast the most famous of planetary nebulae, M27, the Dumbbell Nebula. It was discovered in 1764 by Messier, who compared the nebula's shape to a dumbbell. M27, a "splendid enigma," as Smyth described it in the *Cycle of Celestial Objects*, is one of the closer planetaries (1,200 light-years away), and its physical diameter of 2.8 light-years also makes it one of the larger of them. The gaseous material was blown from the blue dwarf star now at its center during one of the star's death throes some 48,000 years ago (which makes M27 more than twice as old as typical planetaries). M27 is a cylindrical, multiple-shell planetary with different ionization structures seen at its equatorial plane; in fact, M27 is one of the few to have a triple shell. One shell, of doubly ionized oxygen, is expanding at a velocity of 9 miles per second, while another shell, of ionized nitrogen, is expanding at 20 miles per second. The third, outermost shell is faint and extremely complex, and probably represents gases expelled before the formation of the main planetary nebula. From our vantage point, the whole gaseous ring is apparently swelling 6.8″ per century, so the nebula has a mean age of about 3,500 years. The gas shells glow from excitation by ultraviolet radiation emitted by the hot central star, and the nebula glows predominantly in green light.

Stephen McCandliss (Johns Hopkins University) and his colleagues used high-resolution spectroscopy with the Far Ultraviolet Spectroscopic Explorer (FUSE) to study M27 (*The Astrophysical Journal,* vol. 659, p. 1291, 20008). Their results challenge the simple interacting-wind model of a planetary nebula, where a very high speed (~1,000 kilometers per second) stellar wind plows into a slower (~10 kilometers per second), mostly molecular outflow. "Instead," they say, "we find the nebular outflow and ionization balance to be stratified with high ionization states favored at low velocity and low ionization states favored at high velocity. Neutrals and molecules are found at a velocity that marks the transition between these two regimes."

The Hubble Space Telescope image of the Dumbbell Nebula shown here displays a variety of dense gas and dust knots, some looking like fingers, others like nodules or comets. Their sizes typically range from 11 billion to 35 billion miles (17 billion to 56 billion kilometers), which is several times larger than the distance from the Sun to Pluto, and each contains as much mass as three Earths. These knots form in the early stages of a planetary nebula's evolution, and appear to be common among many that HST has imaged. The shape of these features changes over time as the nebula expands.

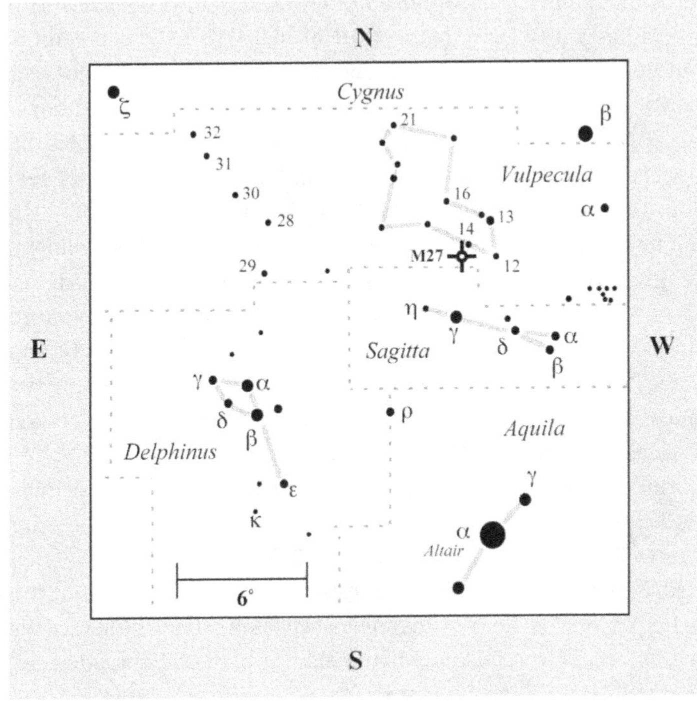

M27 is easily spotted in 7 × 50 binoculars as a roughly 7th-magnitude, 8′ × 6′ glow about 3° due north of magnitude 3.5 Gamma (γ) Sagittae,

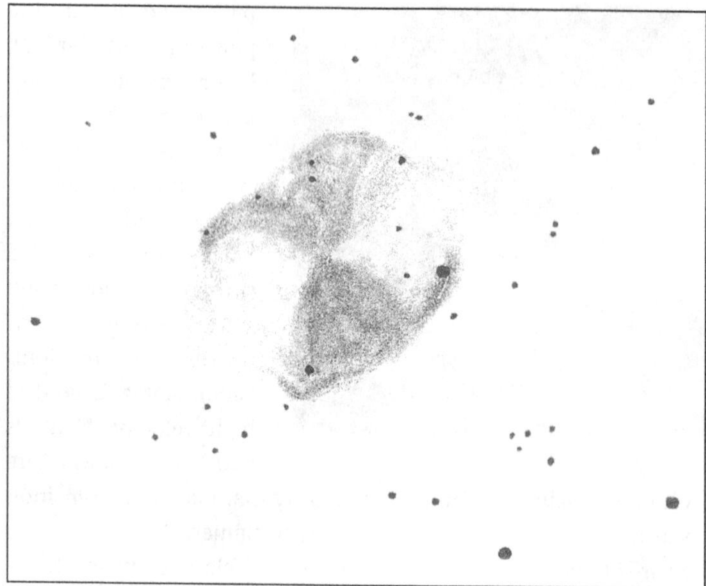

shell of gas oriented 90° from it. There's a definite asymmetry: the southwestern lobe of the dumbbell is brighter than the northwestern lobe, while the northwestern side of the "football" is brighter than the southeastern side.

High magnification should reveal the planetary's turbulent interior. Look for several meandering swirls that create a marble-like texture among a field of foreground stars. Stare at the southwest lobe of the Dumbbell and you may see the clumps of light matter and dark matter coming together to form a wedge-shaped face – devilish almost, with curved, protruding horns.

or less than 1/2° south of the magnitude 5.7 star 14 Vulpeculae. It is one of the few nebulous Messier objects whose visual impression truly matches that of its photographic image.

At 23×, the brightest portion of the nebula shines with a pale green light and has a distinct hourglass shape (oriented northeast to southwest). Faint loops of nebulosity can also be seen extending to the northwest and southeast. In the 4-inch, these faint extensions vanish with increased magnification, because the surface brightness becomes too low for them to stand out against the sky background.

When seen in a 3° field of view, the Dumbbell Nebula appears to be standing upright on a long blanket of stars. Look carefully where the Dumbbell stands, and see if the star blanket seems to be sagging under the Dumbbell's weight. This wonderful illusion is created by clouds of dark nebulosity, which frame the star blanket. You will need a dark sky, though, to see this well. Now return to M27. The dumbbell-shaped gas cloud is part of a more complex structure – namely, the dumbbell sits inside a football-shaped

But M27's central star, a white dwarf with a temperature of 85,000 K, is the real demon. Although spotting stars of 14th magnitude is usually a cinch under pristine skies, glimpsing this one requires keen averted vision. The reason might lie in the difficulty of seeing stars through nebulosity. Using high magnifications usually solves this problem, but it doesn't seem to in this case. It is interesting that drawings of M27 made by Lord Rosse, John Herschel, Admiral Smyth, and Leopold Trouvelot (1827–1895), all using sizable telescopes, do not include the central star! Furthermore, a 1908 Lick Observatory catalogue logs the central star as magnitude 12, while *Sky Catalogue 2000.0* lists the star's magnitude at 13.9. It could be that the earlier magnitude estimates were just off a little, but has anyone investigated the possible variable nature of this star (or of its "17th-magnitude" companion)?

M28

NGC 6626
Type: Globular Cluster
Con: Sagittarius

RA: $18^h24.5^m$
Dec: –24°52′
Mag: 6.9
Diam: 13.8′
Dist: ~17,900 light-years
Disc: Charles Messier, 1764

MESSIER: [Observed July 27, 1764] Nebula discovered in the upper part of the bow of Sagittarius, about one degree from the star λ, and not far from the beautiful nebula [M22] that lies between the head and bow. It does not contain any stars; it is circular and visible only with difficulty in a simple three-and-a-half-foot refractor. Its position was determined relative to λ Sagittarii. Observed again 20 March 1781.

NGC: Remarkable globular cluster, very bright, large, round, gradually extremely compressed in the middle, well resolved, stars from 14th to 16th magnitude.

LIKE M23, M28 IS ANOTHER LOST GEM in the glittering Sagittarius Milky Way. This tiny (10′) but charming globular cluster suffers the misfortune of being too close to the bigger and overpoweringly beautiful globular M22. M28 lies only 1° northwest of magnitude 2.8 Lambda (λ) Sagittarii, the golden K2 star marking the top of the Sagittarius teapot, and can be seen in binoculars shining at 7th magnitude. Messier couldn't resolve any stars in the object (calling it a "nebula") upon discovering it in 1764 or when he reexamined it in 1781. His contemporary William Herschel (who used much larger telescopes than Messier) could, and the object was correctly identified as a globular cluster in Smyth's 1844 *Cycle of Celestial Objects*.

While M28 appears about one-third the size of M22 and is two magnitudes fainter, nearly twice as small, it also lies nearly twice as far away (~17,900 light-years), only about 8,800 light-years from the Galactic center. The cluster is slightly smaller than M22 in true physical extent, however, spanning about 70 light-years of space compared with M22's 100 light-years. The members of M28 have, on average, about 1/20 as much metal per unit hydrogen as the Sun, and the swarm's integrated spectral class is F8. We see it approaching at 10 miles per second.

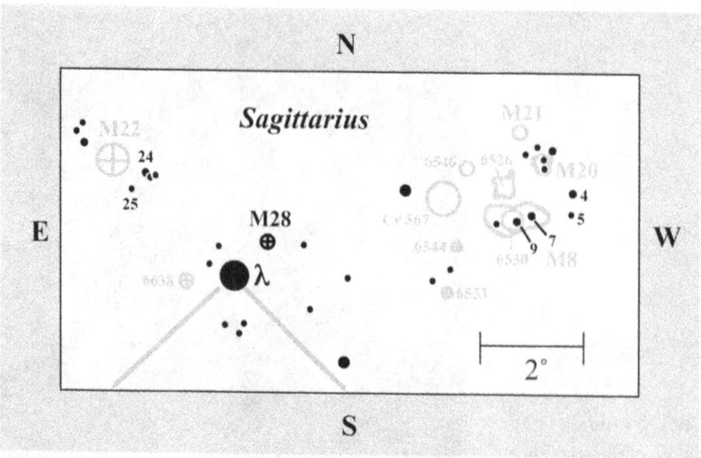

They propose the following scenarios for the possible origins and evolutions of the cluster: "(1) the cluster might have formed in satellite galaxies that were merged and created the Galactic bulge region in the early universe, after which time its dynamical properties were modified by dynamical friction, or (2) the cluster might have formed in primordial and rotationally supported massive clumps in the thick disk of the Galaxy."

The globular is highly concentrated (Shapley-Sawyer Class IV), with a tiny core (30″) of very high surface brightness. The cluster's core has a well-defined giant branch, and 18 variable stars are known. In 1995, T. J. Davidge (Gemini Canada) and colleagues estimated the cluster's age to be in the range of 16 billion to 20 billion years (*The Astronomical Journal*, vol. 110, p. 1177).

In a 2012 paper in *The Astronomical Journal* (vol. 144, p. 26), Sang-Hyum Chun (Yonsei University) and colleagues made a near-infrared study of the stars within the tidal radius of M28 using the Canada-France-Hawaii telescope on Mauna Kea, Hawaii. They found that the spatial density distribution of stars within the tidal radius is asymmetric, with distorted overdensity features, the most prominent of which extend toward the direction of the Galactic plane. The authors suggest that the overdensity features indicate that NGC 6626 experienced a strong tidal interaction with the Galaxy – likely associated with the effects of the dynamic interaction with the Galaxy and the cluster's space motion, and the result of the disk-shock effect of the Galaxy previously experienced by the cluster.

M28 can be seen in 10 × 50 binoculars as a small, condensed glow of diffuse light. At 23× through the 4-inch, the cluster remains much the same but a much sharper and more well-defined glow with a bright center. At moderate power (72×), the cluster appears to have a tightly packed center surrounded by a diffuse halo of stars. The core shows a faint straw tinge, while the halo is pale blue. A "prominent," roughly 12th-magnitude star abuts the halo to the south, but at this magnification the cluster is difficult to resolve in the 4-inch refractor. Still, I can start to resolve some of the brighter members (12th magnitude) populating the halo, which seems significantly elongated to the west. In fact, with averted vision, the distribution of stars in the halo makes the cluster look boxy. Several times I glimpsed a dark "spike" piercing the halo from the southwest. To the north, brushes of starlight flare out from the halo, a characteristic I've seen with many other globulars.

The view is very different at 130×! A tiny, peach-colored nucleus is surrounded by

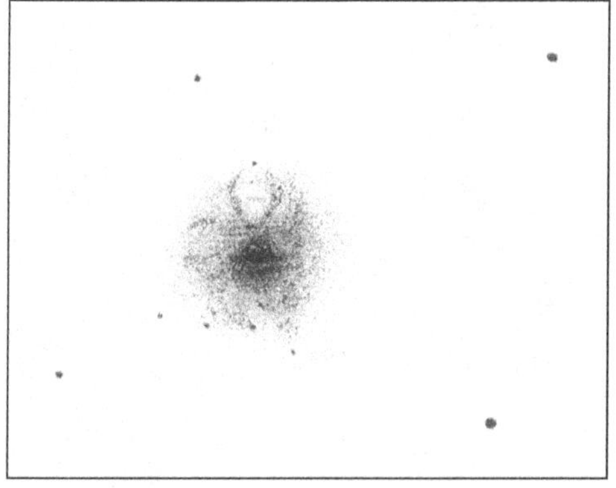

several patches of stars. (Most globulars do not show color at high power; their color is more prominent at low power.) To me the core seems aged – those patches all look very old and weak. With averted vision, the halo teems with innumerable 13th- to 14th-magnitude stars. Through the Lick 36-inch refractor, I found the nucleus to be highly fragmented, with rafts of starlight seeming to drift away from an agitated core. You need to see to magnitude 16.5 to resolve the cluster well.

About 1/2° southeast of Lambda, you'll find the 9th-magnitude globular cluster NGC 6638, which shines like a very condensed comet. At 101×, the cluster looks like a brighter globular star cluster seen at low power. The cluster's brightest stars shine around 14th magnitude, so with keen averted vision they (or clumps of them) can be seen swimming nervously around the cluster's halo.

At 23×, NGC 6638, M28, and M22 can all just barely fit into the same field of view. If your telescope allows it, take some time to look at them together. Here are three globulars of vastly different visual proportions. M22 seems monstrous in comparison with minuscule NGC 6638, which spans 7′ (actually the 4-inch only shows about 2′ of its total diameter); M22 is four times larger in apparent size! And do you see a color difference between M28 and M22?

M29

M29

NGC 6913
Type: Open Cluster
Con: Cygnus

RA: $20^h24.1^m$
Dec: +38°30′
Mag: 6.6
Diam: 10′
Dist: ~5,500 light-years
Disc: Messier, 1764

MESSIER: [Observed July 29, 1764] Cluster of seven or eight very faint stars, which are below γ Cygni, and which look like a nebula in a simple three-and-a-half-foot refractor. Its position was determined from γ Cygni. This cluster is plotted on the chart for the comet of 1779.

NGC: Cluster, poor and little compressed, bright and faint stars.

M29 IS A SMALL PACKET OF 80 STARS contained in a tiny 10′ sphere of sky. Still, this open cluster is visible to the naked eye as a magnitude 6.6 "star" about 2° south-southwest of magnitude 2.2 Gamma (γ) Cygni. It lies in a dense region of Milky Way that runs the length of the celestial Swan (Cygnus) – a region laced with dark lanes of interstellar dust.

This young open star cluster lies in the Cygnus OB1 Association, in a region of high and variable interstellar dust, which makes distance determination difficult. The cluster may contain more than 200 stars of 9th magnitude and fainter, but only about half that number are probable. Studies of its probable members reveal spectral classes ranging from O7 to K3. Only two O-type stars have been identified in this association; 40 percent and 33 percent of M29's total stars are B and A type, respectively, and the F, G, and K stars amount to 25 percent. The cluster is very young, probably between 4 million and 6 million years.

In a 2007 paper in *Astronomy and Astrophysics* (vol. 474, p. 873), N. Schneider (Université Paris Diderot, France) and colleagues note that they found evidence that several molecular clouds, including an H II region known as S106, are directly shaped by the ultraviolet radiation from members of several Cygnus OB clusters, mainly NGC 6913, and are thus located at a distance of ~5,500 light-years in the Cygnus X complex – one of the richest known regions of star formation in the Galaxy. The Cygnus X complex contains as many as 800 distinct H II regions, a number of Wolf-Rayet and O3 stars, and several OB associations; it's largely composed of a single complex 650 light-years wide with a mass of 30 million solar masses. The S106 region

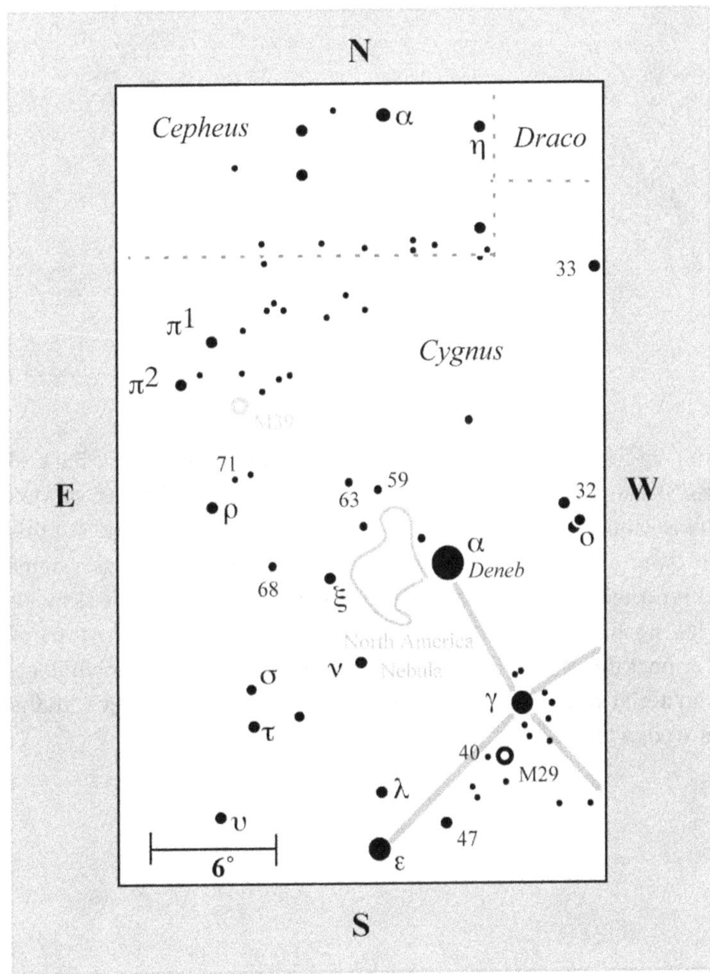

matter that veils the region, dimming the cluster's light by three magnitudes!

Through the 3-inch at medium power (and with time), I counted about two dozen stars in the main cluster. It is ringed by dark nebulosity, making it appear highly asymmetric, with the western side appearing dimmer than the eastern side (this is not as obvious in photographs, which enhance the fainter stars). The ring also appears to separate the core of the cluster from possible outer members, so be sure to let your eye drift beyond the darkness and into the surrounding star fields.

I almost agreed with John Mallas, who said that each increase in magnification reduces the cluster's beauty. But when I used 130× through my 4-inch and let my imagination fly, I saw a stream of 13th-magnitude stars coursing through the banks of brighter stars. Perhaps in larger instruments this stream would not have been so pretty, but it was quite alluring in the 4-inch because it was so faint. Then something interesting happened. As my eye followed that elegant river, my mind drifted to a time when I stood waist high in sawgrass in the Florida Everglades. There my surroundings looked rather bleak, until I caught sight of a single orange wildflower. That tiny splash of color

is located at a Galactic longitude where the local Galactic arm, the Perseus arm, and the outer Galaxy are found along the same line of sight, covering a distance between ~1,000 light-years and 26,000 light-years.

Through 10 × 50 binoculars, M29 is a beautiful tight knot of fuzzy starlight. For small-telescope users at low power, the cluster displays about a dozen stars of 9th to 10th magnitude. The four brightest ones form an obvious box or trapezoid with a couple of dimmer extensions. It would no doubt be a more striking sight were it not for the dense obscuring

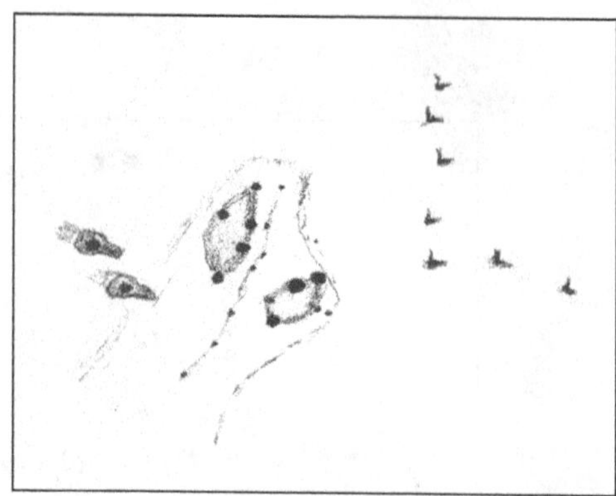

transformed a dull landscape into a grand sensation. Likewise, the delicate stream of stars in M29 turns an otherwise unremarkable cluster into a memorable one.

After plotting these stars, I returned to the eyepiece with the Everglades fresh in my mind. Now my eye played connect the dots, and I saw an entire scene - an aerial view of the glades. I noticed a bright wedge of Milky Way west of M29 and envisioned a flock of flamingos, and two stars to the east of M29 were crocodiles in the salty Atlantic. All this is highly imaginative, but such imagery helps me to unite the wonders I see in the sky with those I've experienced on earth. Such fancy is the very foundation of celestial mythology. Why not create your own mythology with the star patterns in open clusters?

M30

NGC 7099
Type: Globular Cluster
Con: Capricornus

RA: $21^h40.4^m$
Dec: -23°11′
Mag: 6.9
Diam: 12′
Dist: ~26,400 light-years
Disc: Charles Messier, 1764

MESSIER: [Observed August 3, 1764] Nebula discovered below the tail of Capricornus, close to the sixth-magnitude star Flamsteed 41. It is difficult to see with a simple three-and-a half-foot refractor. It is circular and does not contain any stars. Its position was determined relative to ζ Capricorni. M. Messier plotted it on the chart for the comet of 1759, *Mémoires de l'Académie 1760*, plate II.

NGC: Remarkable, globular cluster, bright, large, little extended, gradually, pretty much brighter in the middle, stars from 12th to 16th magnitude.

THE THIRTIETH ENTRY IN MESSIER'S catalogue is easily spotted with binoculars about 1/2° west-northwest of 41 Capricorni – a magnitude 5.5 star 6 1/2° east-southeast of magnitude 3.7 Zeta (ζ) Capricorni. M30 is a fairly large globular cluster with a few hundred thousand stars splashed across 90 light-years of space. We see it whizzing away from us at 115 miles per second.

M30 is one of the Galaxy's most extremely metal-poor globular star clusters, with each of its members, on average, containing about 1/186 as much metal per unit hydrogen as the Sun. The cluster has an integrated spectral type of F3 and an estimated age of 13 billion years. In 2009, the Hubble Space Telescope fully resolved the cluster's sparsely populated central region, which has a post-core collapse ~0.1′ in extent. The HST data revealed a well-defined horizontal branch containing four blue stragglers in the collapsed core. According to an HST press release, the blue stragglers appear to be of two types: (1) those that form in near head-on collisions with one another (2) and those that are in twin (or binary) systems where the less massive star siphons 'life-giving' hydrogen from its more massive companion.

Shining at magnitude 6.9, the moderately condensed glow is surprisingly obvious under a dark sky through 10 × 50 binoculars. Visually compressed, M30 has a tiny core inside a 12′ globular haze, though only about half that size appears dominant through the telescope. Despite the object's brightness, low power does not resolve it at all.

M30

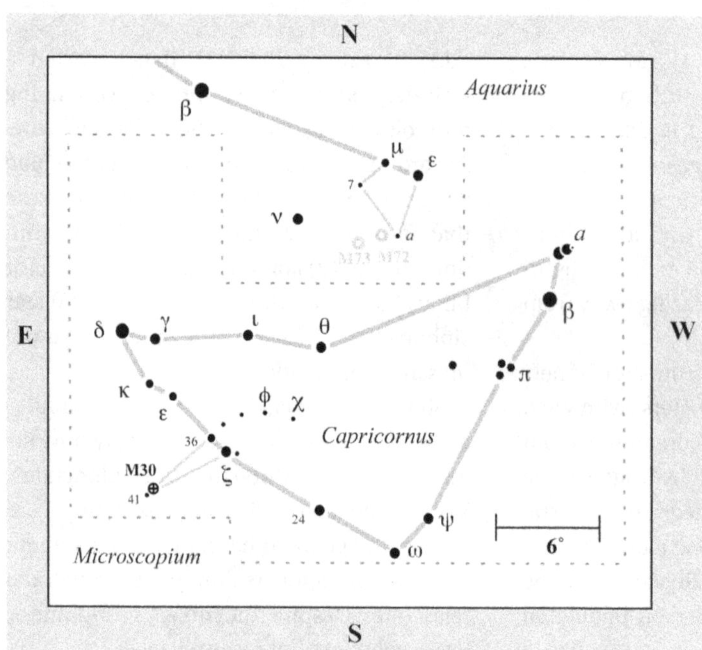

At 72×, the view is most peculiar, as if whoever had started to mold the cluster's shape suddenly decided to stop, never finishing. It has a lopsided appearance; the northern portion displays well-defined fingers of stars, while the southern part looks kneaded but incomplete. Smyth saw this too: "From the straggling streams of stars on its northern verge, [it] has an elliptical aspect, with a central blaze."

At high power, the entire cluster seems to be served on a plate of fainter stars. The northern jets of stars contain bright stellar knots, all between 12th and 13th magnitude, which seem to flow downwind of a gale. More than anything, I'm reminded of the nucleus of a very active comet. (Webb, too, thought M30 was comet-like.) Be sure to spend time concentrating on the core, which is rather corkscrew shaped. In his *Astronomical Objects for Southern Telescopes*, E. J. Hartung writes that M30's "well-resolved centre is compressed and two short straight

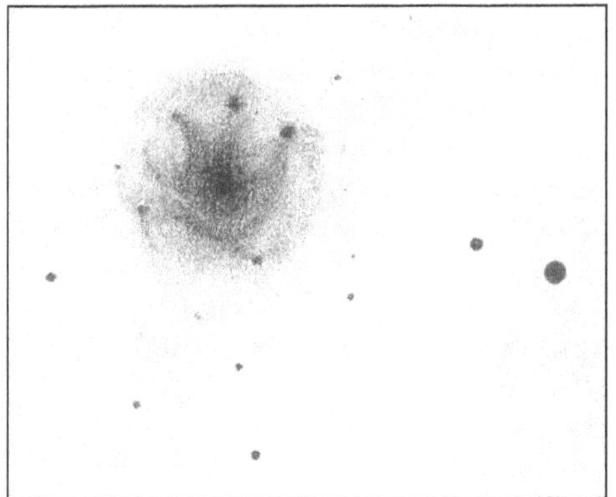

rays of stars emerge [north preceding] while from the N edge irregular streams of stars come out almost spirally." But you'll need to see to 15th magnitude to duplicate this feat.

Return to low power. Now doesn't the cluster look as if it has horns to the north? The challenge is to see a very faint extension of stars on the southeast halo and its very faint semicircular wings.

By the way, while many neutron stars and binary stars are known to populate globular star clusters, observational evidence for cataclysmic variables, as for dwarf novae, are limited to only two good candidates, V101 in M5 and V4 in M30, the latter of which was first discovered by Rosino in 1949. V4 had a photographic magnitude outburst of about magnitude 16.4 and an amplitude of ~2 magnitudes. But G. Machin (Oxford University) and colleagues reported in a 1991 paper in the *Monthly Notices of the Royal Astronomical Society* (vol. 250, p. 602) that their observations with the William Herschel Telescope indicated that the outburst magnitude and radial velocity of V4 show it to be a *foreground* object in the field of M30.

M31

Andromeda Galaxy
NGC 224
Type: Spiral Galaxy
Con: Andromeda

RA: $00^h42.7^m$
Dec: +41°16′
Mag: 3.4
SB: 13.6
Diam: 3° × 1°
Dist: ~2.5 million light-years
Disc: Persian astronomer Al-Sufi, tenth century

MESSIER: [Observed August 3, 1764] The beautiful nebula in the belt of Andromeda, shaped like a spindle. M. Messier examined it with several instruments, but he was not able to detect any stars. It resembles two cones or pyramids of light, joined at their bases, and the axis of which lies northwest to southeast. The two points of light are perhaps some 40 minutes of arc apart. The common base of the two pyramids is about 15 minutes. This nebula was discovered in 1612 by Simon Marius, and has been observed subsequently by various astronomers. M. le Gentil gives a drawing of it in *Mémoires de l'Académie 1759*, page 453. It is plotted in the English *Atlas Céleste*.

NGC: Magnificent object, extremely bright, extremely large, very much extended (Andromeda).

M32

NGC 221
Type: Dwarf Elliptical Galaxy (companion to M31)
Con: Andromeda

RA: $00^h42.7^m$
Dec: +40°52′
Mag: 8.2
SB: 12.7
Dim: 11.0′ × 7.3′
Dist: ~2.5 million light-years
Disc: Guillaume Le Gentil, 1749

MESSIER: [Observed August 3, 1764] Small nebula without stars, below and a few minutes away from the nebula in the belt of Andromeda. This small nebula is circular, its light fainter than that in the belt. M. le Gentil discovered it on 29 October 1749. M. Messier saw it for the first time in 1757, and has not noted any change in its appearance.

NGC: Remarkable, *very* bright, large, round, suddenly much brighter in the middle to a nucleus.

TO THE TRUE ROMANTIC OF ASTRONOMY, M31 will always be known as the "Great Nebula in Andromeda" – a name bestowed on it before spectroscopy revealed that this luminous mist was not the protoplasmic soup of a solar system in formation but a distant island universe like our own Milky Way Galaxy. An enormous pinwheel of dust and gas, the Andromeda Galaxy contains some 300 billion stars spread across 130,000 light-years. It is rushing toward us at 185 miles per second.

M31 is among the largest galaxies known and is by far the largest member of the Local Group of galaxies, which includes our Milky Way and some two dozen smaller systems. The Andromeda and Milky Way galaxies dominate the Local Group with their size, with M31 being twice as massive as the Milky Way. And though we see the Andromeda Galaxy nearly edge-on, astronomers see enough structure to speculate that the Milky Way is similar in shape and structure. If you were in the Andromeda Galaxy looking at the Milky Way, the Milky Way would appear much the same way as M31 does to us.

The Andromeda Galaxy has a black hole of 100 million solar masses at the core of its central hub. In 2012, NASA released a stunning movie that zooms in on a 1992 HST image of the black hole region to provide us with the sharpest visible-light image ever made of the nucleus of an external galaxy. The HST high-resolution image reveals a compact cluster of

blue stars fringing what appears to be a double nucleus – actually enhancements along a ring of old, reddish stars in orbit around the black hole but more distant than the blue stars. "When the stars are at the farthest point in their orbit they move slower, like cars on a crowded freeway," the release explained. "This gives the illusion of a second nucleus."

Astronomers estimate that the massive blue stars surrounding the black hole have an age of no more than 200 million years. Most likely they formed near the black hole in an abrupt burst of star formation, since massive blue stars would not live long enough to migrate to the present position from elsewhere in the Galaxy.

Also in 2012, NASA announced that Hubble Space Telescope measurements of M31's motion through space reveal with certainty that M31 and the Milky Way will have a head-on collision four billion years from now. But it will take an additional two billion years after the encounter for the interacting galaxies to completely merge under the tug of gravity and reshape into a single elliptical galaxy similar to the kind commonly seen in the local universe.

"Although the galaxies will plow into each other, stars inside each galaxy are so far apart that they will not collide with other stars during the encounter," the release explained. "However, the stars will be thrown into different orbits around the new galactic center. Simulations show that our solar system will probably be tossed much farther from the galactic core than it is today."

At 2.5 million light-years distant, M31 is also one of the farthest objects visible to the naked eye. Under reasonably dark skies, it appears as a cocoon of nebulous vapor 1° west of magnitude 4.5 Nu (ν) Andromedae in the belt of Andromeda, the Chained Maiden. M31 stretches 3°, or nearly six times the apparent diameter of the full moon, on the most transparent nights. A good pair of binoculars will show some of the galaxy's subtle detail. Even 7 × 35 binoculars will reveal its elliptical disk, whose surface brightness gradually fades away from a starlike core. The billions of stars in the core are so tightly packed that astronomers believe there may be a black hole at the center.

The galaxy's northwestern rim has a sharp edge to it, which marks the location of a prominent dark lane slicing through that part of the galaxy; in contrast, the galaxy's southeastern rim diffuses gradually into the sky background. Binoculars will also reveal two of M31's companion galaxies: M32 and M110. M32 looks like a slightly swollen 8th-magnitude star on M31's outermost bright rim, roughly 1/2° south and slightly east of the nucleus. M110 is a similarly bright, though larger, elliptical haze 37′ northwest of M31's nucleus.

In the 4-inch at 23×, a bright yellow "star" marks the very center of the Andromeda Galaxy. It lies inside several tightly wrapped pale yellow halos, which start out circular close to the core but become progressively more elliptical and more skewed toward the southwest farther away from the nucleus. This teardrop-shaped patch of golden light

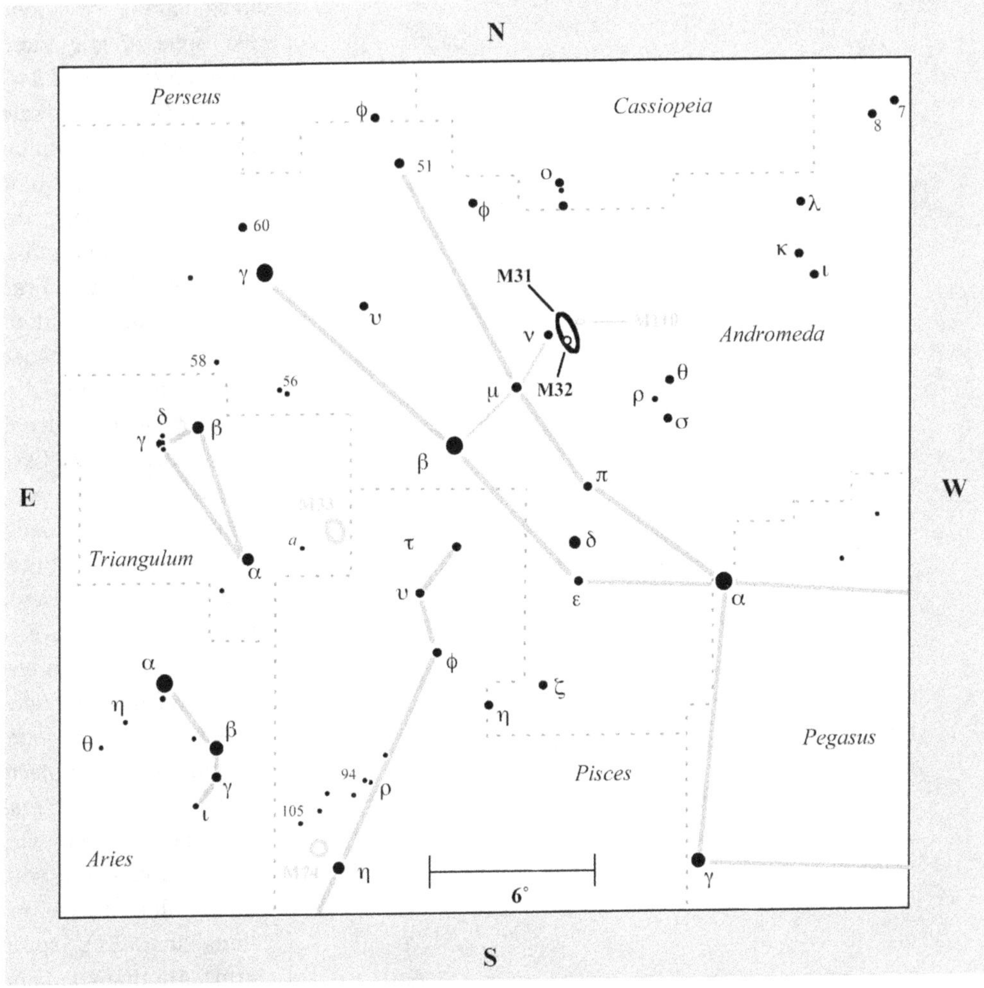

is surrounded by an enormous ashen elliptical halo grooved with faint dust lanes. The detailed drawing shown here is a composite: after spending three hours each night over several nights examining the galaxy, I combined the views made with low, medium, and high powers. I concentrated on the nuclear region the first night, the northeast portion the next night, and the southwest sector last; the two satellite galaxies were viewed on separate nights.

Besides the prominent nuclear region, the most striking feature of M31 is a conspicuous dust lane running along its northwest edge. One night, I followed that lane southwestward for half the galaxy's radius before it looped back in classic spiral fashion. When I used 130× to study the portion of this lane lying closest to the nucleus, it appeared very turbulent. One could spend hours trying to visualize patterns in the lacy swirls and splotches of dust. See if you can detect a leopard-spot pattern there; it almost blends with another dust lane farther to the northwest. Now look just east of the innermost halo surrounding the nucleus. There you should find another lane

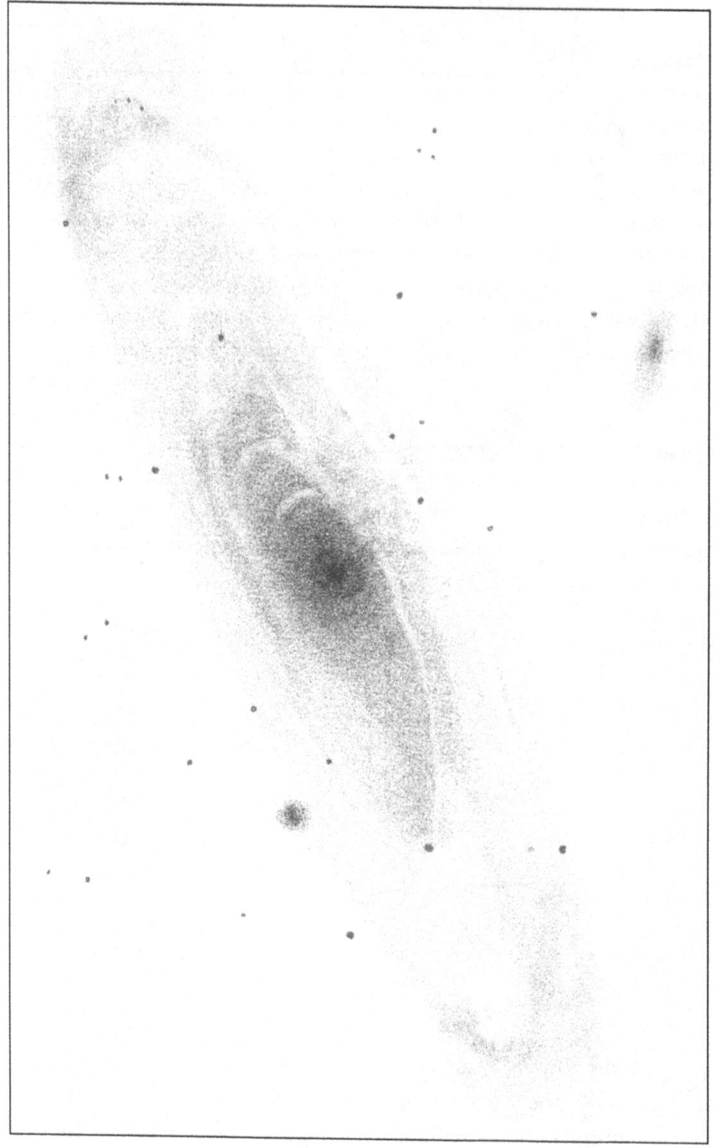

imagination, look like spits of gray sand between streams of dark matter. Glyn Jones said these show best in photographs sensitive to blue light. Interestingly, my eyes appear to be sensitive to blue light, so I can see these concentrations well. The late English nova and comet discoverer George Alcock is also believed to have had blue-sensitive eyes. Other observing friends, such as Michael Mattei of Harvard, Massachusetts, have red-sensitive eyes.

One way you can test your eyes for color sensitivity is to observe astronomical objects with blue or red features and see how you fare. You can start with these blue concentrations in M31's spiral arms. Another good one is NGC 206, close to the galaxy's southwestern rim. This 4′ × 1.5′ patch of fuzzy light is actually an enormous star cloud within the galaxy that

of dust with two faint, dark hooks branching off it to the northeast. If you're having trouble with high magnification, revert to low power because you will condense the galaxy's light and increase the contrast between the bright haze of the galaxy and the silhouettes of dust.

The arms in M31's outer halo contain some bright concentrations, which, with

measures 3,000 × 1,000 light-years. Based on my limiting-magnitude studies, about a dozen globular clusters are within range of a good 4-inch telescope under perfect conditions. By the way, have you ever noticed how the extreme tips of M31's outer spiral arms curve away from the main body? The south-

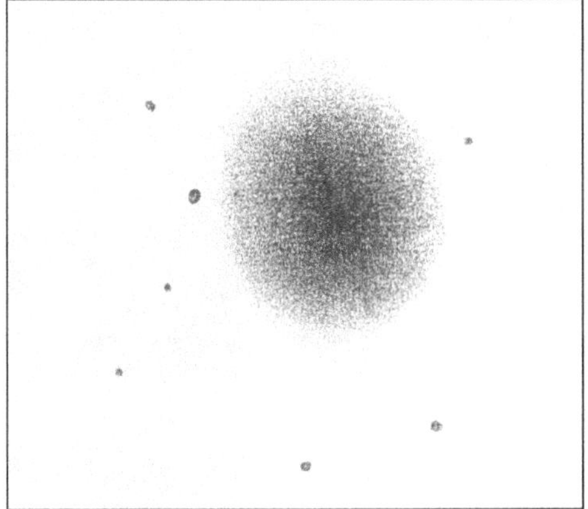

western tip curves to the south, while the northeastern tip curves north.

In 1885, the first and only known supernova to appear in M31 burst onto the scene. Isaac Ward of Belfast, Ireland, first noticed the new star on August 15, and Ernst Hartwig at Dorpat Observatory in Estonia independently recorded it the following evening. Supernova 1885 (S Andromedae) reached naked-eye visibility (6th magnitude) and appeared only 16″ from M31's bright nucleus before fading into near obscurity -- until the Hubble Space Telescope finally revealed the event's optical remnant, appearing only about 4″ across. It also detected a large quantity of cold iron freely expanding at velocities up to 5,000 km/s, which is consistent with its historical Type I classification.

In binoculars, M32 is easy to mistake for a bright star. At low and moderate powers in the telescope, M32 is essentially a featureless circular glow ~7′ wide (~8,000 light-years). But take the time to look for the fainter outer envelope that gives this seemingly round galaxy its larger elliptical shape. At 130×, I could discern a definite starlike core with an odd vertical extension running northeast to southwest through the entire galaxy. Is this a real feature or an illusion created by a peculiar alignment of faint, unresolved foreground stars? The southeastern side of the nucleus seems to be bordered by an arc of bright haze, while at the same distance away to the northeast there appears to be a faint star that flickers in and out of view. I like to see it as a beacon, a friendly message from a distant neighbor.

Like M31, M32 is believed to harbor a black hole, but one perhaps 10 times smaller than the one at the heart of M31. Also like M31, M32's light is blueshifted, but this system appears to be approaching us at a slower speed of 126 miles per second.

M33

Triangulum Galaxy **or** *Pinwheel Galaxy*
NGC 598
Type: Spiral Galaxy
Con: Triangulum

RA: 01^{h}33.8^m
Dec: +30°40′
Mag: 5.7
SB: 14.2
Dim: 67′ × 41.5′
Dist: ~2.8 million light-years
Disc: Messier, 1764

MESSIER: [Observed August 25, 1764] Nebula discovered between the head of the northern Fish [in Pisces] and Triangulum, close to a sixth-magnitude star. The nebula's light is whitish, and almost even in density, but is slightly brighter over the central two-thirds of its diameter, and it does not contain any stars. It is difficult to see with a simple one-foot refractor. Its position was determined relative to α Trianguli. Observed again 27 September 1780.

NGC: Remarkable, extremely bright, extremely large, round, very gradually brighter in the middle to a nucleus.

IN LONG-EXPOSURE PHOTOGRAPHS, M33, another member of the Local Group of galaxies, looks like an enormous spiral with innumerable stars clinging to wildly spinning arms. But with a diameter of 54,000 light-years, you could fit two and a half M33s in the disk of M31. In fact, M33 may be a satellite galaxy of M31, orbiting it just as the Moon does the Earth. The Andromeda Galaxy is also about 15 times more massive than M33, which is about two times smaller and seven times less massive than our Milky Way. As seen from an imaginary planet in the Pinwheel Galaxy, M31 would be an impressive sight – an oblique swarm of faintly glittering stars stretching 6° in that hypothetical sky.

M33 harbors one of the most massive black hole binary systems known, M33 X-7, which has a mass of 16 solar masses and every 3.5 days orbits an O star of 70 solar masses. In a 2010 paper in *Publications of the Astronomical Society of the Pacific* (vol. 435, p. 179), Selma E. de Mink (Argelander Institut fuer Astronomie, Utrecht, The Netherlands) and colleagues suggest that rotational mixing plays an important role in the formation of very massive black holes in a very close orbit with a less evolved massive companion such as M33 X-7. In their new model, helium produced in the center is mixed throughout the envelope. "Instead of expanding during their main-sequence evolution (with the inevitable consequence of mass transfer)," they say, "these stars stay compact, and avoid filling their Roche lobe. They gradually evolve into massive helium stars."

We may get a closer look at the inner dynamics of M33 in about 10 billion years. Roeland P. van der Marel of the Space Telescope Science Institute and his colleagues, in a 2012 paper in *Astrophysical Journal* (vol. 753, p. 9), reported that there is an ~20 percent probability that the Sun will find itself moving through M33 by then. As noted in the M31 discussion, Hubble Space Telescope measurements of M31's motion through space reveal with certainty that M31 and the Milky Way will have a head-on collision four billion years from now, with M33 settling into an orbit around them that may decay toward a merger later. However, the van der Marel team also notes that there is a 9 percent probability that M33 will make a direct hit with the Milky Way at its first pericenter, before M31 gets to or collides with the Milky Way. They also note that there is also a 7 percent probability that M33 will get ejected from the Local Group, temporarily or permanently.

The Pinwheel has long been a naked-eye challenge for amateur astronomers. While some find it easily visible to the naked eye or in binoculars, others cannot see it at all. The problem lies in the galaxy's low surface brightness. Although the total magnitude of M33 is the same as a 6th-magnitude star, the galaxy's light is spread over an area of sky larger than two full moon diameters, making it appear dim. A dark sky, a steady atmosphere, and good vision are required to see M33 with or without optical aid. Its ease of visibility is, as Walter Scott Houston often stated, a barometer for the clarity of one's observing site. The galaxy is completely washed out in urban skies and can be disappointingly dim from suburban locations even through large-aperture telescopes! Try to see it, though, without optical aid, because M33 is one of the farthest objects visible to the naked eye. Some observers claim M33 cannot be seen with the naked eye at all and that the "glow" is caused by some nearby stars. But the brightest naked-eye stars near M33 are 7th magnitude and fainter, so that argument may be specious.

The clockwise-swirling spiral of M33 can be spotted 4° west-northwest of magnitude 3.4 Alpha (α) Trianguli. M33 and M31 lie on opposite sides of, and at nearly equal angular distances from, the 2nd-magnitude red giant Beta (β) Andromedae. Binoculars do a fine job of bringing out the Pinwheel's luster against the black backdrop. Even 7 × 35s show the 54,000-light-year-wide galaxy as a distinct oval glow immediately north of an 8th-magnitude star.

At 23× in the 4-inch, the galaxy's light is compressed into a shimmering disk of optimal contrast, with several faint spiral arms sweeping away from a tight, lens-shaped nuclear region. Take a moment to compare the size of that tiny nucleus with the rest of the galaxy; M33's nucleus contains less than 2 percent of the galaxy's total mass. In comparison, M31's nuclear region is about 1/25 the size of the entire galaxy and about five times larger than M32!

Although John Herschel called M33 unfit for high powers, "being imperceptible from want of contrast with 144×," I find high magnification (130×) perfect for concentrating on its tiny central knot, which appears misshapen by dark matter. Several faint stars or fuzzy kinks lurk in the misty vicinity of the galaxy's core. Using a 12-inch telescope at 225×, Christian

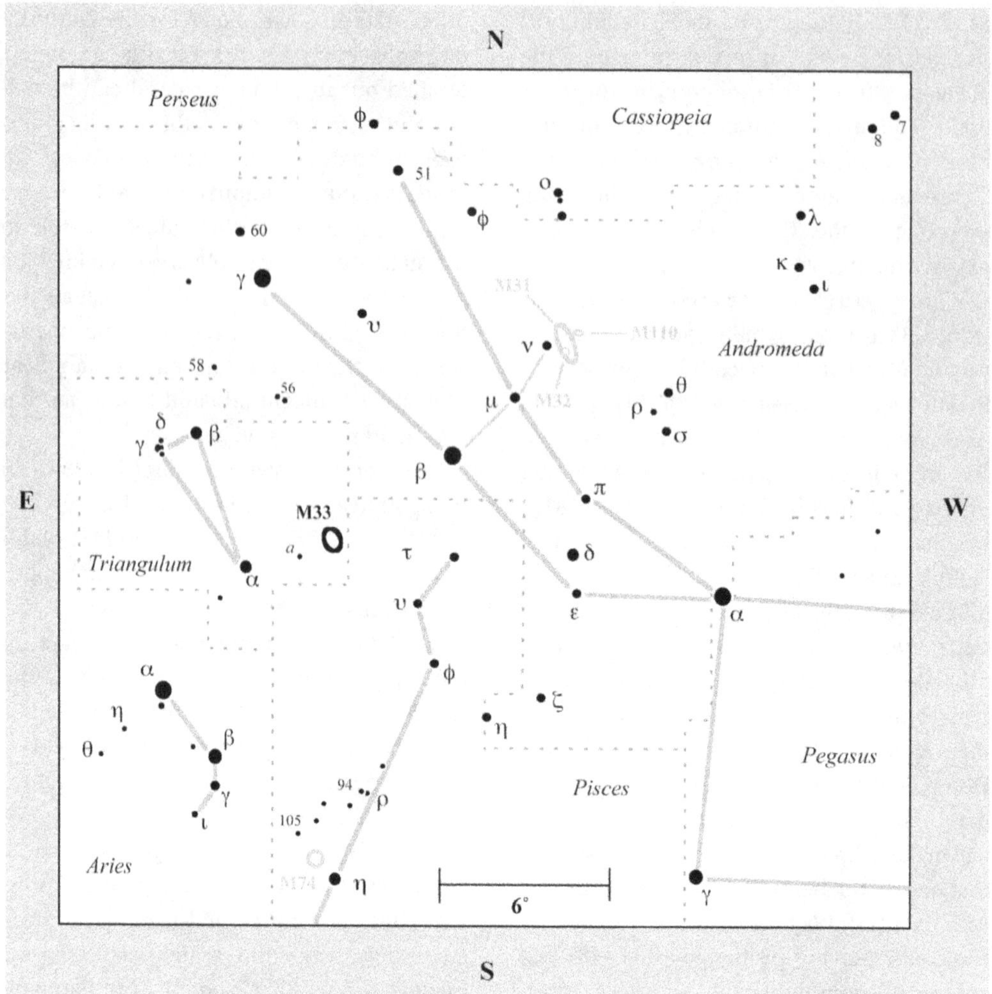

Luginbuhl counted perhaps as many as eight 13th-magnitude stars sprinkled in the southern portion of this region. "The whole surface of the galaxy," he writes, "is covered with faint stellar rings and splotches." Similarly, Lord Rosse saw it full of knots, with two S-shaped curves crossing in the center, though only one long and loose S is conspicuous at low power. But, again, high power will reveal some taut and stubby inner spiral structure.

Without question, M33 is the easiest galaxy beyond the Magellanic Clouds to resolve with large amateur instruments. Skiff notes that many faint clusters, stellar associations, and nebulae embedded in it are within grasp of a 10-inch telescope. I recorded in my observing diary that M33's most prominent arms appear lumpy at 23×, and with 72× the contrast holds up well enough for there to be a suspicion of resolution. M33's largest H II region (NGC 604) is by far the most conspicuous feature in the galaxy besides the nucleus. You'll find it 12′ northeast of the nucleus, just 1′ northwest of a magnitude 10.5 star.

NGC 604 is the second most luminous giant H II region in the Local Group, after 30

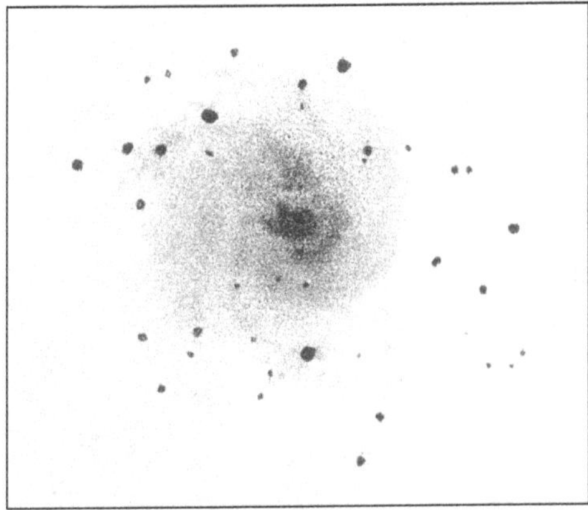

Doradus in the Large Magellanic Cloud, with star formation spread across some 30,000 square light-years. Massive stars form in giant H II regions, but the dusty environments make them a challenge to observe. Recently, Cecilia Fariña (Universidad Nacional de La Plata and IALP-CONICET, Argentina) and her collaborators used the Near-infrared Imager and Spectrometer on Gemini North in Hawaii to penetrate the dust in NGC 604 and identify likely massive young stellar objects (*Astronomical Journal*, vol. 143, p. 43, 2012). The team found 68 candidates with circumstellar material (possibly embedded in nebular emission) that suggests a current generation of star formation, in addition to the older population of the main central cluster of NGC 604.

Luginbuhl spied NGC 604 in a 2.4-inch (60-millimeter) telescope as a concentrated spot in the halo! Can you see it in binoculars?

M34

NGC 1039
Type: Open Cluster
Con: Perseus

RA: $02^h42.1^m$
Dec: +42°45′
Mag: 5.2
Diam: 25′
Dist: 1,530 light-years
Disc: Charles Messier, 1764; though Giovanni Batista Hodierna probably spied it before 1654

MESSIER: [Observed August 25, 1764] Cluster of faint stars, between the head of Medusa [in Perseus] and the left foot of Andromeda, slightly below the parallel of γ. The stars may be detected with a simple three-foot refractor. Its position was determined from β [Persei] in the head of Medusa.

NGC: Cluster, bright, very large, little compressed, scattered 9th-magnitude stars.

CAST YOUR SIGHTS ON THE constellation Perseus and what first catches your eye is the large misty splotch of the Double Cluster (NGC 869 and NGC 884), which in binoculars resolves into a grand pair of bright stellar splashes a half degree apart. While it is curious that the Double Cluster was overlooked by Messier for his catalogue (was it too obviously not a comet?), he did discover another open cluster, now called M34, glimmering nearby, just north-northeast of the midpoint between the famous variable star Algol (Beta [β] Persei) and the beautiful double star Gamma^{1,2} ($\gamma^{1,2}$) Andromedae. It's also probable that the Italian astronomer Giovanni Batista Hodierna (1597–1660) spied it before 1654, listing it as a nebula, meaning nebulous for the naked eye but resolved in a telescope – in this case a simple 20× Galilean refractor.

M34 is a loose aggregation of stars about 225 million years old, which is intermediate in age between the Alpha Persei cluster (80 million years) and the Hyades (~625 million years). Its several hundred members (about 60 of which are obvious through backyard telescopes) are spread across a scant 11 light-years of space. Regardless, Webb called it one of the finest objects of its class.

Shining at magnitude 5.7, M34 is clearly visible to the naked eye under a dark sky. I find it much easier to see than M33. Although they have the same magnitude, M33 is about three times larger in apparent diameter, so it has a lower surface brightness. M34's scattering of bright pearls includes about a dozen stars that shine brighter than 9th magnitude, all of which can be resolved with 7 × 35 binoculars; several of these are white giants. What looks

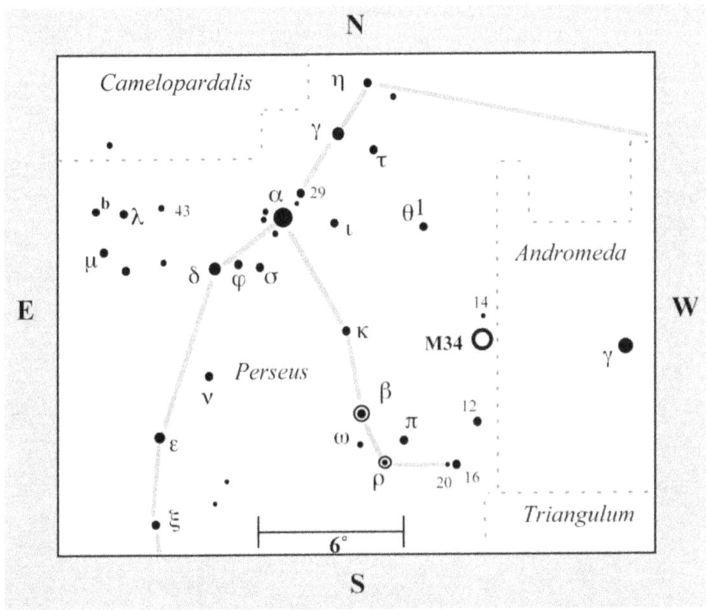

pairs" in this "scattered but elegant group." In fact, he seemed to favor the central double more than the cluster itself, calling M34 a "double star in a cluster."

After probing the depths of the cluster, return to low power and then slightly defocus the view and look for two large stellar arcs abutting the cluster's core to the southwest (see the drawing). If you visually measure the radius from the core to the outer arc and then look about the same distance to the northeast, you may see some hazy outer arcs of starlight. They might show more clearly if you gently sweep the telescope back and forth over this region. Once you see the arcs, use averted vision to stare at them, and they will resolve into individual stars. The region between the faint arcs and the core looks rather vacuous. But this blank area is actually filled with very faint background stars. Algol, the bright star about 5° to the southeast of M34, is one of the most famous variable stars in the night sky. Its placement in the heavens represents the head of Medusa, the serpent-haired Gorgon of classical mythology, held by her slayer Perseus. The "demon star" usually shines at magnitude 2.1, but nearly every three days it mysteriously fades to magnitude 3.4 before brightening again, all in just 10 hours. This periodic dimming occurs because Algol is an eclipsing binary star – a pair of stars orbiting a common center of mass – whose orbital plane lies in our line of sight. Every 2 days 20 hours 48 minutes and 56 seconds, Algol's larger but

like the brightest star in M34 shines at magnitude 7.3, but this star is a foreground star, not a true cluster member. One of the brightest true members is a double star known as Struve 44, whose magnitude 8.4 and 9.1 components are separated by 1.4″.

Although M34 is certainly a first-rate cluster for small telescopes, the late Walter Scott Houston preferred the view through 15 × 65 binoculars. More magnification, he said, merely spreads out the few bright stars that the binoculars show perfectly well. Robert Trumpler classified the cluster as II3r, indicating that it is a rich star cluster with little central concentration, detached from the Milky Way, and composed of bright and faint stars.

Through the 4-inch, the view of the cluster's inner region at 72× is quite dramatic. A spray of double stars, many of similar brightness, appear to be fleeing from two 8th-magnitude type A0 gems separated by 20″ (the double star h1123) at the very heart of the cluster. Smyth also noted the gathering of "coarse

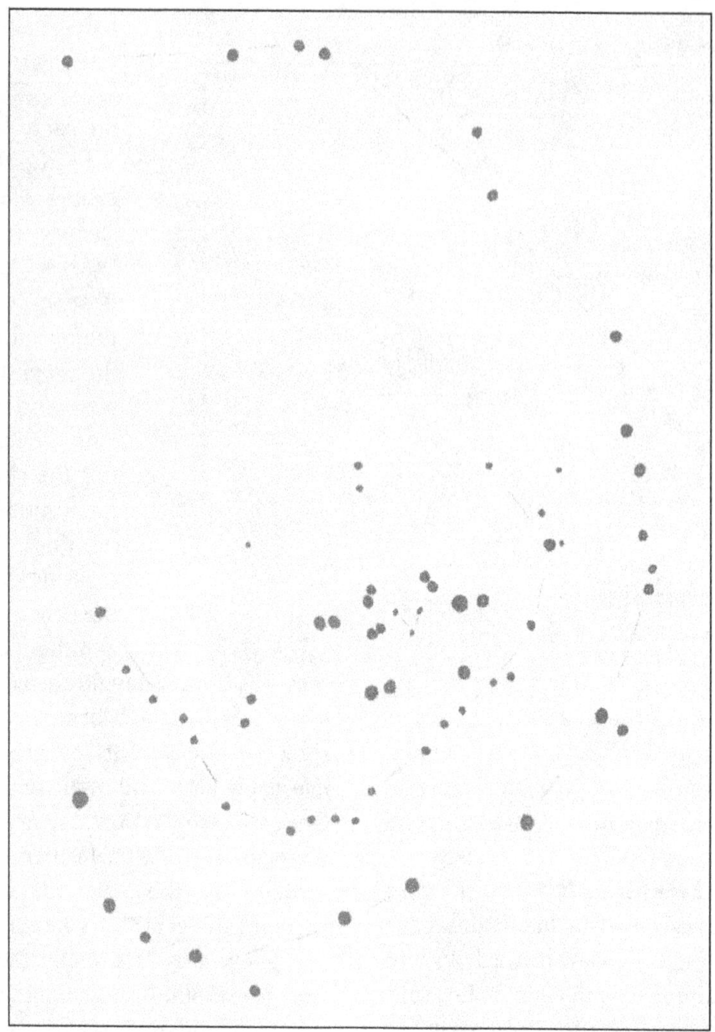

comparatively dim secondary star eclipses its smaller but brighter primary by 79 percent. We see this drama unfold as an apparent winking of the star. By the way, the dimming is not constant, in part because Algol is not a simple binary star but a complicated quadruple system.

NGC 891, a more challenging but worthwhile object in Andromeda, lies 3° to the west-southwest of M34, about midway between it and Gamma Andromedae. This 10th-magnitude spiral galaxy (13′ × 2.8′) is seen exactly edge-on. Furthermore, the system has a diameter of about 160,000 light-years, about 1 1/2 times larger than our Milky Way Galaxy. But we can still imagine NGC 891 as our Galaxy seen edge-on about 43 million light-years distant.

M35

NGC 2168
Type: Open Cluster
Con: Gemini

RA: $06^h09.0^m$
Dec: +24°21′
Mag: 5.1
Diam: 25′
Dist: 2,500 light-years
Disc: Philippe Loys de Chéseaux, 1745

MESSIER: [Observed August 30, 1764] Cluster of very faint stars, close to the left foot of Castor [the western twin of Gemini], not far from the stars μ and η in that constellation. M. Messier plotted its position on the chart for the comet of 1770, *Mémoires de l'Académie 1771*, plate VII. Plotted in the English *Atlas Céleste.*

NGC: Cluster, very large, extremely rich, pretty compressed, stars from 9th to 16th magnitude.

"A MARVELOUSLY STRIKING OBJECT: no one can see it for the first time without exclamation." William Lassell, a nineteenth-century English amateur astronomer, penned this description of M35 in Gemini based on a view through his 24-inch reflector. But this bright (5th-magnitude) open cluster is equally exquisite when seen through small apertures. To the naked eye, it is a mottled splash of hazy light with an apparent diameter nearly that of the full moon (the cluster's true diameter is about 18 light-years). It has a decidedly rectangular shape, with a bright eastern ridge and a seemingly tight core. The mottled naked-eye appearance is almost exasperating, because no amount of time seems enough to resolve it faithfully. The cluster's brightest members shine between 8th and 9th magnitude and, not surprisingly, lie along that conspicuous eastern wall (actually an arc). Someone with keen, young eyes might be able to resolve a few of the stars in the wall from a good site at high elevation. (Although I can convince myself that the cluster is resolved, I cannot pinpoint the location of any given star accurately enough to prove it; and therein lies the challenge.)

The outlying stars fill a 1.5° field and extend east to the 6th-magnitude star 5 Geminorum – a view best appreciated through a rich-field telescope. When I visually connect the brightest stars in this extreme outer halo, they resemble a scallop shell's wavy edges. The sides of the shell are marked by 5 Geminorum to the east-northeast and the tiny, magnitude 8.6 open cluster NGC 2158 to the southwest. A snaking dark river flows between the scallop's halves, making them appear slightly open.

M35 is one of the richest, most compact, and nearby open star clusters located in

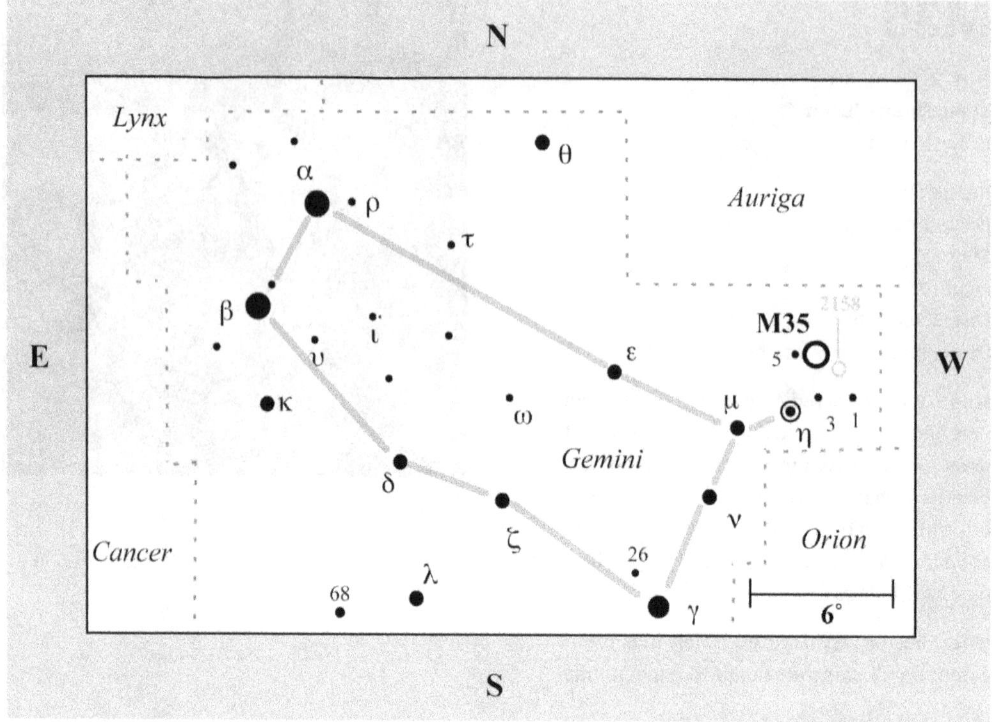

the direction of the Galactic anticenter. Star counts in M35 vary dramatically with the observer. Lord Rosse counted 300 in a field of 26′. Luginbuhl and Skiff tally 200-odd stars in a field of 30′. Åke Wallenquist counted 119 stars in a diameter of 30′, estimating the stellar density there to be about six stars per cubic parsec, which is very loose. (By comparison, the density of M34's core is 21 stars per cubic parsec.)

Recent observations have identified 250 single and 100 binary stars as confirmed members. Other studies have revealed 14 white dwarf stars, some 15 variable star candidates, and 51 x-ray sources.

Hubble Space Telescope observations of the proper motions of 22 stars in M35 provided a radial-velocity dispersion of 0.4 miles per second, a distance of ~2,500 light-years, and an age of 133 million years. In a 2011 paper in *Astronomical Journal* (vol. 142, p. 53), Bernard J. McNamara (New Mexico State University) and his colleagues note, however, that "although this age is consistent with that typically found for M35, the formal error in the dynamical distance of ±19 percent can accommodate ages between 65 million years and 201 million years."

Through the 4-inch at 23×, M35's central stars form a figure eight pattern. A corset of about two dozen stars seems to hold in the cluster's slender waist. This central girdle is flanked by two large voids: one to the south, the other to the northeast. Is this absence of stars caused by obscuring dust in the galactic arm? Look carefully at these voids and see if there is a wavy stream of stars originating at the center of each of them. Both streams flow

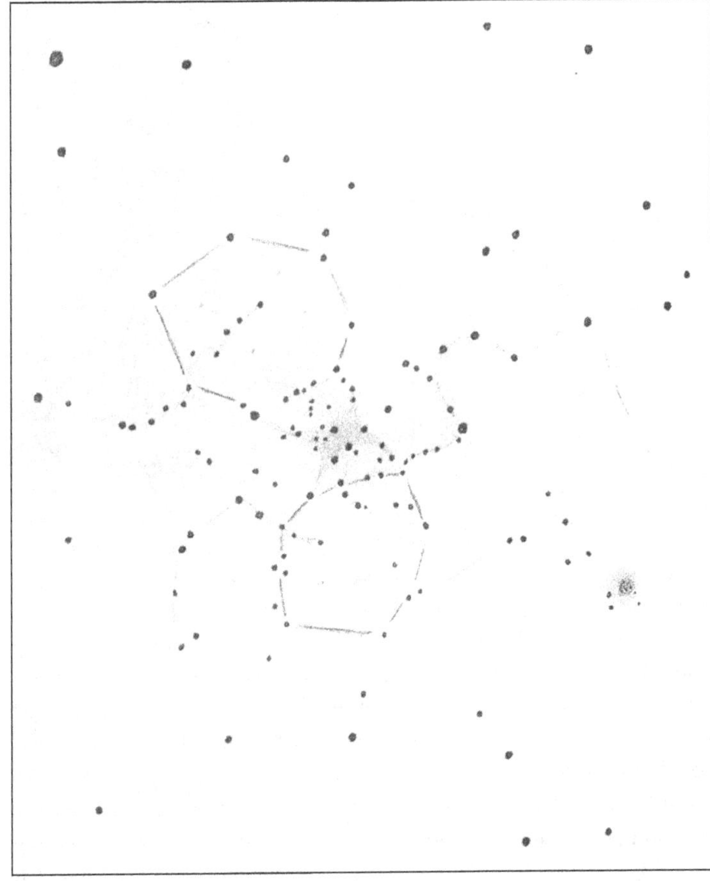

M35, saying its pattern of stars reminded him of a bursting skyrocket.

Earlier I mentioned NGC 2158; it's a pint-sized star cluster only 5′ in diameter that lies 1/2° southwest of M35. Although NGC 2158 appears smaller and dimmer than M35, the two clusters are really very similar in physical size; it's just that NGC 2158 is six times more distant. Visually, NGC 2158 bears some resemblance to NGC 1907 near M38; both clusters appear as milky spots with some stars hovering near the limit of vision.

Early sources severely underestimated the brightnesses of both NGC 2158 and the brightest stars within it, listing the cluster as 11th magnitude, with its brightest members being magnitude 16. I was surprised to find the cluster's milky haze quite easily in the 4-inch, and at 130× I resolved some of its brighter members. Skiff determined the cluster's visual magnitude to be 8.6, and through an 8-inch telescope he resolved some 50 stars of magnitude 13 or fainter. NGC 2158 is very sensitive to light pollution. I never could see this object through Harvard's 9-inch refractor because of urban skyglow.

to the east, though the southern one makes a sudden curve to the south. If you concentrate, you might see these streams coming toward you, an illusion that adds dimension to the otherwise flat cluster.

Star streams also radiate in all directions from the figure eight. They connect to an outer ring of similarly bright stars. This view mimics the smoky remains of a fireworks display faintly illuminated by city lights. The nineteenth-century observer Admiral William Henry Smyth had a similar impression of

M36

NGC 1960
Type: Open Cluster
Con: Auriga

RA: $05^h36.3^m$
Dec: +34°08′
Mag: 6.0
Diam: 10′
Dist: ~4,100 light-years
Disc: Giovanni Batista Hodierna noted it before 1654

MESSIER: [Observed September 2, 1764] Star cluster in Auriga, close to the star φ. With a simple three-and-a-half-foot refractor it is difficult to distinguish the stars. The cluster does not contain any nebulosity. Its position was determined from φ.

NGC: Cluster, bright, very large, very rich, little compressed, with scattered 9th- to 11th-magnitude stars.

M36 IS ONE OF THREE BRIGHT MESSIER clusters decorating a rich band of the Auriga Milky Way (the others being M37 and M38). The cluster is young and metal-poor, with an age of about 40 million years. Italian astronomer Giovanni Batista Hodierna discovered all three before 1654, cataloguing them as nebulae to the naked eye but resolved in his 20× Galilean telescope. Hodierna catalogued these objects in his *De Admirandis Coeli Characteribus* of 1654, though his work went unnoticed until the 1980s. Thus, until recently, Guillaume Le Gentil was often cited as having discovered M36 in 1749.

To find it, look about 8° east of magnitude 2.7 Iota (ι) Aurigae, or about 6° north-northeast of magnitude 1.6 Gamma (γ) Aurigae. M36 and its closest neighbor, M38, appear as hazy "stars," pieces of lint clinging to the dusty veil of the Milky Way. A moderately young cluster, M36 has a true diameter of about 12 light-years, with perhaps 60 telescopic stars visible in a field of 10′ (12 light-years).

Telescopically, M36 appears much as Rev. Thomas W. Webb characterized it, a "beautiful assemblage of stars ... very regularly arranged." Its 15 brightest members (between 9th and 11th magnitude) shimmer against a tight, hazy background caused by the feeble light of dozens of 13th- and 14th-magnitude stars. As with other open clusters embedded in the Milky Way, it's hard to discern the actual outer boundaries of the cluster.

At 23×, M36 displays a loose central concentration of stars with two long, slightly curved arms (extending southwest and northeast) and smaller appendages perpendicular to them. D'Arrest saw the central swirls arranged in three tight spirals. Mallas envisioned a crab, and Jones imagined a rocking chair or a miniature Perseus. I see the central association forming a warped cross – one with an orientation strikingly similar to Webb's Cross

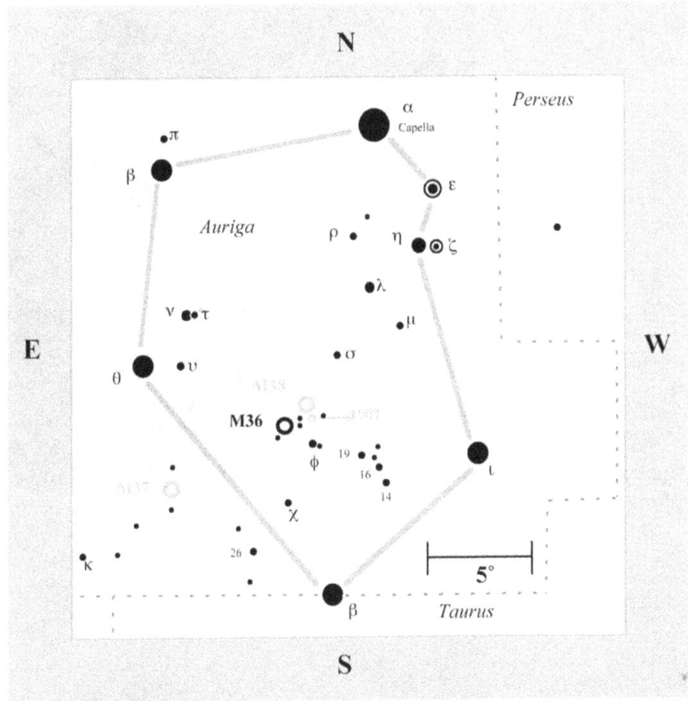

cluster's center. The southerly one is Struve 737, a pair of 9th-magnitude stars separated by 11″. The wider, brighter double to the north is oriented about 90° from Struve 737, and (if you let your mind relax) it looks closer but really is not. It's just smoke and mirrors.

Now return to low power. At 23×, M36 shares the same wide-field view with M38. Place the two clusters so that they are on the northeast edge of a wide-field eyepiece, so you can see 5th-magnitude Phi (ϕ) Aurigae and its attendant stars forming a triangle with them 1 1/2° to the west-southwest. Now sweep the telescope generously back and forth, moving first northeast to southwest, then northwest to southeast. Can you make out the lines of 8th- and 9th-magnitude stars connecting the clusters to one another and to Phi? These stars encompass a delta of Milky Way, which shines with the misty glow of countless faint stars, tiny clusters, and patches of nebulosity. It is one of the most rewarding views for rich-field telescopes in the northern skies. Large binoculars will also do justice to this region. Burnham found the cluster makes its best impression in 6- and 8-inch telescopes at 20× to 60×.

at the heart of M38. Since the two clusters are neighbors, hop from one to the other and see if you can make out the twin crosses.

Switch to medium power to examine the pair of close double stars straddling the

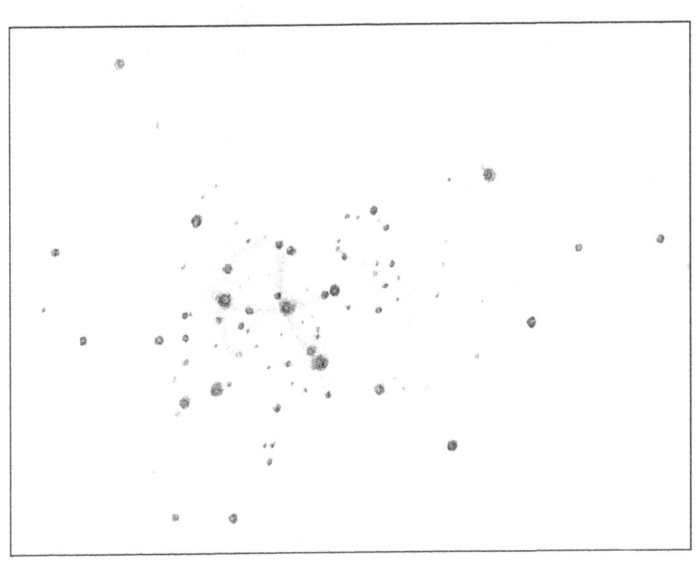

M37

M37

NGC 2099
Type: Open Cluster
Con: Auriga

RA: $05^h52.3^m$
Dec: +32°33
Mag: 5.6
Diam: 15
Dist: ~4,400 light-years
Disc: Giovanni Batista Hodierna, before 1654

MESSIER: [Observed September 2, 1764] Cluster of faint stars, close to the previous one [M36], on the same parallel as χ Aurigae. The stars are fainter, closer together, and are enveloped in nebulosity. It is difficult to see the stars with a simple three-and-a-half-foot refractor. This cluster is plotted on the chart of the second comet of 1771, *Mémoires de l'Académie 1777*.

NGC: Cluster, rich, pretty compressed in the middle, with bright and faint stars.

LOCATE BRIGHT THETA (θ) AURIGAE AND THEN use binoculars to follow a gently meandering stream of 7th- and 8th-magnitude stars flowing about 5° south-southwest of it. The stream pours into M37, a solitary pool of subtle beauty. Put down the binoculars and you should be able to pinpoint the open cluster's magnitude 5.6 glow with the naked eye. It is a small cluster (15′), so its light is compact. Telescopically, M37 is a much more wonderful sight than either M36 or M38. After looking at the two other clusters, M37's grandeur took me by surprise. At 23×, this cluster looks like a finely resolved globular cluster – a wonderful illusion. Indeed, Luginbuhl and Skiff likened it to a "broken down view" of M22 or Omega (ω) Centauri, and these are two of the greatest globulars in the night sky.

M37 is a very rich open star cluster with an age of about 540 million years, making it slightly younger than the Hyades cluster. The cluster may contain nearly 2,000 members. About 500 of these stars are brighter than 15th magnitude, and they are spread across 15′ (about 20 light-years of space). As of 2012, the WIYN Open Cluster Study had observed and identified 545 single and binary members of the cluster's upper main sequence. A total of 24 variables (including nine δ Scuti-type pulsating stars, seven eclipsing binaries, and one peculiar variable star) have also been identified in the cluster.

A deep search for planets transiting stars in M37 conducted at the Multiple Mirror Telescope in 2008–2009 has revealed a total cluster mass of about 3,640 solar masses, light

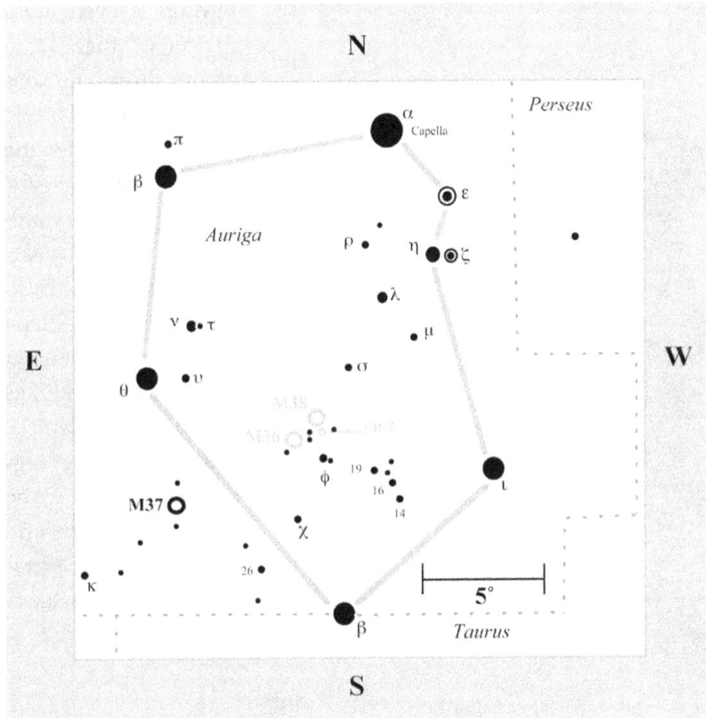

curves for about 500 cluster variables (the bulk of which are most likely rapidly rotating young, low-mass stars), and rotation periods for 575 stars. And while the survey did not detect any transiting planets among the ~1,450 observed cluster members, the researchers, led by J. D. Hartmann of the Harvard-Smithsonian Center for Astrophysics, did, however, identify a Jupiter-sized candidate planet transiting an ~0.8 solar mass Galactic field star with a period of 0.77 days.

Anyone who likes gemstones should appreciate the telescopic view of M37 – a 9th-magnitude topaz jewel surrounded by a pear-shaped cluster of scintillating diamonds. Smyth called it a "magnificent object; the whole field being strewn ... with sparkling gold dust." The color of the central star alone could mesmerize even the most tenured observer, though the degree of color perceived varies with the observer. Some have referred to it as ruby red (another gem!), others as simply red, still others as pale red. Could this star be variable? Someone with a photometer should try monitoring the brightness and hue of this intriguing star over the course of a year.

I am surprised that I haven't come across more references in the literature to the obvious dark lane slicing through the eastern half of the cluster's core. Luginbuhl and Skiff described it as a dark void about 5′ across containing a single 12th-magnitude star. In this respect, the pear-shaped diamond really looks more like a coat of arms. To enhance the dark lane's visual impact, slightly defocus the telescope.

At moderate magnifications, M37 is gloriously detailed. Its stellar richness accounts for only half of the cluster's beauty, though. The other half is the vast network of dark lanes, which, at least from dark skies, becomes visually overpowering the longer I stare at the cluster. When I relax my gaze, the stars seem to fade into the background, while the dark lanes move forward. At that point, I feel as if I'm looking at the silhouettes of leafless trees against a sparkling expanse of Milky Way. If my eye jumps around the field, dark lines appear to slash across the cluster from all directions; it's as if I'm looking at this cluster through a severely fractured eyepiece. The

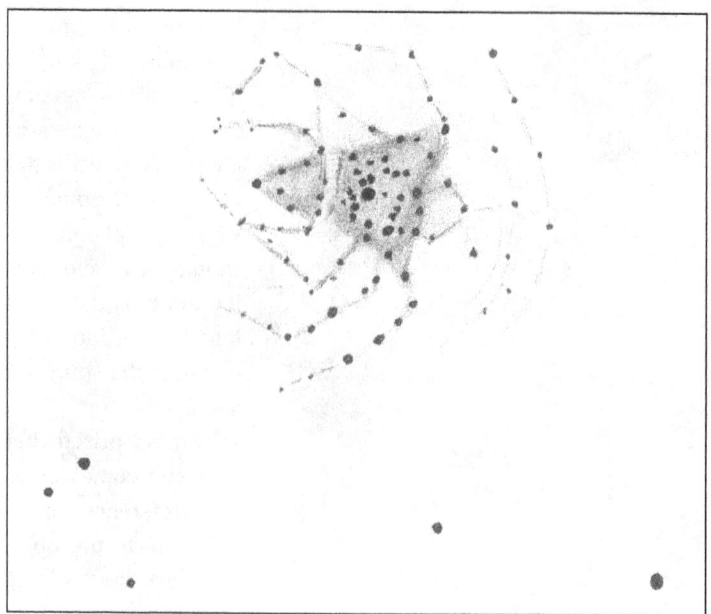

illusion is even more dramatic than that of the dusty cobwebs covering M4.

If you consider the "legs" of stars extending north and south from M37's pear-shaped body, the cluster mimics the shape of a water strider insect in a puddle of rain, as my drawing depicts. I suspect these legs represent part of the "wonderful loops and curved lines of stars" seen by Lord Rosse.

M38

NGC 1912
Type: Open Cluster
Con: Auriga

RA: $05^h28.7^m$
Dec: +35°51′
Mag: 6.4
Diam: 15′
Dist: 4,600 light-years
Disc: Giovanni Batista Hodierna, before 1654

MESSIER: [Observed September 25, 1764] Cluster of faint stars in Auriga, close to the star σ, not far from the two preceding clusters [M36 and M37]. This one is rectangular in shape and contains no nebulosity, if examined carefully with a good telescope. It may extend for about 15 minutes of arc.

NGC: Cluster, bright, very large, very rich, with an irregular figure, bright and faint stars.

ABOUT 2° NORTHWEST OF M36 AND FAINTLY visible to the naked eye is the third Messier open cluster in Auriga, M38. At 15′ (20 light-years) in diameter, M38 is a massive star cluster, one-third larger than M36 in apparent size; it is also a half-magnitude fainter, independent of size. Surprisingly, M38 also has twice the number of stars that M36 has. M38 is an intermediate-age galactic cluster (~375 million years) and contains about 300 stars in an area equal to half the diameter of the full moon. The density of its loosely packed center is about eight stars per cubic parsec – half the density of M36's core and only one-fifth that of M37's.

About a dozen of the brightest stars in M38 are arranged in a distinct cross at the center of the cluster. Webb saw this first as an oblique cross with a pair of brighter stars in each arm, a view that reminds me of a silver crucifix studded with jewels. This is most obvious at high power. Look carefully and you will see a ring of stars at the center of the cross, within which shines a solitary 11th-magnitude star (dimmer stars can be seen, but they are not obvious). A fainter ellipse of stars surrounds the cross, and, with imagination, I can see the arms of the cross wrapped around a hemispherical shell. This evokes a strong image of the bleached shell of a sea urchin seen at a slightly oblique angle, as shown in my drawing.

A ring of bright stars connects M38 to NGC 1907, a compressed 8th-magnitude open cluster 30′ to the south-southwest. (Interestingly, this connection is not obvious at moderate to high powers or in small fields of view.) It doesn't take much imagination to at least contemplate M38 and NGC 1907 as being one cluster with a mighty void at the

M38

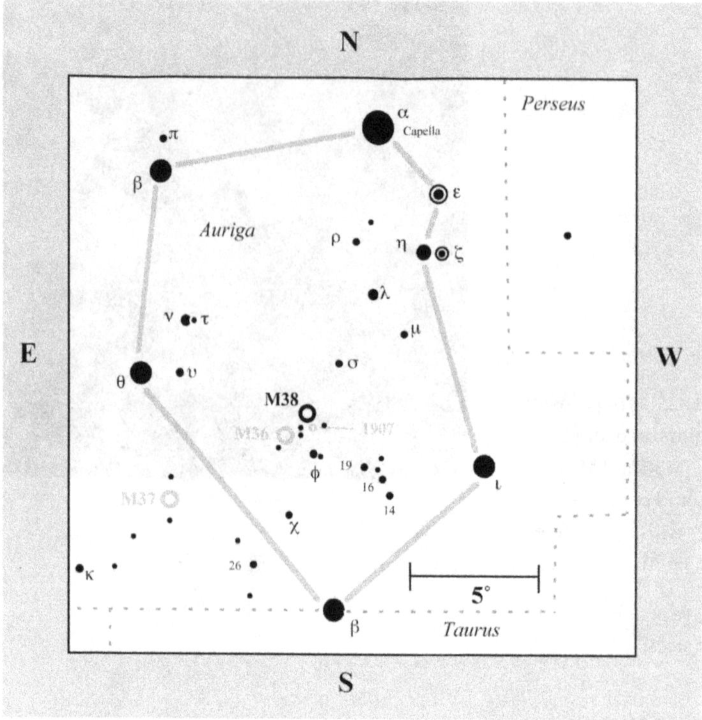

center (remember the black heart of M35?). In a 1995 paper (*Bulletin of the Astronomical Society of India*, vol. 23, p. 449), A. Subramaniam et al. considered M38 and NGC 1907 (a slightly older cluster at 400 million years) as a fairly good candidate for a true physical cluster pair. But the distances of these two clusters have been hard to determine owing to an interstellar cloud in the foreground. But in a 2007 paper in *Publications of the Astronomical Society of the Pacific* (vol. 59, p. 547), Anil K. Pandey and his colleagues state that the two clusters are about 1,200 light-years apart, indicating that in spite of their close locations on the sky, they may have formed in different parts of the Galaxy. Radial-velocity and proper-motion measurements of the cluster in 2002 support that contention, implying that the two clusters are presently experiencing a "fly-by."

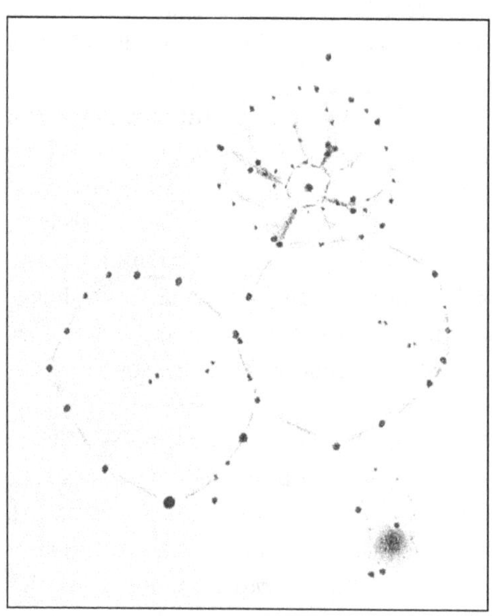

When I view the two together, I like to imagine I am looking at an aerial view of a galactic impact site, as if some hideously large comet once collided obliquely with the Milky Way in Auriga – the ejecta splashing out to the north-northeast to form M38, and a pile of rubble falling to the south-southwest to form NGC 1907. Move the telescope to the southeast, where a second ring of stars adjoins the first.

Inside each ring, you'll find two pairs of tiny doubles (see the drawing for locations).

Examine NGC 1907 at high power. Its dark veins make it appear like a trifid of stars. I really like this cluster because, through the 4-inch, at all powers, it looks like a tiny puff of smoke blowing out of a volcano. But this is an immediate impression. With careful study, it vaguely resolves into a rectangle of equally bright stars. It's rather strange to see such a uniform smattering of stars packed together in so small an area.

M39

NGC 7092
Type: Open Cluster
Con: Cygnus

RA: $21^h31.9^m$
Dec: +48°25.5′
Mag: 4.6
Diam: 31′
Dist: ~950 light-years
Disc: Charles Messier, 1764

MESSIER: [Observed October 24, 1764] Cluster of stars near the tail of Cygnus. They can be seen with a simple three-and-a-half-foot refractor.

NGC: Cluster, very large, very poor, very little compressed, of 7th- to 10th-magnitude stars.

OF THE DOZENS OF BRIGHT CLUSTERS AND nebulae in Cygnus that are available to small telescopes, surprisingly few were recorded by Messier and his contemporaries – only three (M29, M39, and M56). Open cluster M39, near the Swan's tail, is of moderate age and belongs to a class of poorly populated open clusters – only 30 bright telescopic stars span its 31′-wide (~9 light-years) disk.

Yet, shining at magnitude 4.6, M39 does greet the naked eye as a small fuzzy patch of light about 9° east-northeast of Deneb, or 3° north of 4th-magnitude Rho (ρ) Cygni. Charles Messier was the first to provide an accurate position and description of this cluster. The late Kenneth Glyn Jones notes that Bigourdan offered the possibility that Guillaume Le Gentil discovered the object in 1750, recording that it is 6° from the tail of the Swan and that the "cloud" could be seen without a telescope. But he also called it "opaque and very dark," though he did identify several stars through a telescope. It is also possible that around 325 B.C. Aristotle saw M39 without optical aid as a comet-like object.

Actually, under a dark sky, I find that at least three of its brightest members can be distinguished with the unaided eye, and 7 × 35 binoculars easily resolve about a dozen stars in an area of sky the size of the full moon. Smyth seemed pleased with this venerable 300-million-year-old cluster, calling it a "splashy field of stars in a rich vicinity."

The beauty of M39 lies in the uniform brightness of its most prominent members, their diamond blue purity, and the cluster's geometry – a triangle of starlight with a fine double star at its heart. Smyth also noted, as the drawing shows, that many of the stars are in pairs. I do wonder about Flammarion's description: "unusual curved runners of stars, with a compressed cluster of 20 stars, difficult to separate from the rest." My guess is that he based his comment on a binocular view, because I, too, have noticed one significant

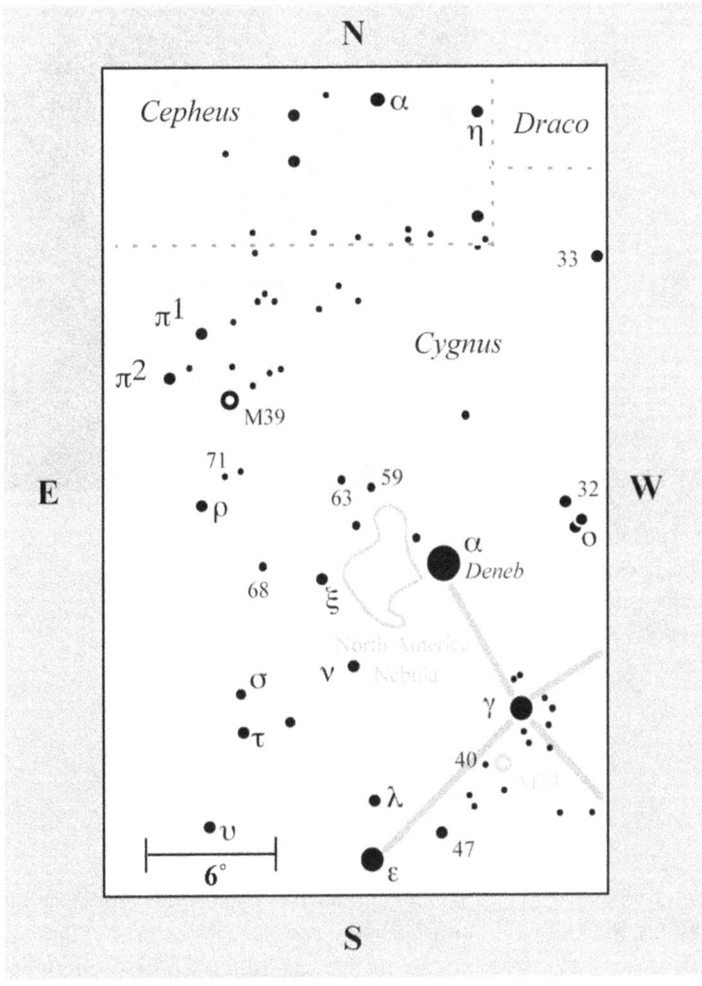

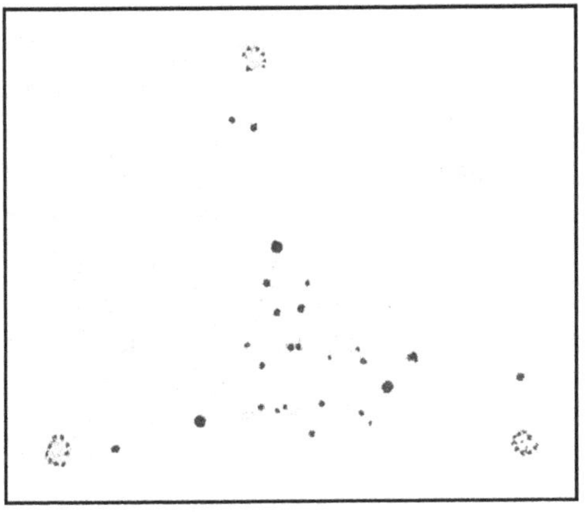

curve of stars running through that triangle in binoculars.

Meanwhile, use low power to survey the area around M39. Do you notice anything curious? Look outward from each of the triangle's three corners and you will find a tiny asterism of stars. Together the three asterisms form a perfect triangular encasement for the smaller triangle of M39.

Sarah Schuff of Brigham Young University and her colleagues photometrically monitored 228 stars in the cluster field to determine their variability. Of these, they found 10 suspected short-period variables. They reported their findings in 2006 in the *Bulletin of the American Astronomical Society* (vol. 38, p. 1135).

M40

Winnecke 4
Type: Double Star
Con: Ursa Major

RA: $12^h22.4^m$
Dec: +58°05′
Mag: 9.0 and 9.6
Sep: 52.8″ (Hipparcos)
Dist: 510 light-years (primary)
Disc: John Hevelius, 1660

MESSIER: [Observed October 24, 1764] Two stars very close to one another and very faint, located at the root of the tail of Ursa Major. They are difficult to detect in a simple six-foot refractor. It was while searching for the nebula that lies above the back of Ursa Major as plotted in the book *Figure des Astres,* and whose right ascension should have been 183°32′41″ and declination +60°20′33″ in 1660 – which M. Messier could not see – that he observed these two stars.

NGC: None.

M40 IS A PAIR OF CLOSE STARS OF NEARLY equal magnitude in Ursa Major that John Hevelius (1611–1687) had described as "a nebula above the back [of Ursa Major]" in an annotation to his *Prodromus Astronomiae* star catalogue, published posthumously in 1690. When Messier saw the object plotted in the second edition of Maupertuis's *Figure des Astres* (1742), he sought it out. Messier tried to find a nebula at the coordinates given by Hevelius, but identified only two stars. After observing it, he decided to include it in his catalogue because "even though Hevelius mistook these two stars for a nebula," he said, "they are difficult to distinguish with an ordinary telescope of six feet." Obviously Messier concluded that users of small apertures could in fact confuse the object with a comet, so he appropriately added it to his list. (Although no one using modern telescopes would have trouble identifying M40 as two close stars, one must consider the poor quality of optics in Messier's day.)

The nebulous appearance of a close double should not be surprising to anyone who has swept the Milky Way with binoculars or a rich-field telescope. Close pairings of stars often look like comets, especially at low magnifications. Ironically, French astronomer Camille Flammarion (1842–1925) observed a different double than the one Messier encountered in his search for the Hevelius nebula. And Smyth did the same! Confusing the matter further is that, when precessed, the position

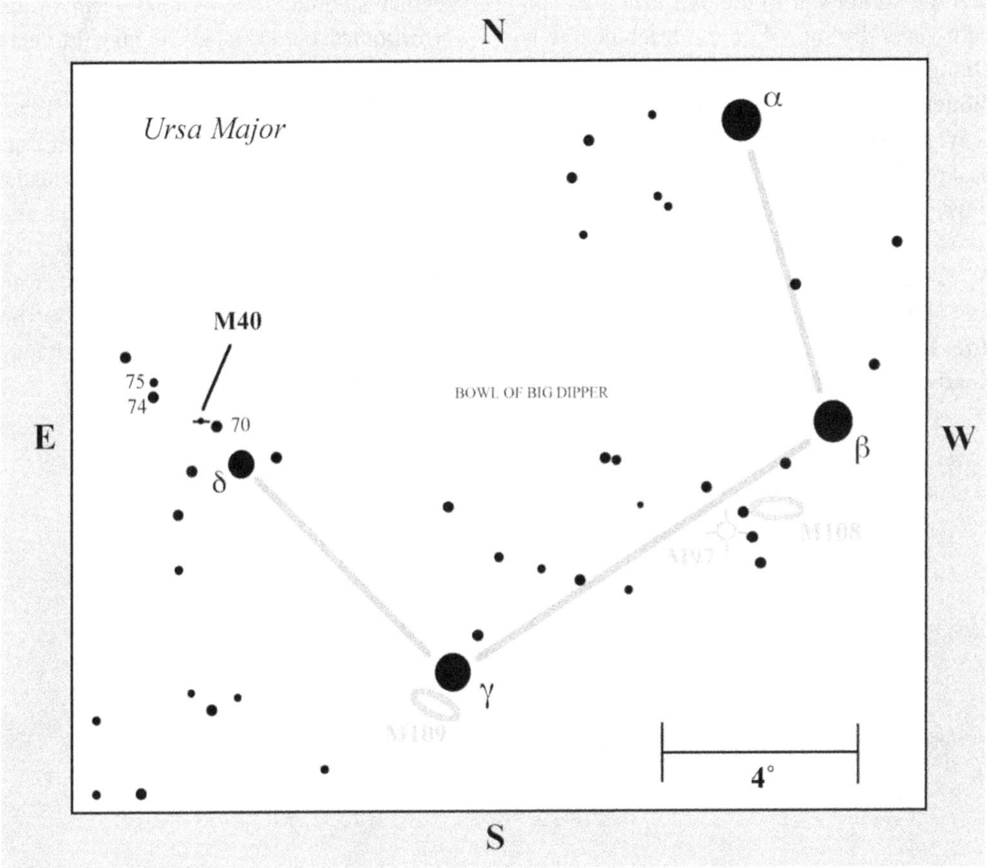

of Hevelius's nebula coincides with the position of the star 74 Ursae Majoris, which is not even double!

Regardless, John Mallas verified that the object Messier observed and catalogued is Winnecke 4 (WNC 4), a double star rediscovered in 1863 by A. Winnecke at Pulkova Observatory in Russia. And because that double is the one recorded in the Messier catalogue, it is the one we will locate. In a 2002 paper in the *Journal of the Royal Astronomical Society of Canada* (vol. 96, p. 63) Richard L. Nugent of the International Occultation Timing Association in Houston, Texas, used Tycho-2 and other data to determine a *minimum* distance to the pair, namely 554±228 light-years. If the pair were gravitationally connected, he says, they would lie 5,000 astronomical units (1 astronomical unit is equal to the distance between the Earth and the Sun) apart and have a minimum period of 232,000 years, which is highly improbable for a true binary star system. Nugent also supplies Hipparcos and Tycho parallaxes, which show that the orbital pair can be explained by the proper motion of two optically aligned stars, indicating no gravitational attraction. Thus, Nugent concludes that WNC 4 is an optical double star and not a true physical pair. The Hipparcos and Tycho data verify that the fainter star in the pair is *at least* 490 light-years distant, while the brighter one is about 1,860 light-years distant, making the fainter star the closer of the two.

Winnecke 4 is easily found about 1/2° northeast of 70 Ursae Majoris, near Delta (δ) in the bowl of the Big Dipper. Its stars have magnitudes of 9.0 and 9.6, and are separated by 49". At 23×, Delta, 70, 74, and 75 Ursae Majoris, and M40, all fit in the same low-power field. The barred spiral galaxy NGC 4290 lies nearby and is easily visible in the 4-inch as an irregular splotch. *The Deep-Sky Field Guide to Uranometria 2000.0* lists its magnitude as 11.8, though I estimate it to be 11.5.

When I push my vision to the limit, I can see a bar, which looks like a detached segment, almost making the galaxy appear like a double nebula. More challenging and at the limit of visibility is yet another galaxy, NGC 4284, immediately west of NGC 4290. Can you see it? Several times I suspected it. The galaxy's magnitude is listed as 13.5, but it may be brighter.

M41

Little Beehive
NGC 2287
Type: Open Cluster
Con: Canis Major

RA: $06^h46.1^m$
Dec: -20°46′
Mag: 4.5
Diam: 39′
Dist: 2,100 light-years
Disc: John Flamsteed, 1702; Giovanni Batista Hodierna recorded it before 1654, and Aristotle noted it about 325 B.C.

MESSIER: [Observed January 16, 1765] Cluster of stars below Sirius, close to ρ Canis Majoris. This cluster appears nebulous in a simple one-foot refractor. It is no more than a cluster of faint stars.

NGC: Cluster, very large, bright, little compressed, stars of 8th magnitude and fainter. (Note: The *New General Catalogue* incorrectly recorded this cluster as M14.)

As Canis Major, the Great Dog of Orion, rises above cool winter landscapes, open cluster M41 hangs below its collar like an ice-covered tag reflecting moonlight. One of the faintest objects recorded in classical history, M41 was seen by Aristotle as a star with a tail. With the naked eye, I find it surprisingly similar to the Beehive Cluster (M44), only smaller. Virtually sitting on 6th-magnitude 12 Canis Majoris, about 4° south of Sirius (the brightest star in the night sky at magnitude -1.5), M41 shines at magnitude 4.5 and fills an area of sky 30 percent larger than the full moon. From nineteenth-century England, Reverend Webb hailed it as a "superb group visible to the naked eye."

Unfortunately for many of today's observers, the naked-eye view is not as superb, or even possible, thanks to light pollution. But if you do enjoy dark skies, M41 looks to the naked eye like a ghost image of Sirius, or a falling clump of snow illuminated by Sirius's fire-fused crystal blaze. In 7 × 35 binoculars, both Sirius and M41 fit in the field of view, and binoculars resolve the cluster quite well. But even with the unaided eye, I get the distinct impression of resolution – a provocative sight because of the 80 or so stars in this 200-million-year-old cluster, about 50 shining between 7th and 13th magnitude; at least three of them are 7th magnitude, and seven others are 8th magnitude. I think that after several observations over time, an astute observer could resolve some individual stars with the unaided eye. The stars in

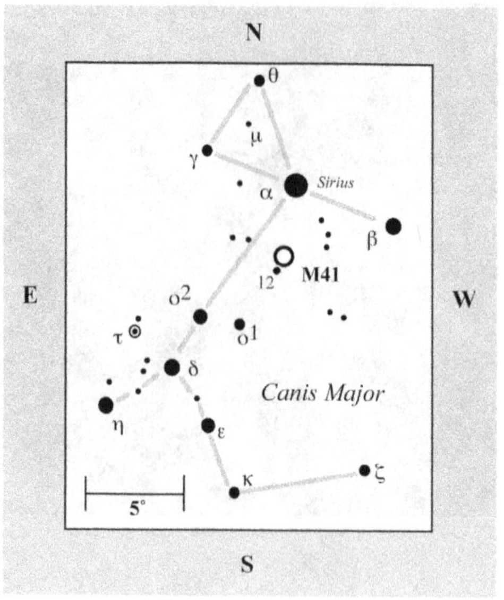

M41 are of solar type and spread across about 24 light-years; the cluster's central density is six stars per cubic parsec. Its Trumpler class is I3r, meaning it is a rich cluster of bright and faint stars with a strong central concentration that's detached from the Milky Way.

In a 2012 paper in the *Monthly Notices of the Royal Astronomical Society* (vol. 423, p. 2815), P. D. Dobbie (University of Tasmania, Hobart, Australia) and colleagues note that their CCD imaging survey of M41 revealed a white dwarf star (WD J0643-203) with an estimated mass of 1.02–1.16 solar masses, which is potentially the most massive white dwarf so far identified within an open cluster. "Guided by the predictions of modern theoretical models of the late-stage evolution of heavy-weight intermediate-mass stars, we conclude that there is a distinct possibility that it has a core composed of oxygen and neon," they say.

The first thing I notice in the 4-inch is a pair of equally bright red stars in the center of the cluster. But this observation is rather enigmatic, considering that most accounts refer to a single red star there. Webb saw "larger stars in a curve with [a] ruddy star near [the] centre." Glyn Jones described a "central, slightly orange star," and Mallas referred to it as the "famous red star at the center." Burnham notes that the "central reddish star" is a K-type giant, and that several other K giants are known in the cluster. The central star may also be variable. I compared my drawing of M41 (with two central red stars) with the drawing by Glyn Jones (with one central red star) in *Messier's Nebulae and Star Clusters.* Interestingly, I can identify my second red star due east of Glyn Jones's single red star. But he drew it markedly fainter!

Curiously, Smyth writes of M41: "A double star in a scattered cluster. ... A9, lucid white, B10, pale white." He could have been referring to a double star on the western edge of the cluster, as Glyn Jones believed (see Mario Motta's close-up image of M41). But, if so, then the prosaic Smyth, who often refers to spectacularly colored stars in clusters, apparently did not notice any predominantly red stars in M41.

Using my imagination, I can see the pattern of bright stars in M41 outlining a fruit bat, or

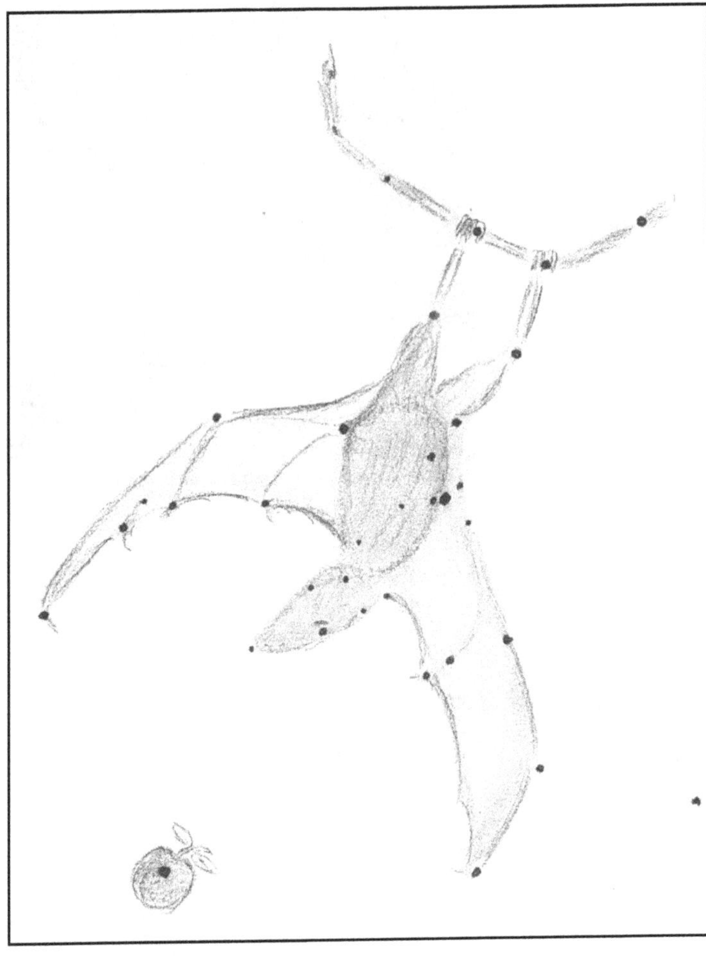

to eye. This pattern can also be made out in 7 × 35 binoculars. At 72×, the entire bat asterism fits perfectly in the field of view. But because of its larger apparent size, the asterism loses some of its impact.

The stars do stand out more boldly though, and the dark rifts between the stars seem to slither across like black snakes. The cluster suddenly appears loosely gathered into groups of stars that form triangles, arcs, lines, and other geometrical patterns, as if in a cataclysmic explosion, with celestial debris flying off in various directions.

flying fox. The bat's wings are slightly opened in a curve. The bat is hanging by a branch (which flying foxes do), and 12 Canis Majoris makes a tempting piece of fruit for the bat

M42

The Great Orion Nebula
NGC 1976
Type: Emission Nebula and Cluster
Con: Orion

RA: $05^h35.4^m$
Dec: –05°27′
Mag: 3.7
Dim: 1.5° × 1.0°
Dist: 1,500 light-years
Disc: Nicholas Peiresc, 1610

MESSIER: [Observed March 4, 1769] The position of the beautiful nebula in Orion's sword, around the star θ, which lies within it together with three other, fainter stars, which can be seen only with good instruments. M. Messier went into great detail about this great nebula; he gives a drawing, made with the greatest care, which may be found in the *Mémoires de l'Académie 1771*, plate VIII. Huygens discovered it in 1656, and many astronomers have observed it subsequently. Plotted in the English *Atlas Céleste*.

NGC: Magnificent Theta Orionis and the Great Nebula.

M43

NGC 1982
Type: Emission Nebula
Con: Orion

RA: 05^{h}35.6^{m}
Dec: -05°16′
Mag: 6.8
Dim: 20′ × 15′
Dist: 1,500 light-years
Disc: Jean-Jacques Dortous de Mairan, before 1750

MESSIER: [Observed March 4, 1769] Position of the faint star that is surrounded by nebulosity and that lies below the nebula [viewed with south up] in the sword of Orion. M. Messier depicted it on the drawing of the large nebula [M42].

NGC: Remarkable, very bright, very large, round with a tail, much brighter in the middle, contains a star of magnitude 8 or 9.

ONE OF THE GREATEST PARADOXES IN VISUAL astronomy is that Galileo (who paid great attention to Orion, according to Reverend Webb) apparently never noticed the Great Nebula – though in 1617 he did record three stars in the Trapezium through his telescope. Furthermore, there appears to be no mention of the nebula in medieval records. Yet here is one of the grandest naked-eye nebulae in the heavens, dangling from one of the best-known asterisms (Orion's belt) in one of the most famous and brilliant constellations. That Galileo would have missed seeing the nebula is especially puzzling given that Al-Sufi and others had noticed the fainter "nebula" in Andromeda (M31) as early as A.D. 905 with their unaided eyes!

Messier credits Huygens with the discovery of the Orion Nebula. But Kenneth Glyn Jones says that its discovery is generally attributed to Nicolas Peiresc in 1610, the year after Galileo turned his telescope to the stars. One certain discovery was made independently in 1618, when, according to Rudolf Wolf of Zurich, Johan Baptist Cysat (1588–1657) of Luzern published an account of it that was later mentioned by Friedrich Bessell (1784–1846) in the *Berliner Jahrbuch* for 1808, p. 122. Before 1654, G. B. Hodierna made the first known drawing of the nebula. Huygens added the fourth star to the Trapezium on January 8, 1684.

Nevertheless, Huygens did independently discover the nebula, and his description of it should interest modern-day observers: "There is one phenomenon among the fixed stars worthy of mention, which as far as I know, has hitherto been unnoticed by no one and indeed, *cannot be well observed except with large telescopes."* (Emphasis is mine.) He found it while examining the stars in Orion's sword in 1656 with his 2 1/3-inch telescopes with focal lengths of 12 and 23 feet and magnifications from 48× to 92×.

Yet, in how many New England winters have I looked across snow-laden fields to see this nebula, in the middle of Orion's sword, looking like angel's breath against a frosted sky? Even today, M42 is visible with the naked eye from the heart of downtown Boston, Massachusetts!

The Orion Nebula is an enormous cloud of fluorescent gas, predominantly hydrogen – with traces of helium, carbon, nitrogen, and oxygen – ~ 40 light-years in diameter. It glows by the ultraviolet radiation streaming from Theta (θ) Orionis, a bright grouping of four massive stars, whose light energy is burning a hole through the mackerel sky of the nebula's sharply defined interior. Commonly called the Trapezium, these four jewels shine between 5th and 8th magnitude, and it is the brightest member that produces 99 percent of the energy that illuminates the clouds of gas we see through our telescopes. All the hot, young Trapezium stars probably began shining a mere 1 million years ago (the Sun is 4.5 billion years old). And there are about a thousand unseen members (300,000 to 1 million years old) hiding in the dense cloudscapes. Recent Hubble Space Telescope (HST) images (see page 174) show the expansive nebula in unprecedented detail: tumultuous swirls of colorful churning gas; 40-billion-mile-long "comets" of dust and gas, whose comas enshroud newborn stars; dark protoplanetary disks silhouetted against the nebula's spectral glow. The HST has confirmed what astronomers have long suspected: M42 has all the ingredients for solar and planetary creation.

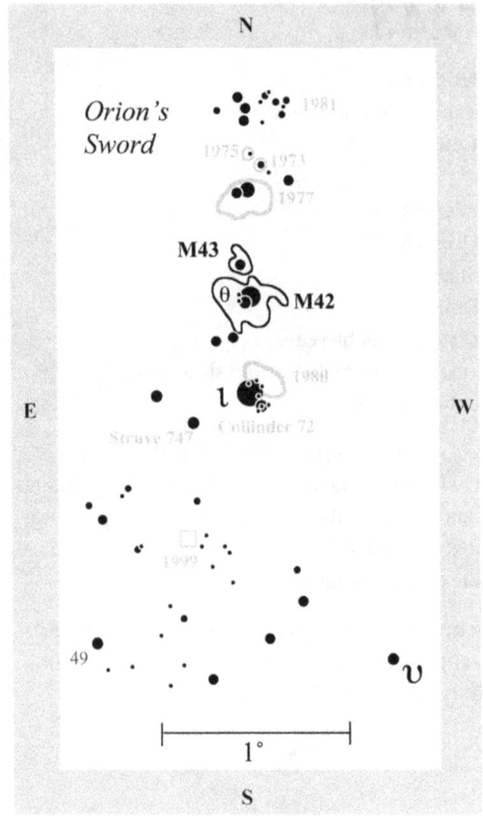

As NASA announced in 2006, among the stars HST spotted for the first time in visible light were young brown dwarfs and a population of possible binary dwarfs orbiting each other. The nebula is a cauldron of star formation – a perfect laboratory to study star formation owing to its proximity. Massimo Roberto of the Space Telescope Science Institute says that a new large-scale HST image of the Orion Nebula (a mosaic containing a billion pixels, shown here) shows the "entire formation history" of Orion printed into the nebula's features, which includes arcs, blobs, pillars, and rings of dust that resemble cigar smoke. "Each one tells a story of stellar winds from young stars that impact the environment and

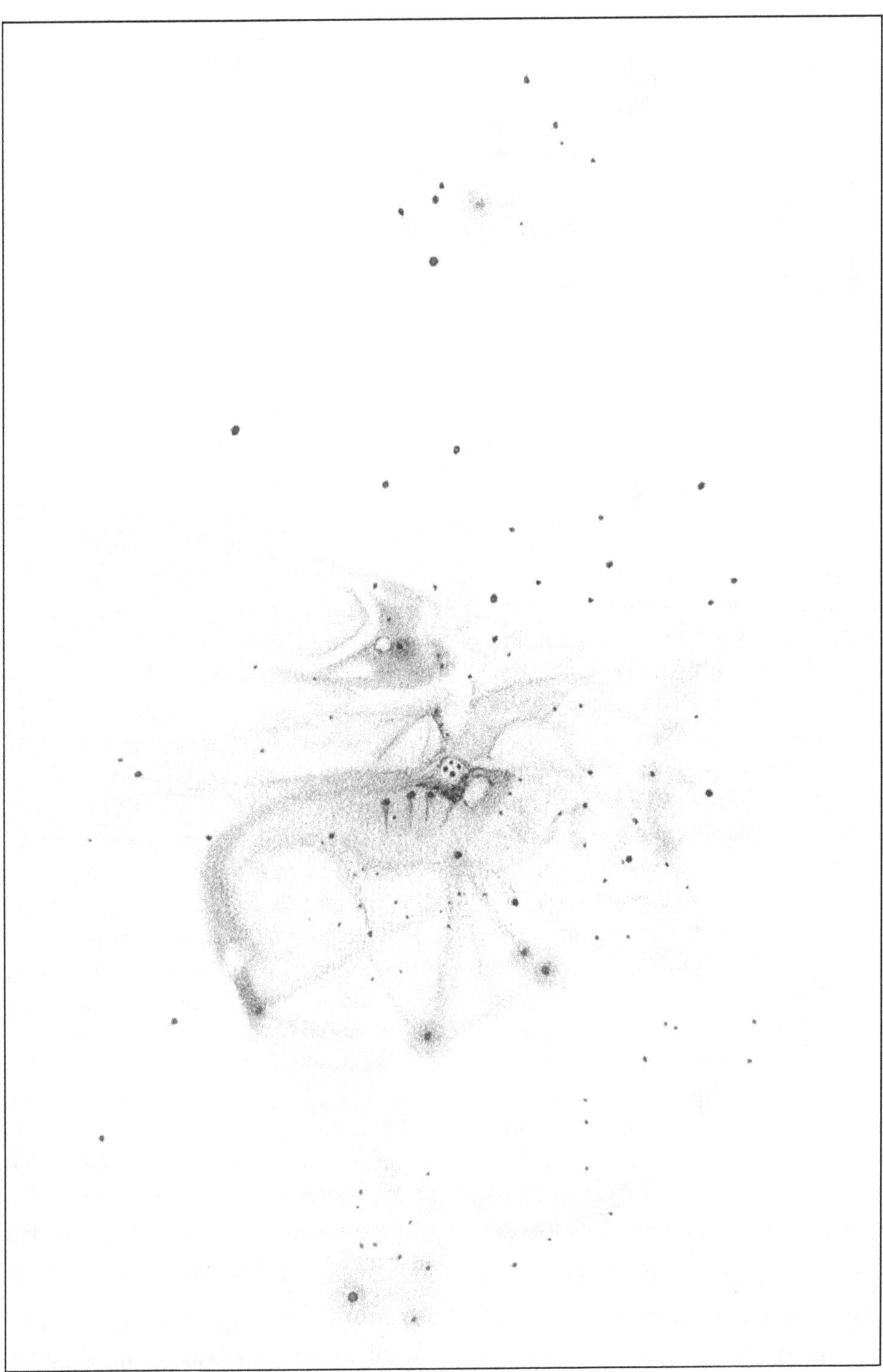

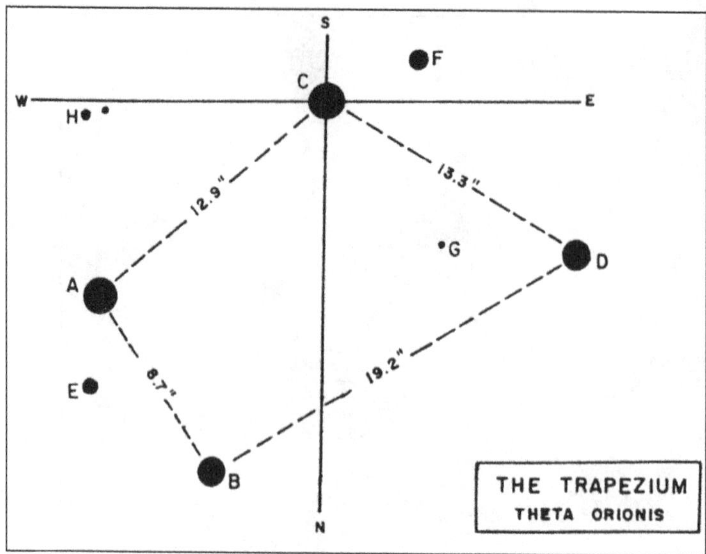

Nebula, however, fits nicely in most low-power eyepieces.

How can anyone draw the Orion Nebula and do it justice? The inner square of nebulosity surrounding the Trapezium is a chaotic witch's brew. This stellar grouping, under very steady conditions, looks like an emerging embryo cradled in a soft womb of nebulosity. The Trapezium actually contains several visible fainter stars in and around it. Two of them are commonly seen by amateurs with moderate-sized instruments. Use the chart on page 172 to search for these more difficult stars. The easiest is the 11th-magnitude star "E," about 4″ due north of "A." Also shining at 11th magnitude is star "F," 4″ southeast of "C," but I find this a more difficult object and believe it to be fainter than "E." Robert Burnham Jr. writes that both "E" and "F" have been seen with apertures smaller than 3 inches! The two other members shine at 16th magnitude and are challenges for the largest amateur instruments. Texas deep-sky wizard Barbara Wilson and I glimpsed "G" through a 36-inch Dobsonian at the 1994 Winter Star Party, though "H" remained elusive. The "X" on the Trapezium chart marks the location of a star that E. E. Barnard saw at the limit of vision through the 36-inch refractor; he saw it on two nights in 1888 and 1889, but to my knowledge it has not been recorded visually by anyone since.

Look carefully at the inner square at high magnification and try to locate the four sharp

the material ejected from other stars," Roberto says. "This appears to be a typical star-forming environment. Our Sun may have been born 4.5 billion years ago in a cloud like this one."

At 23×, I have no trouble seeing the overall shape of the nebula as a blazing comet, the kind that invoked fear and superstition in the impressionable populations of days gone by. Leading to this cometary image is the broad, sweeping gaseous mass to the southwest and the curved jets streaming from the fractured nucleus. (Remember Comet West and how its nucleus split into four fragments when it rounded the Sun in 1976?) Low power is also required to follow the large bubble of vapors rising above that turbulent central region; the thick northern arms of the bubble are the bat wings E. E. Barnard saw beating against the darkness to the south. The overwhelming structural detail of this nebula makes it appear larger than it is. M42 is about 1.5° at its widest, which is three times smaller than M31, the Andromeda Galaxy, and roughly the same size as open cluster M44. The entire Orion

flares shooting off to the south of it and the large, hedgerow "prominence" to the southwest. Now study the dark bay northeast of the Trapezium (Smyth called this the "fish's mouth"); use averted vision and you should see several herringbone bridges crossing it. If you concentrate solely on the dark bay and forget the nebulosity, a powerfully dark image appears. It is a mighty silhouette, and you can follow its course throughout the entire region. The faint nebulosity forming the bay's northern bank appears to have turbulent edges near the Trapezium, and its pale gray color and soft texture are so unlike the stronger green wreath-like nebulosity in the main body of M42.

The Orion Nebula is one of the few gaseous clouds that shows color through amateur telescopes. Imagers have no problem recording vivid reds and greens in it with proper exposures. Of course, Hubble images depict a full spectrum of vibrant colors throughout the nebula. Our eyes, on the other hand, are largely insensitive to color at night, so visual observers have to settle for subtle shades of green or red. Interestingly, when we do see these colors, pale as they might be, our mind tends to exaggerate their intensity, and we say the color is "strong," as I did earlier.

At the 1991 Winter Star Party in the Florida Keys, the view through Tippy D'Auria's 18-inch Dobsonian reflector was breathtaking. The center of the nebula, called Huygen's Region, filled the field with brilliant gas clouds mingling with dark swirls of dust. The region around the Trapezium looked like massing storm clouds. The clouds shone weakly with green light, while the outer filaments of gas were tinged ever so lightly with a pink color. A slight push of the tube brought me to the graceful outer arc of nebulosity that passes near Iota (ι) Orionis, which looked like a ballerina's arms held in high 5th position – gentle spits of light scalloped as if by the action of waves.

M43 is a wedge of nebulosity surrounding a 7th-magnitude star (called Bond's star) 10′ northeast of the Trapezium. Indeed, when French scientist Jean-Jacques Dortous de Mairan (1678–1771) discovered it, he penned, "one of the stars included in M. [Huygens's] drawing is seen to be surrounded by a bright glow very similar to that which, I believe, may produce the atmosphere of our Sun, if it were dense and extensive enough to be visible in telescopes at a similar distance."

Commonly overlooked because of its proximity and association with the Orion Nebula, M43 harbors a wealth of detail that is well worth becoming acquainted with, so plan an evening with it. You will have to block out the commanding presence of its larger neighbor, but one way to seclude M43 is to use high magnification. See if you can resolve the tiny, dark pool due east of Bond's star. One night, a short but strong earthquake struck while I was examining M43. The telescope began to shake and, when it did, my eye caught sight of some extremely faint extensions looping to the west and a long wedge of material flowing to the east. Instead of waiting for an earthquake, try tapping your telescope tube and see if you can make out these subtle features. Also, be sure to move north of M43 to NGC 1977, more nebulosity whose wavy texture is delightful at low powers. Then move farther north to the loose open cluster NGC 1981 to complete your tour of Orion's sword.

By the way, returning to M42 and the subject of challenges, I have also spent considerable time peering at the region just inside the Great Nebula's arcing western rim and searching for "GOD" – a series of

dark, nebulous swirls that spell out the word GOD in capital letters. Although I have seen portions of these letters, I have yet to piece together the details in a single strong view. This challenge might be a bit too difficult for the 4-inch, but I do urge those with larger telescopes to look for GOD in the Orion Nebula.

M44

Praesepe **or** *Beehive Cluster*
NGC 2632
Type: Open Cluster
Con: Cancer

RA: $08^h40.4^m$
Dec: +19°40′
Mag: 3.1
Diam: 70′
Dist: 577 light-years
Disc: Known since antiquity

MESSIER: [Observed March 4, 1769] Cluster of stars known as the nebula in Cancer [Praesepe]. The position given is that of star C.

NGC: Praesepe Cancri

CANCER IS THE ONLY CONSTELLATION WHOSE brightest stars are fainter than a Messier object within its boundaries. In fact, if it weren't for the mystifying cloudy appearance of open cluster M44, which draws your gaze to the surrounding 4th-magnitude stars, it is conceivable that dim Cancer might have either gone unnoticed by ancient stargazers or have been envisioned differently. To the naked eye, the 3rd-magnitude glow of M44 looks like the bearded head of a tailless comet passing between the 4th-magnitude stars Gamma (γ) and Delta (δ) Cancri.

The Beehive Cluster has been known since antiquity. Aratus and Pliny both noted that whenever the object's misty form vanished from naked-eye view, foul weather was on the way – an observation most likely related to the approach of thin, high cirrus clouds, which can dim the appearance of the night sky before an impending storm. The misty glow of M44 had a more macabre significance in ancient China, where it was seen as Tseih She Ke, "exhalation of piled-up corpses." Observing from Alexandria in ancient Egypt, Ptolemy (ca. A.D. 83–161) recorded the celestial mist as the "center of the cloud-shaped convolutions in the breast [of Cancer], called Praesepe."

Praesepe is a derivative of the Latin verb *praespire* (meaning to enclose), which is also the root word for *presepio* (nativity scene). The term appears to be linked to an early Christian church on the Esquiline Hill in Rome, known since the seventh century as Sancta Maria ad praesepe – the place, according to tradition, where the remains of the holy manger were brought for safe keeping. The Praesepe we see in the night sky, then, is the fuzzy, straw-like cloud of unresolved starlight visible to the unaided eye between the 4th-magnitude stars Gamma (γ) and Delta (δ) Cancri; seen through ancient eyes, these stars were the two Aselli, or donkeys, guarding Christ's crib.

The nature of the "cloud" remained a mystery until Galileo turned his tiny telescope on it and saw a "mass of more than 40 small stars."

To my knowledge, English observer John Herschel was the first to call M44 the Beehive. In his 1833 *Treatise on Astronomy*, Herschel writes, "In the constellation Cancer, there is ... a luminous spot, called Praesepe, or the beehive, which a very moderate telescope, - an ordinary night glass, for instance, - resolves entirely into stars." The description makes sense because in the nineteenth century, the traditional beehive, called a skep, was made of straw. So, if you're a purist, you should call the misty glow visible to the unaided eye the Praesepe (a heap of straw) and reserve the Beehive moniker for the binocular or telescopic view, which shows the individual stars as the swarming bees.

There is an interesting parallel between early observations of the Praesepe and those of bees. "If Praesepe is not visible in a clear sky," Pliny said, "it is a presage of a violent storm." As Daphne Moore explained in 1976 in *The Bee Book*, "bees disappear in bad weather and reappear with the Sun and flowers." In other words, the disappearance and reappearance of bees is tied not only to the natural cycle of death and rebirth as reflected in the seasons but also to the whims of weather - "a vital element," Moore says, "in Earth-Mother worship."

Elsewhere in her book, in an interesting parallel, Moore explains that a Breton legend "tells us how the falling tears of the crucified Christ turned into bees and flew away to bring sweetness to the world." You can see the falling tears of Christ in the mind's eye. Just take the time to examine the Praesepe's shape with your unaided eye. At first, the glow will appear round. With time and keen averted vision, however, you should see the cluster's core elongated north to south. Now relax. Take a moment to walk around while breathing regularly. When you feel refreshed, look at M44's elongated core once again, but this time see if it breaks apart into three distinct misty patches - the "falling tears of Christ." This triple aspect has been observed by sky observers since before the invention of the telescope.

M44 did not go unnoticed by Galileo, who described its telescopic appearance as "not one only but a mass of more than 40 small stars." In modern telescopes, about 200 confirmed cluster members ranging from magnitude 6 to 14 are visible, 80 of which are brighter than magnitude 10. The age of the cluster is estimated to be about 650 million years, and its core measures 12 light-years in true physical extent.

At first glance, M44 appears round to the naked eye, but, as mentioned, if you concentrate, it should look elongated roughly north to south. You might also begin to suspect arms extending in various directions and stars shining within the central haze. Is M44 resolvable to the unaided eye? Yes, and it's not difficult to do! In fact, the cluster contains 15 stars between magnitudes 6.3 and 7.5. One night, while observing at 14,000 feet, I gazed at the Beehive without optical aid while occasionally breathing oxygen through an oxygen mask. To my amazement, the haze of unresolved stars vanished and the individual cluster members stood out boldly as rock-steady pearls, each a tiny, bright disk surrounded by an intense black rim. I did not count the stars that night because they seemed so numerous. I used the oxygen in brief spurts to help keep me alert because, at 14,000 feet, the atmosphere is 40 percent thinner than at sea level. Lack of oxygen can also decrease the sensitivity of the eyes.

For example, without supplemental oxygen at 14,000 feet, I had to use averted vision

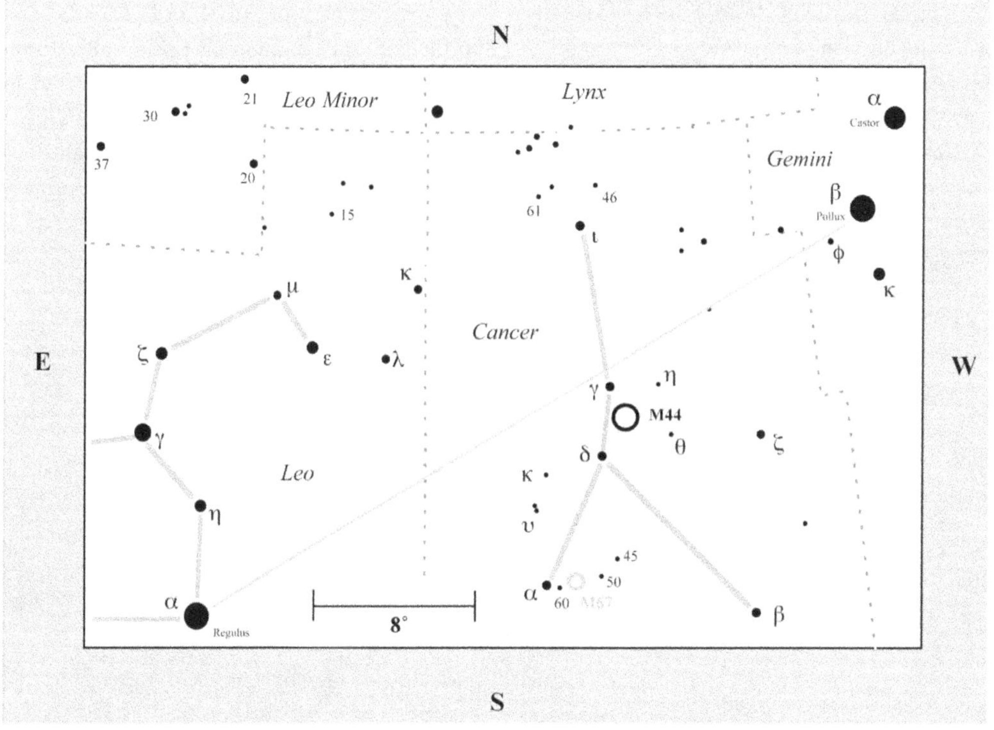

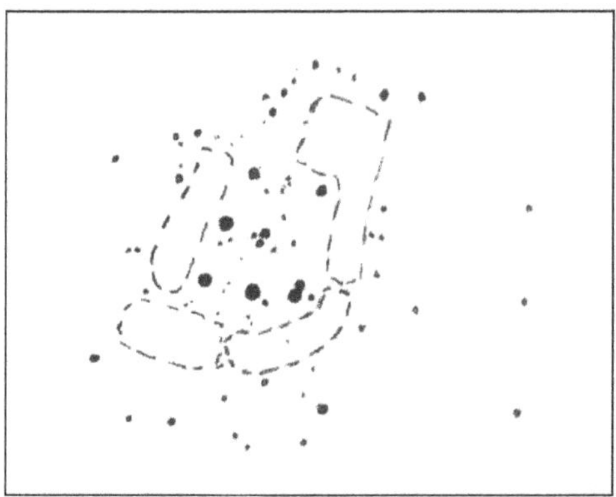

to see the Andromeda Galaxy well; 10,000 feet lower, however, the galaxy was an incredible naked-eye sight with direct vision. There is no reason for sea-level observers to use additional oxygen because you're working in an oxygen-rich environment (hyperventilating, however, can increase the immediate supply); too much oxygen will make you sick. I did count the stars in M44 from 9,000 feet and 4,000 feet without oxygen. With a long, dedicated effort, I recorded a dozen cluster members without optical aid – that's twice the number of stars most observers casually see in the Pleiades (M45)!

Through the 4-inch at 23×, the entire 70′-wide cluster is visible, and the stars seem to swarm with sparkling madness – truly a beehive of nervous starlight. At least one-fifth of all the stars are doubles! The main body of stars is sharply outlined by several strong dark lanes (reminiscent of the naked-eye view from 14,000 feet), with a tiny coalsack to the northwest. High-power or large telescopes will bring out the faint stars within the dark lanes, but why destroy such a visually stunning illusion of emptiness?

I would be curious to know how the celestial backdrop affects brightness estimates of open clusters in the Milky Way band. For example, one reason M44 appears so visually distinct is that its brilliant orb of starlight is seen against the star-poor region of Cancer, which is off the main stream of the Milky Way. It is seen projected against a darker background than, say, either M6 or M7, which are nestled in the most brilliant region of the Milky Way. I can only imagine how stunningly bright M7 would appear if it were seen against the stars of Cancer and how dim M44 would appear if seen against the heart of the Milky Way!

M45

Pleiades **or** *Seven Sisters*
NGC (not listed)
Type: Open Cluster
Con: Taurus

RA: $03^h47.5^m$
Dec: +24°06′
Mag: 1.5
Diam: 2°
Dist: ~440 light-years
Disc: Known since antiquity

MESSIER: [Observed March 4, 1769] Star cluster known as the Pleiades. The position given is that of the star Alcyone.

NGC: None.

ON CRISP WINTER EVENINGS, THIS BRILLIANT 1st-magnitude open cluster rides high on the shoulder of Taurus, the Bull, whose V-shaped face itself (minus Aldebaran) is another, brighter open cluster – the Hyades. Together these clusters are among the most alluring sights in the heavens. But the larger, more open, Hyades pales in comparison to magnificent M45. With the naked eye, the Pleiades looks like a tiny dipper, a forest of starlight bathed in moonlit mist, or a distant gathering of veiled brides. Indeed, photographs show the entire cluster swaddled in ice-blue nebulosity, which reflects the light of these young, hot stars. The stars of the Pleiades emerged from their dusty cocoon some 20 million years ago. And some astronomers have conjectured that rather than the surrounding nebula belonging to them, the stars may just be passing through a nebulous region in Taurus.

Just how many stars (Pleiads) in the Pleiades are truly visible with the naked eye is the subject of some debate. Traditionally, the number has been seven. But that count stems from ancient Greek mythology and refers to the seven doves that carried ambrosia to the infant Zeus, or to the seven sisters who were placed in the heavens so that they might forget their grief over the fate of their father, Atlas, condemned to support the sky on his shoulders. Although largely symbolic, the age-old association of the Pleiades with the number seven remains fixed to this day – to the point that some observers swear they cannot see more than seven members, even though the Pleiades contains 10 stars brighter than 6th magnitude. Some observers question how it is possible to see 10 Pleiads in the Seven Sisters (a demonstration of the power of words or, as Clerke inferred, the power of the Pleiades over human affairs). The fact is that almost three times that magic number of stars can be seen without magnification by an astute observer under dark skies. Clerke notes that Johannes Kepler's tutor, Maestlin, saw 14, and he mapped 11 before the invention of the telescope. Archinal reports routinely seeing 12 Pleiads and sometimes 14 under good skies. Houston counted 18, and about 20 years ago I logged 17 from Cambridge, Massachusetts! The trick is to spend a lot of time looking and plotting.

Another lingering debate concerns whether the nebulosity associated with the cluster can be seen with the naked eye. Some argue that the haze is just an illusion, a vision created by the tight gathering of stars whose unresolved light simply appears fuzzy – just as close doubles do at low power (like M40). Others, however, claim that the cluster appears just as Tennyson described it in Locksley Hall:

> Many a night I saw the Pleiades,
> rising thro' the mellow shade,
> glitter like a swarm of fireflies
> tangled in a silver braid.

Houston once scoffed, "Why does everyone fuss over the Pleiades nebulosity? Of course you can see it with the naked eye!" I did not reply, and he didn't expect a response; he knew he was preaching to the choir. However, Skiff points out that some of the perceived nebulosity may be scattered starlight. He suggests testing this hypothesis by screening Alcyone, the brightest Pleiad, behind the edge of a building. Try it.

The 125-million-year-old cluster contains about 100 bright telescopic stars in a sphere 15 light-years in true physical extent, though it is sailing through space at about 25 miles per second, and it will take them about 30,000 years to move an apparent distance of one Moon diameter. The brightest Pleiads are all rapidly rotating, with Pleione spinning about 100 times faster than the Sun. And Alcyone, the centralmost and brightest Pleiad, is a thousand times more luminous than the Sun. In a 2010 study (*Monthly Notices of the Royal Astronomical Society*, vol. 405, pp. 666–680) of the dynamical evolution of the Pleiades, Joseph M. Converse and Steven W. Stahler of the University of California, Berkeley, found that the original cluster, newly stripped of gas, already had a radius of 13 light-years. Over time, they say, the cluster expanded further and the central surface density fell by about a factor of two. They attribute both effects to the liberation of energy from tightening binaries of short period. They found that the ancient Pleiades also had significant mass segregation, which persists in the cluster today. "In the future," they say, "the central density of the Pleiades will continue to fall. For the first few hundred million years, the cluster as a whole will expand because of dynamical heating by binaries. The expansion process is aided by mass loss through stellar evolution, which weakens the system's gravitational binding. At later times, the Galactic tidal field begins to heavily deplete the cluster mass. It is believed that most open clusters are eventually destroyed by close passage of a giant molecular cloud. Barring that eventuality, the density falloff will continue for as long as 1 billion years, by which time most of the cluster mass will have been tidally stripped away by the Galactic field."

Through a telescope, the Pleiades and its attendant nebulosity fill a low-power field, where they are best seen. The nine brightest members fit well in a 1° field (a true diameter of 7 light-years). From dark skies, the cluster looks like a cobwebbed coffin filled with glistening jewels. The individual Pleiads are interconnected by gauze-like veils of nebulosity, the brightest of which surrounds the star Merope (23 Tauri) and is known as "Tempel's Nebula" or the Merope Nebula; what follows is but a highlight.

While in Venice, Italy, on October 19, 1859, German-born Wilhelm Tempel turned his 4-inch Steinheil comet seeker to the Pleiades.

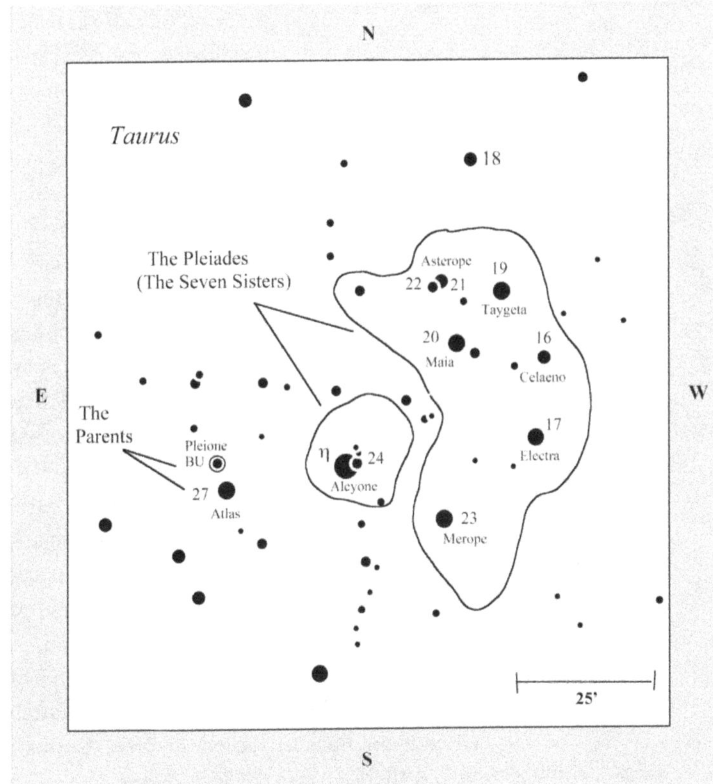

at Arcetri, Italy. Through these telescopes, he was "greatly surprised to find it so *plain, large,* and (in one direction at least) so *sharply defined.*"

Skepticism arose, however, when several renowned observers using great refractors and reflectors could not see the nebula. Heinrich d'Arrest was among Tempel's biggest critics. After failing to detect the nebula through his 11-inch refractor at Copenhagen, Tempel jousted, "I have hitherto been able positively to see nothing. ... I therefore, even yet, am of opinion that this nebula is variable, otherwise the original announcement of the discovery ... must be looked upon as being greatly exaggerated."

He hadn't looked at the group in six months, he said, and consequently found a "large bright nebula," about 30′ in length, just south of Merope. Except for an uncertain observation by French astronomer Edme-Sébastien Jeaurat in 1786, Tempel's was the first well-documented sighting of nebulosity in the Pleaides.

In a letter to Christian A. F. Peters, editor of the *Astronomische Nachrichten*, dated December 23, 1860, Tempel said that at first he mistook the Pleiades to be a "fine large comet, but convinced myself on the following night, the 20th, that its position remained unchanged." He went on to observe the nebula repeatedly through a variety of instruments, including the 8-inch and 11-inch refractors at the Royal Florentine Observatory

Others, including E. Schönfeld, disagreed. Observing with a 6 1/2-inch refractor at Mannheim, Germany, in 1862, Schönfeld argued that the nebula "instantly stuck out in the telescope," adding later that it had been seen "very clearly" and was "immediately conspicuous, without accurate knowledge of the position."

In response to mounting swipes, d'Arrest admitted that he had "after a long effort ... actually set eyes on Tempel's Nebula," but, in support of his own argument that Tempel's claim was exaggerated, added the following verbal parry: "This is the faintest object which I remember ever having seen in the refractor." Adding fuel to the fire was the fact that even those observers

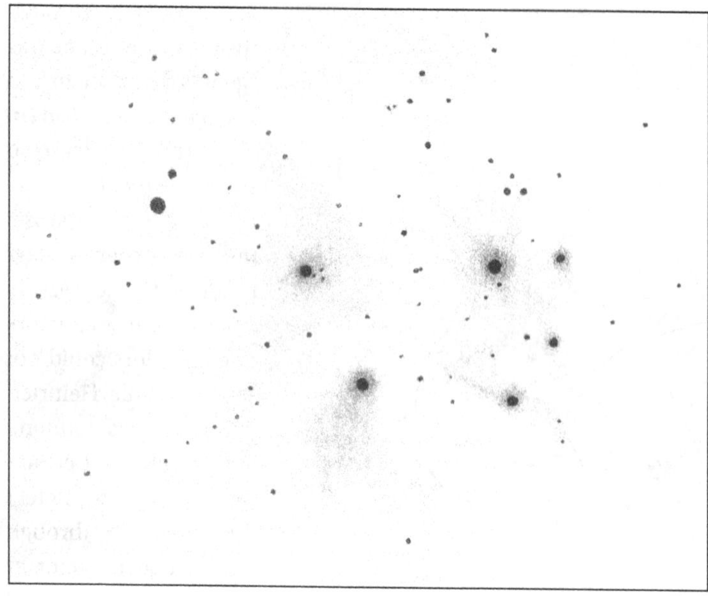

who rallied support for Tempel could not agree on the nebula's exact form, casting more doubt on the veracity of the claims.

In his "Great Nebula in the Pleaides," which appeared in an 1888 issue of *Knowledge*, A. Coper Ranyard opined, "The observations are worth examining, as they throw some light on the differences which are always likely to exist when observations are pushed into the border-land of vision, where by reason of the extreme faintness or minuteness of the objects examined, the eye begins to fail, and the imagination begins to play a larger and larger part in filling up the gaps where the senses of the eye-straining observer fail him."

In fact, in 1863, British observer Rev. Thomas W. Webb spied it through a 2-inch telescope, while finding it "invisible" through an 11-inch. Even d'Arrest finally admitted, "Here are nebulae, invisible or barely seen in great telescopes which can easily be seen in their finders." Indeed, what you see will depend greatly on the size of your telescope, the clarity of the night, the darkness of the sky, the magnification used, the size of your field of view, and your vision.

The visual observations of the Merope Nebula and other veils of nebulosity associated with the Pleiades were ultimately vindicated in October 1886, when Isaac Roberts made a 3-hour exposure with the 20-inch Grubb reflector that showed the nebulosity extending "in streamers and fleecy masses, till it seems to almost fill the spaces between the stars, and to extend far beyond them."

The only visual detail that went "unnoticed" in the photographs was IC 349, which eagle-eyed Barnard spied in 1890 through the 36-inch refractor at Lick Observatory. He called it the "new Merope nebula" because the 30″-wide cometary glow (the brightest part of the nebula) lies only 36″ southeast of Merope and thus hides in the star's glare.

At first glance, through the 4-inch at 23×, the Merope Nebula network looks like a tapered comet tail. But if you take the time to sweep the telescope back and forth over its surroundings, you might notice that this patch is but part of a larger fan of material that sweeps westward. Using low magnification, just stare at the cluster for a while and occasionally tap the telescope tube. You should start to see wide dark lanes running amid the nebulosity, especially at the cluster's center, where it drapes around a bridal veil of nebulosity adorning Alcyone. Study the drawing and photograph and look for prominent streaks of nebulosity, especially around Maia and Electra.

M46

NGC 2437
Type: Open Cluster
Con: Puppis

RA: $07^h41.8^m$
Dec: -14°49′
Mag: 6.1
Diam: 20′
Dist: ~5,300 light-years
Disc: Charles Messier, 1771

MESSIER: [Observed February 19, 1771] Cluster of very faint stars between the head of Canis Major and the two rear hoofs of Monoceros, determined by comparing the cluster with the sixth-magnitude star Flamsteed 2 in Argo Navis [now 2 Puppis]. These stars can be seen only with a good telescope. The cluster contains some nebulosity.

NGC: Remarkable, cluster, very rich, very bright, very large, involving a planetary nebula.

M47

NGC 2422 = NGC 2478
Type: Open Cluster
Con: Puppis

RA: $07^h36.6^m$
Dec: –14°29′
Mag: 4.4
Diam: 25′
Dist: ~1,500 light-years
Disc: Giovanni Batista Hodierna, before 1654

MESSIER: [Observed February 19, 1771] Cluster of stars not far from the previous one [M46]. The stars are brighter. The center of the cluster was determined relative to the same star, Flamsteed 2 Argo Navis [now 2 Puppis]. The cluster contains no nebulosity.

NGC: Cluster, bright, very large, pretty rich, with bright and faint stars.

USE YOUR BINOCULARS AND SCAN ABOUT 15° (a fist- and two finger-widths) east of Sirius in a line due south of Procyon and Alpha (α) Monocerotis. There, hiding among the monotonously faint star fields of the Puppis Milky Way, are the curious Messier open clusters M46 and M47. They lie a mere 1 1/2° (a finger width) apart but look markedly different. The farther east of the two, M46, is a relatively young galactic star cluster with an age of 245 million years; it appears as a round, uniform 6th-magnitude glow. Gawky M47, on the other hand, is an irregular gathering of reasonably bright but dissimilar stars; the cluster's age is about 70 million years, comparable to the Pleiades. And though both clusters have impressive apparent diameters three-quarters the size of the full moon, their wildly different appearances can distract you from that realization. What is immediately obvious, however, even in binoculars, is that whatever M46 lacks in visual grandeur, M47 lacks in visual grace, and vice versa. It's like trying to compare a flower with a rock.

At 23×, M46 is a bright sphere of finely resolved starlight with faint, clockwise spiral structure. If you sweep the telescope to the west, you might notice that M46 appears to be on the eastern end of a wide, V-shaped string of 9th-magnitude stars connecting it to M47. But this is deceptive: M46 is about 3,800 light-years more distant than M47 and is actually somewhat larger than its closer neighbor! M46 contains 186 stars ranging from magnitude 10 to 13, distributed rather evenly over an area of 20′ with a loose central gathering. But this central gathering is also somewhat illusory, because moderate to high magnification will reveal that M46 has a dark, not a bright, core. With imagination, I see that

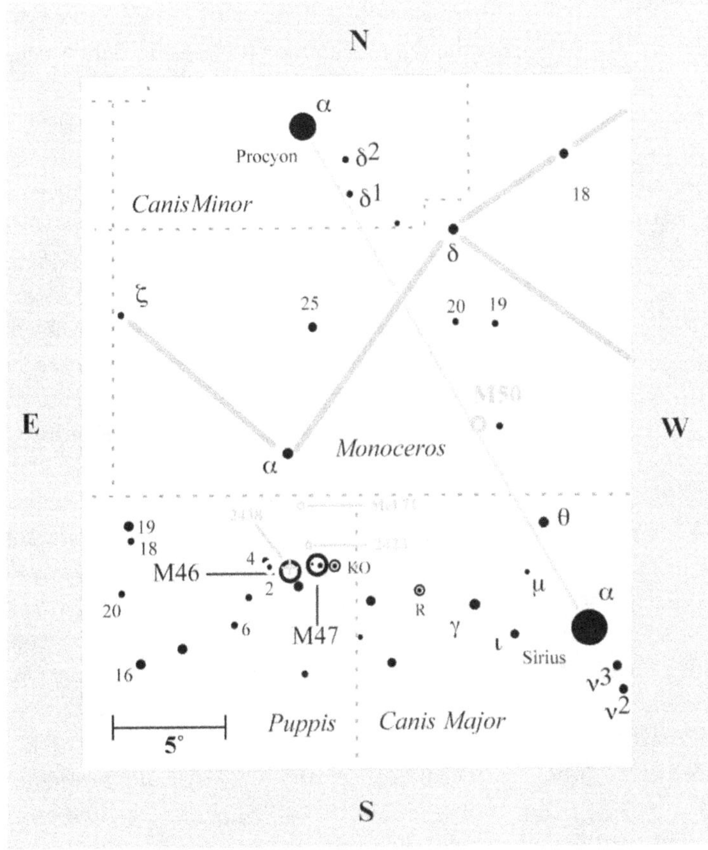

void as an atoll, with a 2′-diameter black lake rimmed by 10th-magnitude coral specks that break a wave of stars flowing around it from the south.

There is yet another illusion associated with M46. It appears to contain a tiny planetary nebula, NGC 2438, which is distinctly visible in the accompanying photograph. But the cluster and nebula are not physically associated, because the cluster is 5,300 light-years distant, whereas the nebula's distance has been a matter of debate. Estimates have placed it from about 6,520 light-years (making it farther away than the cluster) to about 3,000 light-years away, placing it roughly halfway between M46 and M47. NGC 2438 started to form about 45,000 years ago.

In a 1996 paper in *Astronomische Nachrichten* (vol. 317, no. 6, p. 413), however, R. Pauls and L. Kohoutek write that they determined the recessional velocity of NGC 2438 to be about 37 miles per second, which is in agreement with the recessional velocity of four stars in M46 (38 miles per second). "We conclude," the authors write, "that contrary to earlier statements the nebula is probably associated with the cluster." Positioned just a few arcminutes north of the cluster's center, this 11th-magnitude planetary measures only about 1′ in diameter. I suspected it at 23×, but 72× shows it clearly as a ghostly mote among the multitude. With averted vision at high power, it looks like the glow from the flame of a distant candle. Try, if you must, for the planetary's central star, but the task might be daunting, given that it shines at about 16th magnitude! Finally, turn your gaze north of the planetary. Do you see the wide canal in the cluster's outer halo of faint stars? It runs southwest to northeast, and its southern bank abuts the planetary.

M47, an open star cluster in Puppis, proved problematical to identify with Messier's catalogued position: right ascension $7^h44^m16^s$; declination −14°50′08″ (epoch 1771.0). When precessed to epoch 2000.0

M47

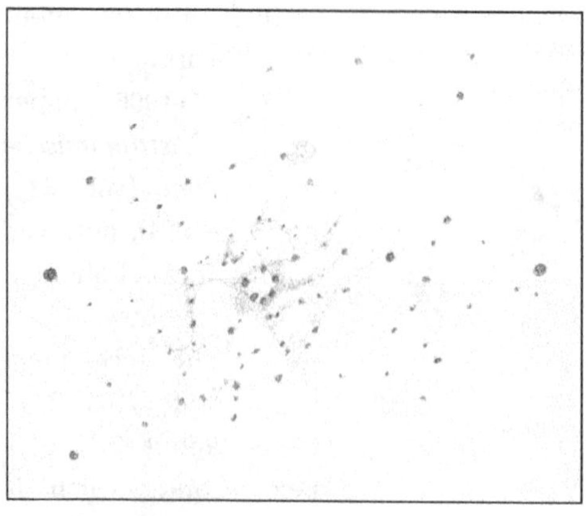

($7^h54^m13^s$; –15°23′24″), the coordinates place M47 some 3° *southeast* of M46, in a region of sky devoid of any deep-sky objects.

Confusing the matter, this spot in the rich Puppis Milky Way received the NGC designation 2478. As no observer since Messier had been able to identify a cluster of stars at or near Messier's position, M47 (NGC 2478) became popularly known as one of Messier's "missing" or "nonexistent" objects.

But German astronomer and astronomy popularizer Oswald Thomas (1882–1963), the man who gave us the term Summer Triangle to refer to the pattern formed by the bright stars Vega, Altair, and Deneb, offered a viable solution in his 1934 book *Astronomie*, where he identifies M47 with the bright and obvious open star cluster NGC 2422. Twenty-five years later, T. F. Morris of the Royal Astronomical Society of Canada's Montreal Centre arrived at the same conclusion, adding that Messier appears to have made an error in computation.

To determine an object's position, Messier would simply find a catalogued star of the same declination near the object and measure the difference. He would then record the times that the object and the known comparison star transited a wire centered in his field of view; the difference in transit times provided him with the difference in right ascension between the star and the object.

In the case of M47, Messier had used 2 Navis (now 2 Puppis) as the comparison star. Morris noticed that if you simply reverse Messier's offset positions (both east to west and north to south) from 2 Navis, you arrive at NGC 2422. Clearly M47.

But the history is even more involved. William Herschel discovered NGC 2422 in 1785, not knowing, of course, that this was in fact what Messier was referring to as his forty-seventh object and what Dreyer had listed in his *New General Catalogue* as a new object (2422). Curiously, as astronomical historian Michael Hoskin wrote in 2005 in the *Journal for the History of Astronomy* (vol. 36, p. 373), William's sister Caroline had twice observed, recorded, and identified M47 as the object for which William would later claim discovery credit in 1785. The first occasion was on the evening of February 26, 1783, and the second was on March 4, 1783, when Caroline erroneously catalogued M46 as her No. 3. One might speculate whether this error might be related to Messier's wrong position for M47.

M47 is an elaborate mix of bright and faint stars that look like they've been tossed together by someone in great haste. The sight makes me feel as if I have opened yet another treasure chest and am looking into a tangle of secret riches. M47 contains at least 117 members, some as bright as 5th and 6th magnitude, in an area 25′ in diameter. Thus, M47 measures a mere 14 light-years in diameter, less than half that of M46. I resolved "three" stars in M47 with the unaided eye. I use quotation marks here because the central one is actually the double star Struve 1121, which consists of two opalescent 7th-magnitude stars separated by 7″. Three nicely spaced doubles reside in the cluster, as do two 8th-magnitude orange stars

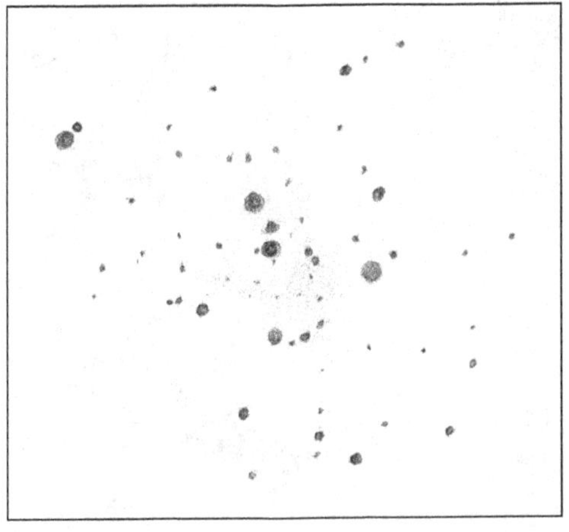

(one north of the brightest star, the other to the south). I am quite surprised at my magnitude estimate of 5.7, which is fainter by more than a magnitude than that in Luginbuhl and Skiff's *Observing Handbook* and by 0.5 magnitude from the one offered by Kenneth Glyn Jones.

Return to low power and follow a chain of 8th- to 9th-magnitude stars off M47 north to the smaller, magnitude 6.7 NGC 2423. Then look to the northwest and you should see another chain of 8th-magnitude and brighter stars curving to yet another, still smaller patch of stars: open cluster Melotte 71. Together, all these clusters and strings form a pattern that reminds me of the alien craft depicted in the 1953 film *The War of the Worlds*.

M48

M48

NGC 2548
Type: Open Cluster
Con: Hydra

RA: $08^h13.7^m$
Dec: -05°45′
Mag: 5.8
Diam: 30′
Dist: ~2,400 light-years
Disc: Charles Messier, 1771

MESSIER: [Observed February 19, 1771] Cluster of very faint stars, without nebulosity. This cluster is close to three stars that lie at the root of the tail of Monoceros.

NGC: Cluster, very large, pretty rich, pretty much compressed, 9th to 13th magnitude stars.

THE MOST INTRIGUING "MISSING" Messier object, M48 is now believed to be NGC 2548 – a knitted stellar gathering of about 80 stars between magnitude 8 and 13 located in a rather inconspicuous region of Hydra where it borders Monoceros. M48 marks the southern tip of a nearly equilateral triangle with magnitude 3.4 Epsilon (ε) Hydrae, in the head of the Serpent, and magnitude 0.4 Alpha (α) Canis Majoris, and lies only 3 1/2° southwest of the obvious naked-eye grouping of 1, 2, and χ Hydrae.

Messier's recorded position for M48 is off by about 5° in declination, placing it due north of NGC 2548 (where there are no clusters). Astronomical historian Owen Gingerich, in his chapter in the 1978 classic *The Messier Album*, explained:

> Messier determined the declination of a nebula or cluster by measuring the difference between the object and a comparison star of known declination. ... Since no conspicuous star is located 2 1/2° away in declination, we cannot account for this position by another error in sign [as was the case with M47]. ... Messier did not publish the name of the [comparison] star he used, and his original records are apparently no longer extant. ... Thus, a careful survey of the region described by Messier leads to the conclusion that NGC 2548 is the cluster that the French observer intended as his 48th object, for lack of any other cluster nearby that fits the description.

On March 8, 1873, Caroline Herschel (1750–1848), William's sister, swept up "a cluster of scattered stars ... not in [Messier's] catalogue." Her brother added the object to his catalogue as H VI-22, which later became NGC 2548, the missing M48. It is only through William's inclusion of his sister's find in his catalogue that astronomers became aware of its existence.

While NGC 2548 is certainly the most likely candidate for M48, I had an interesting experience when I went out to make my

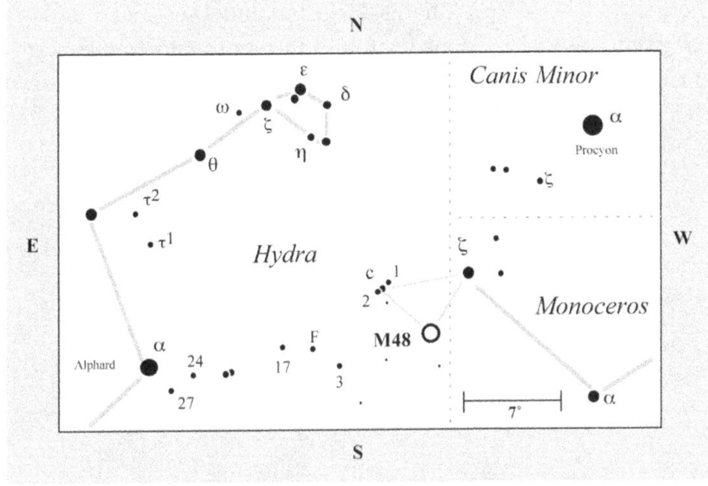

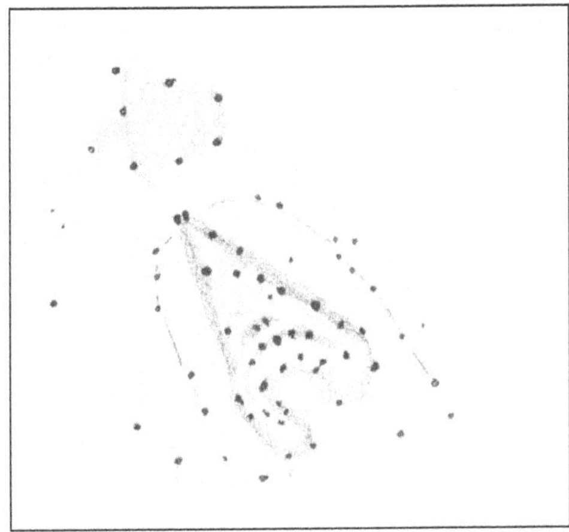

first observation of it for this book. I began by looking for the cluster with the unaided eye. I easily located the line of stars 1, 2, and χ Hydrae, and then immediately noticed a diffuse patch south of them. I swung the telescope to this region and swept the area at 23×. To my amazement, I could not find the cluster. I looked up again, located the hazy patch, and tried to find the cluster, but without success. Then I realized my mistake: I was looking too far to the southeast. I moved the telescope about 5° northwest and – bang – there was M48.

I didn't think anything of this experience until I repeated that very same mistake the next time I went out! I finally viewed this mystery cloud through the 4-inch at 23× and found it to be a loose collection of stars, some of similar brightness. When I placed it in my 7 × 35 binoculars and slightly defocused the view, I saw a small, "fuzzy" object (the unresolved loose gathering of stars in the mystery cloud) surrounded by a brighter circlet of stars (giving the cloud brightness and breadth). Nevertheless, this object does not fit Messier's description of M48, so it was just a curiosity. As the late Kenneth Glyn Jones pens in his book *Messier's Nebulae and Star Clusters*, "The cluster is a little difficult to find as there are many small aggregations of stars in the neighbourhood which look like clusters."

In a CCD photometric study of M48, L. Balaguer-Núñez (University of Barcelona, Spain) and colleagues determined several properties of the cluster, including its distance (~2,400 light-years), its age (~400 million years), and a near solar metallicity. Their results were reported in a 2005 paper in *Astronomy and Astrophysics* (vol. 437, p. 457).

Through the 4-inch, M48 is a tremendously pleasing cluster, a perfect arrowhead of bright stars with a tight, elliptical, off-axis core. Like Lord Rosse, I find the cluster riddled with dark lanes and openings. One dark feature I call the "keyhole" – it's a notch in the base of the arrow to the southwest. Actually, I like to call M48 the "alligator" because it reminds me of an aerial view of a stalking alligator (when all one sees is its triangular head and snout). With this image in mind, I see a wake of stars flowing around the snout and, to the northeast, a circle of stars representing a turtle – an alligator's special meal.

M49

NGC 4472
Type: Elliptical Galaxy (E2)
Con: Virgo

RA: $12^h29.8^m$
Dec: +08°00′
Mag: 8.4
Dim: 8.1′ × 7.1′
Dist: ~52.5 million light-years
Disc: Charles Messier, 1771

MESSIER: [Observed February 19, 1771] Nebula discovered close to the star ρ Virginis. This is rather difficult to see with a simple three-and-a-half-foot refractor. M. Messier compared the comet of 1779 with this cluster on 22 and 23 April: the comet and the nebula had the same luminosity. M. Messier plotted this nebula on the chart showing the path of this comet, which will appear in the Academy volume for the same year, 1779. Observed again 10 April 1781.

NGC: Very bright, large, round, much brighter in the middle, mottled.

M49 IS AN ENORMOUS ELLIPTICAL SYSTEM that lurks at the center of a subcluster of galaxies, called the Virgo Cloud, in the heart of the vast Virgo Cluster – a rich concentration of galaxies spanning some 20 million light-years of space. In the early 1990s, ROSAT (an earth-orbiting x-ray observatory) imaged M49 and revealed that it possesses a halo of hot (10,000,000 K) gas. Observations with previous x-ray telescopes have shown that the hot gas in galaxy clusters cannot be confined by the combined gravity of the gas and the galaxies alone.

Thus, the ROSAT data imply that perhaps 75–95 percent of the mass of small galaxy clusters, such as the M49 subcluster, could be in the form of dark matter. It has also been discovered that elliptical galaxies that lie at the center of galaxy clusters lack the raw materials from which new stars form, so the ellipticals are excessively red. Modeling of the x-ray emission from M49 indicates a mass of about 3×10^{11} solar masses. Indeed, M49 is the most luminous galaxy in the Virgo Cluster. It spans 124,000 light-years and is receding from us at about 630 miles per second. It's estimated that some 6,300 globular clusters orbit the galaxy.

M49's core is a low-ionization nuclear emission region (LINER) and possesses a two-sided weak radio jet. In 2004, Catherine Boisson of the Paris Observatory and her colleagues reported in *Astronomy and Astrophysics* (vol.

M49

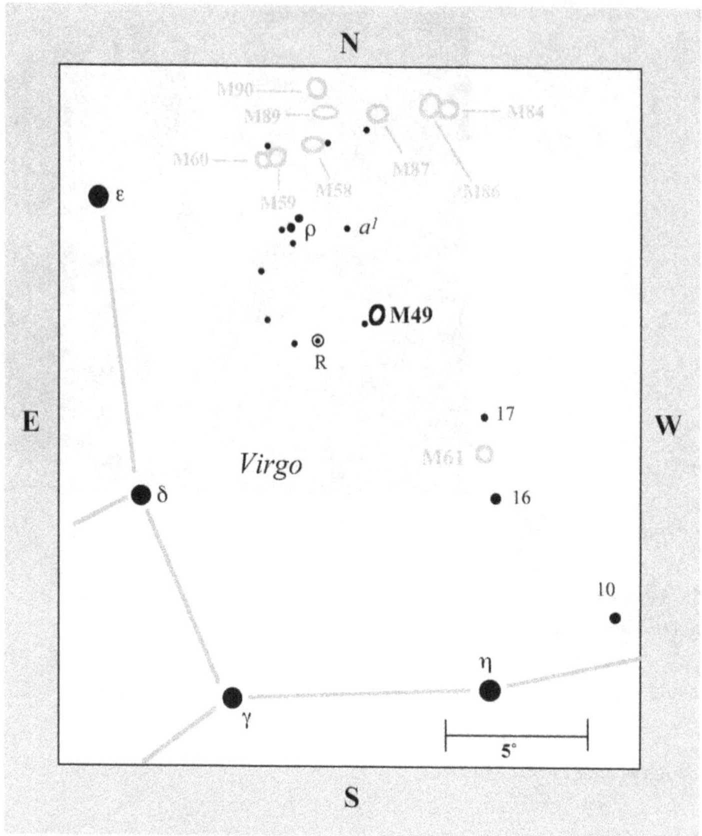

and her team had also made a study in 2003 (*Astrophysical Journal*, vol. 599, p. 1302) of novae in M49 with the HST. Their 55-day observing campaign led to the discovery of nine of them. They suggest that about 100 novae erupt in the galaxy every year.

To find M49, start by using the chart at the back of the book to find 3rd-magnitude Epsilon (ε) Virginis. Then use the chart here to seek out 5th-magnitude Rho (ρ) Virginis about 5° to the west-southwest. The star is easy to identify in binoculars since it is the central bright star in a 30′-wide upside-down Y of stars (oriented north to south). Now look for 6th-magnitude 20 Virginis a^1 due west; use binoculars first, then center Rho Virginis in your telescope and make that large sweep to 20 Virginis. M49 is about 2 1/2° southwest of 20 Virginis, appearing as a very compact, 8th-magnitude glow nestled between two magnitude 6.5 stars. The galaxy shows up well in binoculars from a dark-sky site.

Through the 4-inch at moderate power, M49 reveals a bright nucleus surrounded by a tight inner core and a diffuse halo. A magnitude 12.5 star punctuates the galaxy's bright eastern fringe like a splendid supernova. The overall sight reminds me of an unresolved globular cluster or the head of a comet. Barnabus Oriani, an eighteenth-century Italian observer,

428, p. 373) that the galaxy has an old and metallic stellar population that's homogeneous in its inner region, with the bulk of the stars being older than a few billion years and somewhat metallic. They also found that a population of less than 20 percent young stars (A dwarfs and K supergiants) also make a poorly defined contribution to this region.

In a 2006 paper in *Astrophysical Journal Supplement* (vol. 164, p. 334), Laura Ferrarse of Rutgers University and her colleagues say they used Hubble Space Telescope (HST) images to discover a thin, boomerang-shaped dust lane crossing the galaxy's center at a position angle of roughly 45°, extending ~1.5″ on either side of the galaxy's core; they could not, however, detect a nucleus. Ferrarse

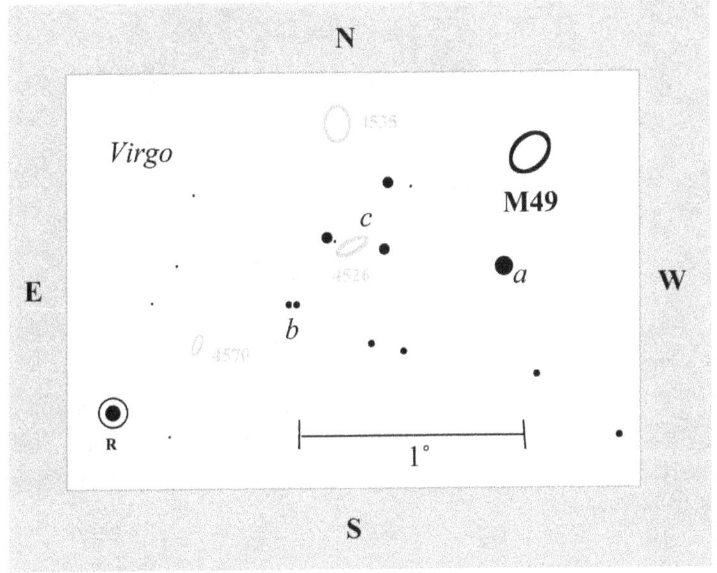

appearing as a mottled nuclear region and two arcs of light at the outer edge of the inner halo, one to the northeast and the other to the southwest. They also appear mottled and cause me to suspect spiral structure, though this may be an illusion. D'Arrest, however, believed he resolved the misty glow of M49 into stars of 13th to 14th magnitude near the edges. Speaking of illusions, do you see a cross of faint knots or stars bordering the bright inner region? It resembles the gravitational-lensing effect known as Einstein's Cross. The visible outer envelope itself is very deceiving: at high power, it makes the galaxy appear round, but moderate to low magnifications reveal a broad shield of diffuseness that transforms this sphere into an elliptical glow oriented northeast to southwest.

When using the telescope, refer to the detailed finder chart here to scan the area, because, as I quickly discovered, the low-power field is chock full of galaxies, and they're all pleasantly distracting. The most prominent of them is NGC 4526, another elliptical about 1° southeast of M49, between two 7th-magnitude stars.

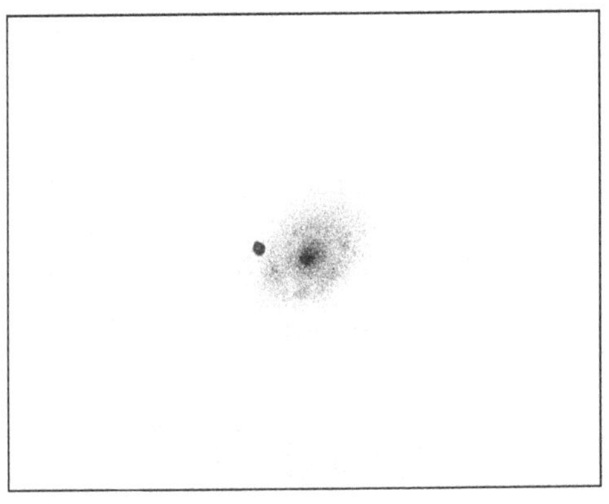

thought the haze looked exactly like the Comet of 1779 (apparently, so did Messier).

But look carefully at the nucleus with high power. I get hints of inner structure,

M50

M50

NGC 2323
Type: Open Cluster
Con: Monoceros

RA: $07^h02.8^m$
Dec: –08°23′
Mag: 5.9
Diam: 15′
Dist: ~3,300 light-years
Disc: Probably Giovanni Domenico Cassini, before 1711; Charles Messier rediscovered it in 1772

MESSIER: [Observed April 5, 1772] Cluster of faint stars of differing brightness, below the right thigh of Monoceros, above the star θ in the ear of Canis Major, and close to a seventh-magnitude star. It was while observing the comet of 1772 that M. Messier observed this cluster. It has been plotted on the chart for that comet that he prepared, *Mémoires de l'Académie 1772.*

NGC: Remarkable cluster, very large, rich, pretty compressed, elongated, stars ranging from 12th to 16th magnitude.

M50, AN OBSCURE OPEN CLUSTER IN AN equally obscure constellation (Monoceros), has an aura of enigma about it. Giovanni Domenico Cassini (Cassini I, 1625–1712), father of Jean Dominique Cassini (who discovered the first gap in Saturn's rings in 1675), appears to have chanced upon this object before 1711. But this fact was only related by Giovanni's son, Jacques Cassini (Cassini II, 1677–1756), in his 1740 work *Elements of Astronomy*, where he mentions his father sighting a nebula in the area between Canis Major and Canis Minor. Messier looked for the object in 1771 but did not find it, concluding that Giovanni Cassini had probably witnessed a passing comet. One year later, however, Messier did sweep up an open cluster in Monoceros, which would become the fiftieth entry in his catalogue.

A rich young cluster, M50 lies in the region of the stellar association CMa OB1, though apparently M50 is not physically connected to it. The cluster contains about 2,100 stars brighter than magnitude 23 and spans about 14 light-years in true physical extent. Jasonjot Singh Kalirai of the University of British Columbia and colleagues reported in a 2003 paper in *Astronomical Journal* (vol. 126, p. 1402) that they used the Canada-France-Hawaii Telescope atop Mauna Kea to study

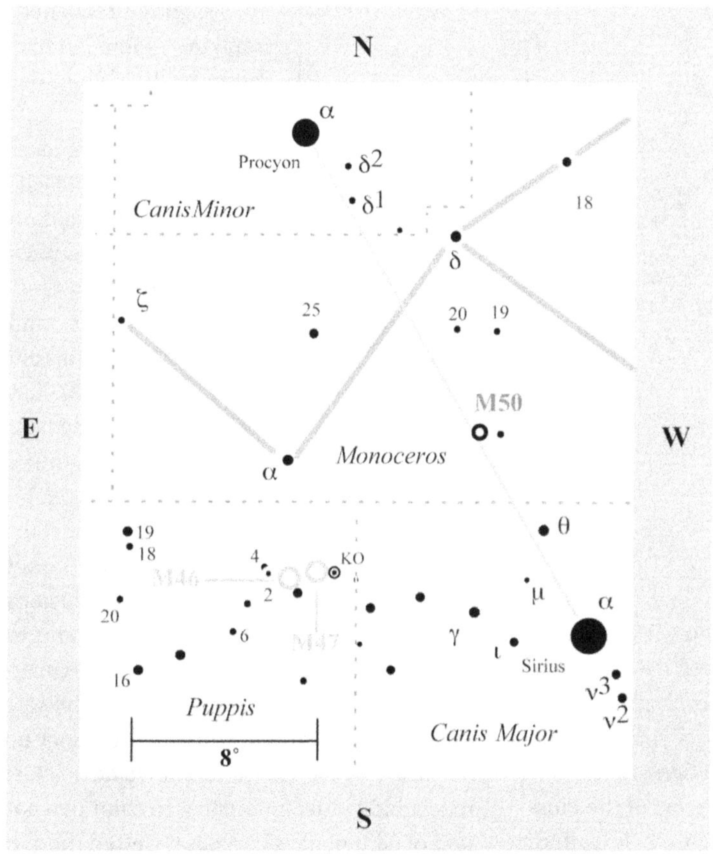

M50 and found that the cluster has a clear main sequence. Their analysis of the luminosity and mass function of M50 suggests that, despite its young age (130 million years), the cluster is "somewhat dynamically relaxed, exhibiting signs of mass segregation."

In a 2012 paper in *Astronomy Letters* (vol. 38, p. 74), Russian astronomers V. N. Frolov (Pulkovo Astronomical Observatory) and colleagues reported results of a comprehensive proper-motion and photometric study of the cluster using six plates from the Pulkova Observatory's astrograph that span 60 years. They found 508 true members on these plates and determined the cluster's age to be about 140 million years.

You can use binoculars to locate M50. It will appear as a bright, 6th-magnitude glow bathed in a rich Milky Way field just 7° north of Gamma (γ) Canis Majoris – a 4th-magnitude star about 5° (two finger widths) east of Sirius. Once you've found this snowy blur, lower the binoculars and try to see it with the naked eye. This might be a challenge if your skies are at all light polluted.

At low power through the 4-inch, the cluster's true extent is rather hard to make out because the Milky Way surrounding M50 is bursting with stellar clumps and asterisms. Indeed, Skiff says it is hard even to determine M50's declination because of irregular concentrations of stars in the immediate vicinity. Wallenquist found at least 50 members in the central 14′ of the cluster, and Skiff counts 150 members across 25′ in a 10-inch telescope. Archinal tallies only 80 cluster members in a 15′ diameter, though. The question is, Where does the cluster really end?

With moderate magnification, the main pattern of stars forms a crude cross whose broken arms extend nearly 1° on a side. These prominent but star-poor arms radiate from the cluster's bleak center; they would make a striking spiral pattern, except that one of the arms is "turning" the wrong way. But, again, are these long stellar extensions part of the true cluster? If you cut back to low power and look at

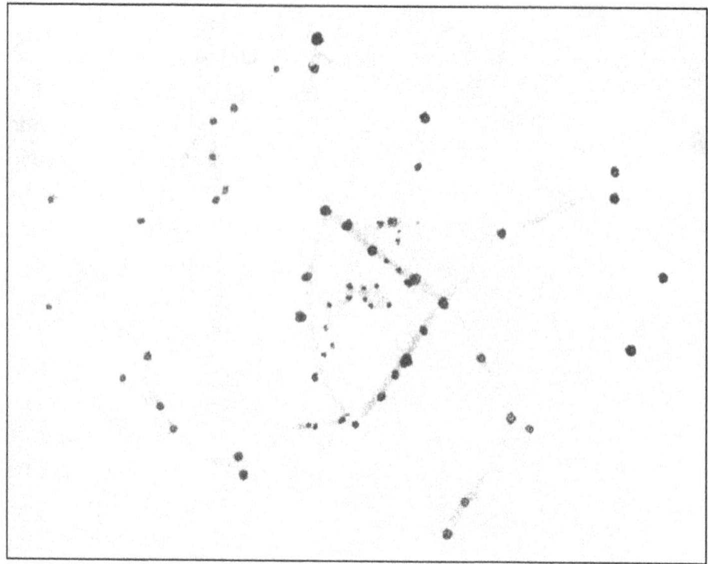

are slightly reddened, implying some obscuration by dust along our line of sight.

Many observers over the past century and earlier have commented on a red star in the southern portion of the cluster; Smyth, d'Arrest, and Webb all noted it. But I did not see it, nor did Mallas, who wrote, "A red star mentioned by Admiral Smyth and T. W. Webb is presumably the 8th-magnitude object about 7′ south of the cluster's center, but the color was not conspicuous to [me]." Skiff notes that photometry of this magnitude 7.8 star on the southern edge shows it to be definitely red. Could the discrepancy be because this star is a variable? Or, does it just reflect individual differences in color perception of faint objects? Archinal viewed the star with a 20-inch f/7 reflector and described it as "particularly bright orange." He suspects the reported variations in the star's color are caused by variations in observers' eyes, particularly at low light levels.

Use low power to follow a meandering river of 9th- to 10th-magnitude stars flowing south of M50. It is even more striking in binoculars, looking like a fuzzy strand of DNA. It winds its way toward another weak and loose open cluster, NGC 2335, 2° to the southwest.

the cluster's overall shape, you will see fainter stars filling in the gaps between the arms, giving the cluster a distinctly oval appearance, though it is just slightly out of round. M50's shape reminds me of Omega Centauri's.

Now concentrate on the heart of the cluster. Do you see a distinct right-angle pattern of stars with a curved hypotenuse to the east? A smaller cluster of stars resides in the northern section of that warped right triangle. This minicluster contains about a dozen members that form a scoop-like asterism. Apparently, Luginbuhl and Skiff also noticed this grouping, observing that "the middle of the cluster is relatively empty except for one small group." They also recorded a dark patch near the cluster's center, which is probably why M50's heart looks so empty. The stars of M50

M51

Whirlpool Galaxy
NGC 5194
Type: Spiral Galaxy (SAbc)
Con: Canes Venatici

RA: $13^h30.0^m$
Dec: +47°16′
Mag: 8.4
SB: 13.1
Dim: 11.2′ × 6.9′
Dist: ~27 million light-years
Disc: Charles Messier, 1773

MESSIER: [Observed January 11, 1774] Very faint nebula without stars, near the more northerly ear of Canes Venatici, below the second-magnitude star η in the tail of Ursa Major. M. Messier discovered this nebula on 13 October 1773, when observing the comet that appeared in that year. It may be seen only with difficulty with a simple three-and-a-half-foot refractor. Nearby there is an eighth-magnitude star. M. Messier plotted its position on the chart of the comet observed in 1773 and 1774, *Mémoires de l'Académie 1774*, plate III. It is double: both parts have bright centers, and they are 4′35″ apart. The two atmospheres are in contact; one is much fainter than the other. Observed several times.

NGC: A magnificent object, great spiral nebula.

JUST AS THE CRAB NEBULA, M1, CAUSED Messier to undertake his catalogue, the Whirlpool Galaxy, M51, caused me to undertake my own version of his catalogue. My reason for starting, though, was different from Messier's. He would have preferred that you avoid the objects in his list, whereas I'm encouraging you to seek them out!

Messier discovered M51 on October 13, 1773, "when observing the comet that appeared in that year." After making a refined observation of it on January 11, 1774, he described the object in his expanded catalogue, which appeared in the French almanac *Connaissance des Temps* for 1783 (published in 1780). He called it a "very faint nebula without stars," noting that it could be seen "only with difficulty with a simple three-and-a-half-foot refractor. Nearby there is an eighth-magnitude star." Messier makes no mention of any other "nebula" in the field, and a 7th-magnitude star does lie roughly 20′ to the east.

Harvard historian Owen Gingerich notes, however, that in the 1784 *Connaissance des Temps* Messier appended his original catalogue description of M51 to include these words: "It [M51] is double: both parts have bright centers, and they are 4′35″ apart. The two atmospheres are in contact; one is much fainter than the other. Observed several times."

Despite what may be falsely implied, Messier didn't discover M51's companion; Méchain did – a fact unearthed by French astronomy popularizer Camille Flammarion (1842–1925), who owned an original copy of Messier's 1784 manuscript of his catalogue. In it, Flammarion found a margin note – in Messier's handwriting (accompanied with "a little sketch") – stating that "M. Méchain" saw M51 as "double." He then transcribed Méchain's original notes: "March 20th, 1781, saw this nebula; effectively it is double. The center of each is brilliant and clear; distinct and the light of each touches each other."

After receiving Méchain's discovery notes, Messier must have gone out and measured the separation of the two objects, thus explaining the noted separation and his comment that he observed them "several times." Some might argue that, like most astronomers of his day, Messier believed that nebulae were clouds of matter from which stars were born – so it would not be surprising that, as with M78 in Orion, he would treat M51 and its companion as a single object.

But Messier didn't see M78 as a "double" nebula. He and Méchain (who discovered the object) described it as a cluster of stars involved with nebulosity, which had two bright nuclei – a very different description than what was penned for M51 and its companion. *Each* of these objects, Méchain said, has a "distinct" center. In other words, he did not see two nuclei embedded in a single nebulosity but two objects, clearly identifiable, yet touching.

When William Herschel noticed M51's companion during a sweep of the heavens on May 12, 1787, he might not have been aware of its existence (especially if he was using the 1783 *Connaissance des Temps*), which would explain why he felt justified in giving the companion an independent catalog number: H I-186 (now NGC 5195).

Gingerich concludes that, "While it's safe to assume that Herschel believed he had discovered a new object, Méchain is NGC 5195's rightful discoverer, and Messier was the first to include it in a catalog."

Thus, it seems reasonable that, if you'd like, you could "tick off" NGC 5195 as an unsung Méchain missive, thereby adding an additional object to the list. It's an object mentioned in Messier's original catalogue; it was discovered by Pierre Méchain; it's a distinct object (a peculiar barred spiral galaxy, well worth one's attention); and no one can argue the credibility of its discovery or of its existence.

Astronomers once believed M51 to be a great swirling nebula. Rosse detected the nebula's "spiral convulsions" in 1845 with his 72-inch speculum-mirror reflector at Birr Castle in Ireland, making M51 the first galaxy shown to have spiral structure.

But some eighteenth- and nineteenth-century observers, such as John Herschel, were nearly clairvoyant about the object's nature. Herschel's drawing of M51 shows a split ring surrounding a central condensation, a view that closely resembled his concept of the Milky Way's structure. Admiral Smyth went so far as to describe the nebula as a "stellar universe, similar to that to which we belong, whose vast amplitudes are in no doubt peopled with countless numbers of percipient beings." But these statements were based on the knowledge of the day; the astronomers did not envision a galaxy of countless stars but rather a hazy vortex that verified the cosmology established by the French mathematician Pierre Simon de Laplace, who postulated that our solar system condensed out of a rotating gaseous nebula. This notion of M51 being a solar system in formation was not shattered until 1923, when astronomers discovered the true nature of the mysterious spiral nebulae.

We now know that M51 is the finest example of a face-on spiral galaxy. It is a highly metallic system with well-defined spiral structure. A near neighbor of our own Galaxy, just 27 million light-years away, this graceful pinwheel of stars, dust, and gas measures about 87,000 light-years across and shines with a luminosity about 10 billion times the Sun's. According to scientists at the Space Telescope Science Institute, M51's central region is about 400 million years old and has a mass 40 million times larger than the Sun's: "The concentration of stars is about 5,000 times higher than in our solar neighborhood, the Milky Way Galaxy. We would see a continuously bright sky if we lived near the bright center."

The bright central region is surrounded by an elderly population of stars, at least 8 billion and perhaps 13 billion years old, or the age of the universe itself. Beyond these ancient stars lies a "necklace" of very young star clusters comprised of infant stars, younger than 10 million years, which are about 700 light-years away from the galaxy's core. Since astronomers usually find such extremely youthful inhabitants thousands of light-years away from galactic centers, they suggest that perhaps M51's compact core formed about 400 million years ago when a dwarf companion galaxy passed close to it, stirring up dust and material that triggered the unusually high star-formation activity in the bright necklace of young stars.

Until recently, astronomers had not determined an accurate distance to M51. But as J. Vinkó (University of Szeged, Hungary) and colleagues report in a 2012 paper in *Astronomy and Astrophysics* (vol. 540, p. 93), their studies of Hubble Space Telescope (HST) data taken of two recent supernovae in the galaxy (SN 2005cs and SN 2011dh) provided a distance of 27 million light-years, which is in good agreement with other recent values.

The importance of giving some spotlight to NGC 5195 is evidenced by the many careful visual observations of it. In images taken with large telescopes, the galaxy is a marvel unto itself, especially as it is interacting with M51 (Arp 85) and is clearly distorted by it, with distinct stellar plumes characteristic of tidal interaction.

NGC 5195 most likely passed closest to M51 some 70 million years ago and is now on M51's far side, receding from us with a radial velocity of 290 miles per second. Indeed, high-resolution images reveal a vast swath of dust in M51's northeast arm slicing across the eastern face of NGC 5195, lending visual proof of the pair's relative placement with respect to us.

Otherwise, NGC 5195 appears as a diffuse disk galaxy, inclined 46° from face-on, with numerous internal dust lanes that appear on the galaxy's west side. The galaxy is small, measuring only about 55,000 light-years in extent. If you could take a broom and sweep away that dust, we might see NGC 5195 as an SB0 type galaxy, with peculiarities induced by tidal interactions with M51. But no one is certain of the true classification of the system (thus the question mark in the data table for this object); for now it's all guesswork. The description in the *Hubble Atlas* suggests

that the infrared emission most likely results from dust heated by the evolved starburst population. "It is clear that NGC 5195 has undergone a starburst," the Boulade team argues and that "this starburst has ceased." The team also notes that very few stars younger than B5 exist in NGC 5195, thus leading the researchers to believe that the thermal emission comes from the evolved stellar population.

In 2006, Korean astronomers Narae Hwang and Myung Gyoon Lee of Seoul National University announced (*Astrophysical Journal*, vol. 638, p. 79) how they had used mosaic images taken by the Hubble Space Telescope to discover about 50 faint, fuzzy star clusters around NGC 5195 that, with effective diameters of about 46 light-years, are larger than typical globular star clusters. The clusters are about 100,000 times more massive than the Sun and older than 1 billion years. Most of these new clusters are scattered in an elongated region almost perpendicular to the northern spiral arm of M51, slightly north of NGC 5195's nucleus. In contrast, NGC 5195's normal compact red clusters are located around the bright optical body of the host galaxy. Hwang and Lee suggest that at least some faint fuzzy

that the morphology has similarities to the Amorphous class, including M82.

Like M82, NGC 5195 appears to be a starburst galaxy, as evidenced by its warm thermal emission deduced from IRAS (Infrared Astronomical Satellite) measures. Curiously, no star-formation regions have yet been found. In a 1996 paper in *Astronomy and Astrophysics* (vol. 315, p. L85), Olivier Boulade (CEA/DSM/DAPNIA/Service d'Astrophysique, Dedex France) and his colleagues suggest clusters are experiencing tidal interactions with the companion galaxy M51 and must be associated with the tidal debris in the western halo of NGC 5195.

To find these interacting marvels, use the chart at the back of the book to locate 2nd-magnitude Eta (η) Ursae Majoris, the easternmost star in the Big Dipper's handle. Now use your unaided eye or binoculars to find 5th-magnitude 24 Canum Venaticorum, which is 2 1/4° west-southwest of Ursae Majoris. M51 and NGC 5195 lie about 3° to the south-southwest.

As my drawing indicates, M51 does not reveal much detail in small telescopes. But the features I did record are impressive for a 4-inch telescope. For example, the galaxy's main spiral arms are often stated as being difficult for telescopes smaller than 10 inches. Much depends on a practiced eye and very favorable viewing conditions. Still, had I been an observer in the nineteenth century, armed with the telescope I own today, I would not have shied away from at least suggesting the possibility that M51 had spiral structure. The brightest portions of the arms do appear mottled and are clearly separated from the stellar nucleus and its tightly wound core by dusky patches (this is the "ring" Herschel and others saw). I would have agreed with Smyth, who saw M51 as "resembling the ghost of Saturn with its rings in a vertical position."

Admittedly, the most surprising detail in my drawing is the dim bridge of gas connecting M51 to NGC 5195. The visibility of this bridge in small apertures has long been debated. Some argue that it cannot be seen except from under a dark sky and with at least a 12-inch telescope. Others say it can be seen in a 4-inch. Could the view of the bridge in small apertures be illusory – a manifestation created by the proximity of the two galaxies coupled with the knowledge that the bridge exists? I did an experiment to find out.

As I mentioned in Chapter 2, the nineteenth-century astronomer George Bond recorded the outer extensions of a diffuse object by placing the object outside the field of view and then letting it drift back in until he detected a change in the brightness of the sky background. I tried a modified version of this technique with M51. Using 130×, I moved the telescope toward M51 from the west and found a clear gap between M51 and its companion until I reached the eastern boundaries of the two, where I encountered a slight increase in brightness that ended the canal. I repeated this sweep several times. Each time, soon after my mental canoe went sailing down that dark canal, it collided with this phantom bridge. Then, I reversed the sweep, moving from east to west toward the two galaxies. There was no "entrance" to the canal; I was stopped immediately by the bridge. Try this sweeping technique yourself to discover your own truth about the visibility of the phantom bridge.

In addition to the two supernovae mentioned earlier, M51 hosted another, SN 1994I, bringing to three the total number of supernovae known in the galaxy. Tim Puckett and Jerry Armstrong of Atlanta, Georgia, discovered SN 1994I, which appeared at magnitude 13.5 some 14″ east and 12″ south of the nucleus; it rose to about magnitude 12.8 at its brightest. Wolfgang Kloehr of Schweinfurt, Germany, found SN 2005cs, a Type II supernova – a giant star whose massive core collapsed to become a neutron star or black hole – that exploded 78″ south of the galaxy's nucleus and rose to 14th magnitude. SN 2011dh, another Type II supernova, was a multiple discovery (first by Tom Reiland of Glenshaw, Pennsylvania); it appeared 126″ east and 92″ south of the

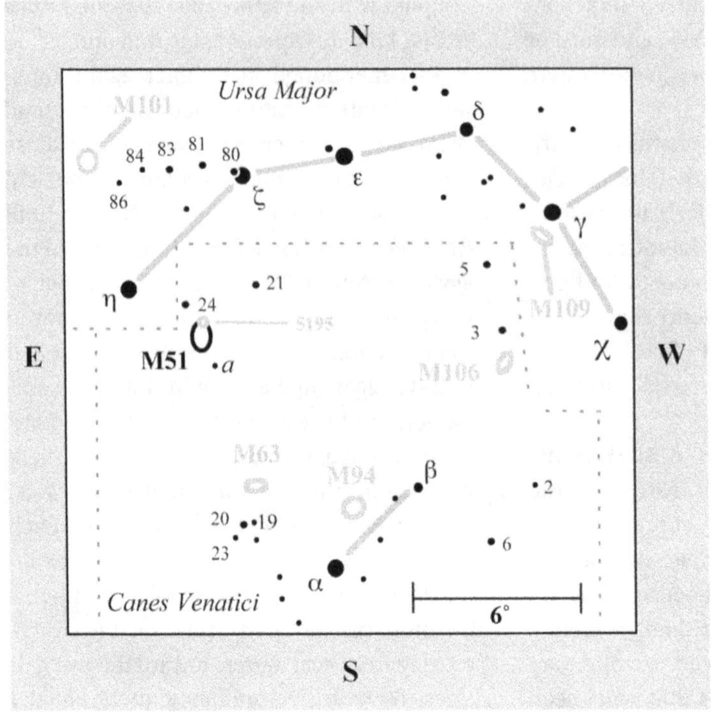

outer coma. With averted vision, the view at 60× shows NGC 5195 appearing more oval shaped, oriented east to west. A bright, nebulous "star" shines at the galaxy's core, which is surrounded by a circular collar of snowy white light. This in turn is nestled in a diffuse and slightly elliptical outer halo.

At 94×, the galaxy looks irregular, with faint extensions bleeding off to the west from the galaxy's northern and southern tips; the core (like a stubby bar oriented north to south) also appears to bulge in that direction, as if it were impregnated with matter. The galaxy's overall appearance is that of a capital Sigma (Σ). With time, much attention, and averted vision (almost to the point of straining), I can see some diffuse material just north of NGC 5195, appearing as a dim patch of feeble light separate from NGC 5195. The views were consistent enough for me to put faith in the fact that this was light from the tidal plumes so apparent in larger telescopes.

It's hard to say just how beautiful the M51-NGC 5195 pair appeared when I observed it from the Texas Star Party one year through Larry Mitchell's 36-inch reflector. Seeing it in the sky required standing in peril high atop a 20-foot ladder near the top rungs and holding onto the telescope's truss supports as a precaution against demonic wind gusts that continually threatened my existence. Feet braced, hands clutching, heart

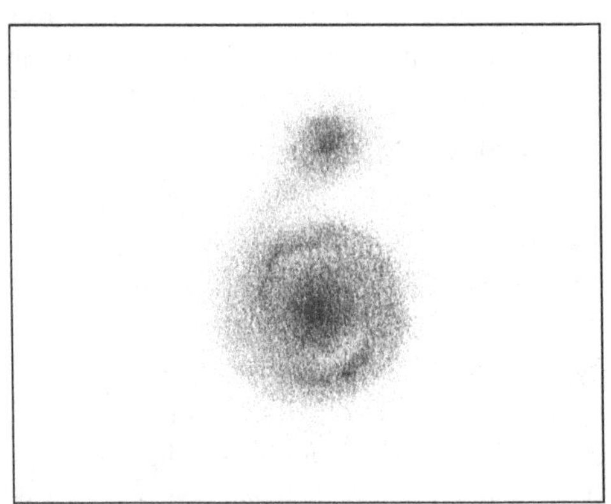

nucleus and achieved magnitude 12.7 at maximum brightness.

As for NGC 5195, at 33× in the 4-inch, it is a brilliant 3′-wide circular glow, with an intense core that's brighter than M51's! It shines like a 9th-magnitude comet (sans tail) with a bright nucleus, intense inner coma, and diffuse

pounding, I brought my eye to the eyepiece and saw M51's spiral wonder with one long arm embracing NGC 5195. Defaced with dust, "arms" contorted into wide-sweeping nebulous arcs on the galaxy's northern and southern edges, NGC 5195 burned forth with imperial light. And, without much need for averted vision, I saw wide plumes of gaseous matter "sweeping off" the galaxy to the north and spreading into a glorious wash of wispy tendrils that seemed to flow away like ash blown by a gust of wind – the beauty that is the hobby of astronomy captured in a single view.

M52

The Scorpion
NGC 7654
Type: Open Cluster
Con: Cassiopeia

RA: $23^h24.8^m$
Dec: +61°36′
Mag: 6.9
Diam: 16′
Dist: ~5,100 light-years
Disc: Charles Messier, 1774

MESSIER: [Observed September 7, 1774] Cluster of very faint stars, mingled with nebulosity, which may be seen only with an achromatic refractor. It was while observing the comet that appeared in that year that M. Messier saw this cluster, which was close to the comet on 7 September 1774. It is below the star *d* [Flamsteed 4] in Cassiopeia. Star *d* was used to determine the positions of the star cluster and the comet.

NGC: Cluster, large, rich, much compressed in the middle, round, stars from 9th to 13th magnitude.

M52 IS ONE OF MY FAVORITE OPEN CLUSTERS for binoculars – not because it trumpets starlight but because it doesn't. The 7th-magnitude cluster is merely a large, uniform glow; subconsciously, I must find that simplicity soothing. Although I tried very hard, I could not convince myself of seeing it with the unaided eye. The cluster lies about 1/2° south of 4 Cassiopeiae, a reddish 5th-magnitude star about 6° northwest of Beta (β) Cassiopeiae. Through 10 × 50 binoculars, the cluster looks like a double star in a haze.

The binocular view belies the fact that M52 is a very rich telescopic cluster. About 60 million years young, it measures 24 light-years in true linear extent and contains nearly 200 members brighter than 15th magnitude. The cluster's tidal radius, however, spans about 43 light-years, and has a total mass of about 1,200 solar masses (C. Bonatto et al., *Astronomy and Astrophysics*, vol. 455, p. 931, 2006). The cluster's computed central density is 56 stars per cubic parsec; in Messier's catalogue, only M11 and M67 are denser open clusters. The cluster's color–magnitude diagrams show a wide age spread. Star formation then may have been biased toward higher masses early in the cluster's history, with low-mass stars forming later. Photometric observations of M52 by Y. P. Luo (National Astronomical Observatories) and colleagues, reported in

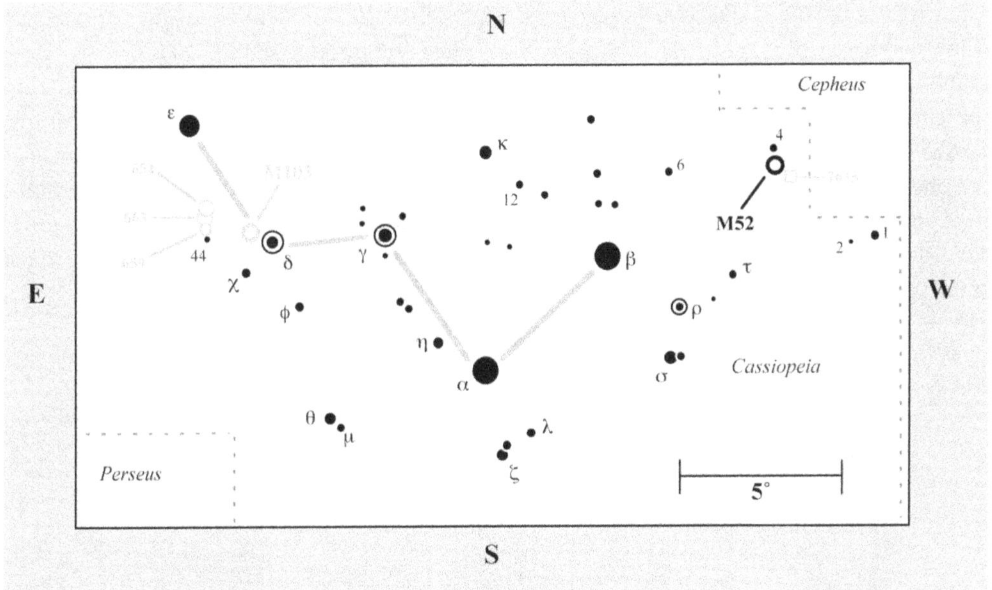

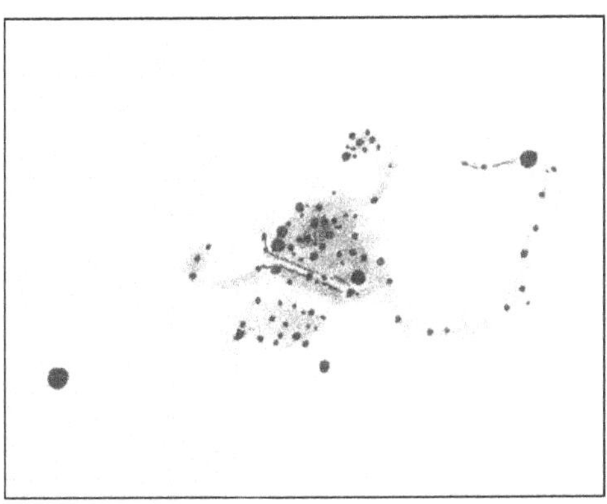

2012 in *Astrophysical Journal* (vol. 746, p. 7), revealed 18 slowly pulsating B (SPB) stars (14 newly discovered), of which 12 are probable members. This makes M52 the richest galactic cluster in terms of SPB content.

Telescopically, you will be immediately struck by the forceful presence of an 8th-magnitude topaz field star on the cluster's southwestern edge. This star all but leaps out at you, as if trying to steal the show. It is an impostor, however, not an actual member of the cluster. Interestingly, when I first looked at the star and then at the cluster, which has an overall bluish hue, the topaz color of the star seemed even more pronounced. Webb and Smyth saw this star as orange, and Mallas recorded it as a conspicuous reddish star. Webb described M52 as "irregular, with orange star, as is frequently the case," an apparent indication that usually the most outstanding member of an open cluster shines with a ruddy hue.

With a quick glance at low power, M52 looks like a tight ball of tiny crystal chips reflecting blue light. A longer view will reveal a little isolated patch of starlight just to the northwest of a heart-shaped central body; a much larger patch lies to the southeast. With 72×, the shape of the cluster's stars looks rather arachnid, like a scorpion. Thin wisps of faint stars jut from the body like tiny legs, and patches of starlight form the claws. The scorpion's swooping tail curves to the north, where it joins the topaz star. Two stinger stars follow to the east.

The main, heart-shaped body of starlight contains many uniformly bright stars caging a haze of fainter members, which requires peripheral vision to resolve. When you don't stare directly at this cage, do you see the thin, dark rift running along the southeastern side of the scorpion's body?

While in the vicinity, try to glimpse the elusive Bubble Nebula (NGC 7635). Only 35′ southwest of M52, this very dim ring of gas, whose brightest section surrounds a magnitude 8.5 star, is only weakly visible from dark skies. Although it resembles a planetary nebula, the Bubble is a fairly ordinary H II region, a vast cloud of ionized hydrogen.

M53

NGC 5024
Type: Globular Cluster
Con: Coma Berenices

RA: $13^h12.9^m$
Dec: +18°10′
Mag: 7.7
Diam: 13′
Dist: ~58,400 light-years
Disc: Johann Elert Bode, 1775

MESSIER: [Observed February 26, 1777] Nebula without stars discovered below and close to Coma Berenices, close to the star Flamsteed 42 in that constellation. This nebula is circular and conspicuous. The comet of 1779 was directly compared with this nebula, and M. Messier plotted it on the chart for that comet, which will be published in the Academy volume of 1779. Observed again 13 April 1781. It resembles the nebula that lies below Lepus.

NGC: Remarkable, globular cluster, bright, very compressed, irregularly round, very much brighter in the middle; contains 12th-magnitude stars.

WILLIAM HERSCHEL DESCRIBED M53 AS one of the most beautiful sights in the heavens. Still, this whopping 220-light-year-wide sphere of stars is an unsung hero among globulars for small telescopes. Perhaps that is because it sits on the fringe of the Coma–Virgo cluster of galaxies, like a flower outside a forest of redwoods. But it is a bold glow, and even binoculars will reveal its magnitude 7.5, 13′ disk only about 1° north-northeast of the 4th-magnitude binary Alpha (α) Comae Berenices (Diadem).

German astronomer Johann Elert Bode (1747–1826) discovered the object on February 3, 1775, describing it as a "new nebula, appearing through the telescope as round and pretty lively." Through his larger reflectors, William Herschel found it had the "form of a solid ball, consisting of small stars quite compressed into a blaze of light with a great number of loose ones surrounding it and distinctly visible in the general mass." William's son John added that it looked like "curved appendages of stars, like the short claws of a crab."

M53 is currently about 58,400 light-years from the Galactic center, close to its perigalacticon of ~51,200 light-years. The cluster has a spectral type of F6, and each of its members has about 1/125 as much metal per unit hydrogen as the Sun. Astronomers are still debating

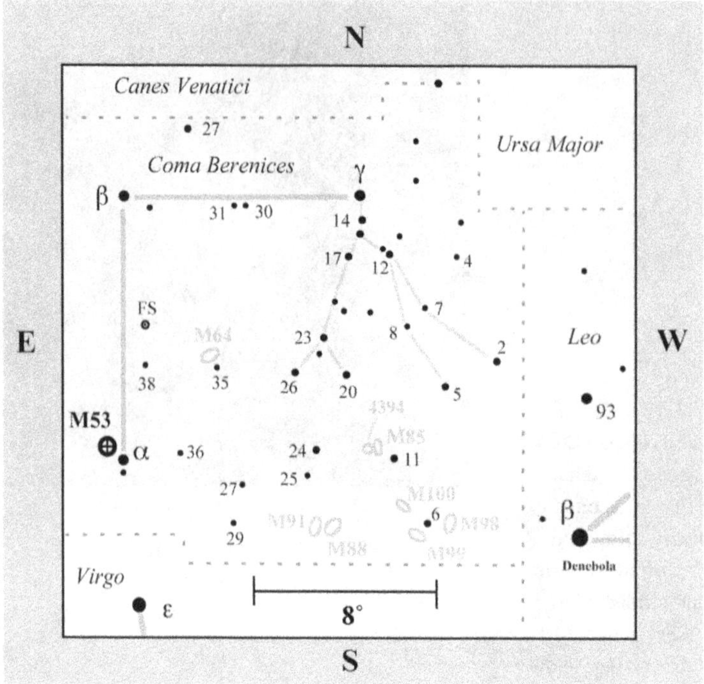

the two appear very close to one another in the sky, they are actually separated by about 3,300 light-years. Chun et al. (*Astronomical Journal*, vol. 139, p. 606, 2010) detected a tidal bridge-like feature around the two globulars as well as tidal features that may result from dynamical interaction between the two clusters. Katrin and Jordi, on the other hand, found none.

whether M53 belongs to the Sagittarius tidal stream – the tidal tail of the Sagittarius dwarf spheroidal galaxy, of which globular star cluster M54 in Sagittarius is the nucleus. Katrin Jordi and Eva K. Grebel of the Astronomisches Rechen-Institut in Heidelberg, Germany, in a 2010 paper in *Astronomy and Astrophysics* (vol. 522, p. 71), note that their study of M53 found an obvious large-scale structure connecting it to that stream, but they did not draw any conclusion.

What a haunting field M53 is in telescopically. Whenever I peer at the globular, my averted gaze continually picks up a multitude of dim spectral shapes, which seem to flitter across the low-power field. While most are close doubles, one pale ashen sheen belongs to the loose globular NGC 5053, which looms about 1° southeast of M53 like the departed soul of its more brilliant neighbor. Though

With a glance at low power, M53 appears as a tight, unresolved ball of light with a faint outer halo. Use averted vision, however, and the inner halo, which surrounds a crisp, star-like nucleus with a tangerine hue, begins to look boxy. At 72×, the inner region reveals its beauty: look for a pale gemstone inside a tiny jeweled chest, which is encased inside a larger jeweled chest; the chests are aligned east to west. High power shatters the jewel, whose opalescent fragments seem to sparkle through a thinning cloud of dust. Thus, the core no longer looks so tight, though I still see a tiny "nucleus" flanked by a wall of stars to the south. Use keen averted vision and relax your gaze. Do you see the dark rift abutting the south wall (refer to the drawing)?

Now return to 72×. The cluster appears twisted, as if someone had tried to wring out the stars in the northern half. This strange sight is created by four claws of stars ripping through the outer hood to the north and west; the three northern extensions appear to

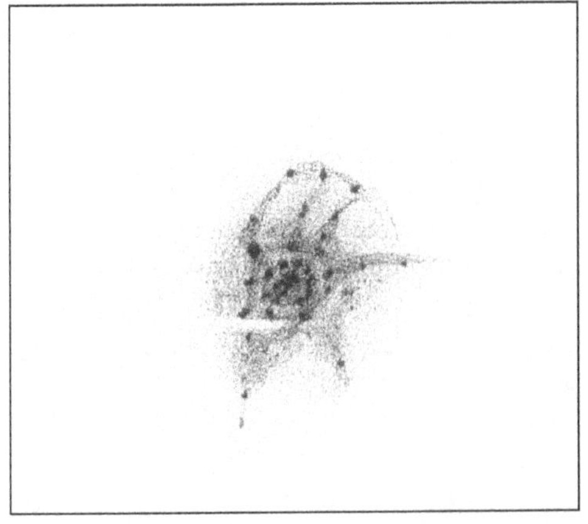

be topped by a faint arc of starlight emanating from the westernmost finger. Meanwhile, the southern half of the cluster looks triangular, with the apex of the triangle pointing due west. I wonder if this is what Luginbuhl and Skiff were referring to when they wrote that "in poor seeing the core of the cluster looks roughly triangular."

Although M53 can easily be seen with binoculars, I could not positively identify it with the naked eye. I do believe someone with eyes younger than mine could. The problem is separating it from Alpha Comae Berenices. Also through binoculars, look just to the northeast of M53 for a beautiful 1°-long asterism, which I call "the Rail," as it reminds me of one of those sleek dragsters that tore up the strips in the 1960s.

M54

NGC 6715
Type: Globular Cluster
Con: Sagittarius

RA: $18^h55.1^m$
Dec: -30°29′
Mag: 7.7; 7.2 (O'Meara)
Diam: 12′
Dist: ~86,400 light-years
Disc: Charles Messier, 1778

MESSIER: [Observed July 24, 1778] Very faint nebula discovered in Sagittarius. The center is bright, and it does not contain any stars; seen with an achromatic three-and-a-half-foot refractor. Its position was determined relative to third-magnitude ζ Sagittarii.

NGC: Globular cluster, very bright, large, round, gradually, then suddenly much brighter in the middle, well resolved, with 15th-magnitude stars.

BEAMING FROM A DISTANCE OF 86,400 LIGHT-years from the Sun and 61,000 light-years from the Galactic center, M54 is the farthest of all the Messier globulars and the second most massive known. It's an impressive system 300 light-years wide that's approaching us from the deep beyond at 460 miles per second.

Despite its distance, M54 is no problem for 7 × 35 binoculars, appearing as a bright, star-like object 1 1/2° west-southwest of Zeta (ζ) Sagittarii, at the base of the teapot. But being so distant, the cluster does not easily reveal its secrets to telescopic viewers. For instance, James Dunlop observed the object four times in 1826 and not once suspected that it consisted of stars. But it's no wonder, given that the cluster's brightest stars shine at a dim magnitude 15.2, while its horizontal-branch magnitude is a dim 18.2. Therefore, you have to be able to see to beyond 18th magnitude to fully resolve the cluster.

In 1784, William Herschel did resolve M54. Of it, he wrote: "240 power shows two pretty large stars in the faint part of the nebulosity, but I rather suppose them to have no connection with the nebula. I believe it to be no other than a miniature cluster of very compressed stars resembling that near the 42nd Comae [M53]."

In 1994, astronomers discovered a large, faint dwarf galaxy in Sagittarius 81,000 light-years from the Sun, orbiting within our Galaxy's halo with a period of about 1 billion years. Furthermore, the new object was superposed on the globular cluster M54. Since then, evidence has mounted to prove almost certainly that M54 belongs to that Sagittarius dwarf spheroidal galaxy (SGR

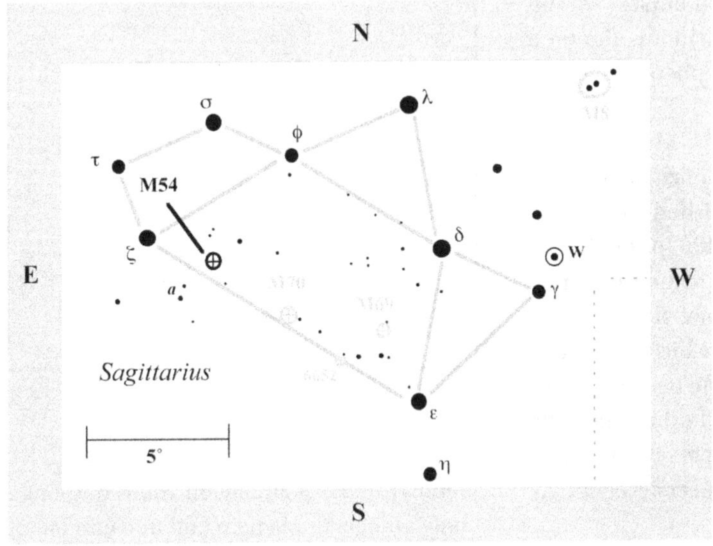

monolithic collapse, it's estimated that the Milky Way began swallowing the Sagittarius dwarf spheroidal about 10 billion years ago.

The Sagittarius dwarf spheroidal and its stream now constitute the most dramatic evidence that hierarchical merging processes have contributed heavily to the buildup of the Milky Way and continue to do so today.

In a 2007 paper in *Astrophysical Journal* (vol. 667, p. L57), Michael H. Siegel of McDonald Observatory and his colleagues presented new Hubble Space Telescope photometry of M54 and the superposed core of the Sagittarius dwarf spheroidal galaxy, which yields an unprecedentedly detailed color–magnitude diagram showing the extended blue horizontal branch and multiple main sequences of the combined system. Their preliminary assessment finds the M54+Sgr field to be dominated by the old metal-poor populations of the dwarf galaxy and the globular cluster.

Multiple turnoffs indicate the presence of at least two intermediate-aged star-formation epochs with ages of 4 billion and 6 billion years and metallicities ranging from one-half to one-quarter solar values. "We also clearly show, for the first time, a prominent, ~2.3 [billion-year-old Sagittarius dwarf galaxy] population of near-solar abundance," Siegel and his team say. They also found a trace population of even younger (~0.1 billion to 0.8 billion years old), more metal-rich (1/4 as much iron per hydrogen atom as the Sun) stars. They say that this is, in part, consistent with

dSph), which our Milky Way Galaxy is now cannibalizing and tidally destroying. Deep, wide-sky surveys have detected the dwarf spheroidal stellar content in the halo along a massive tidal stream. The stream is in fact the most prominent halo substructure yet detected. M54 appears not only to belong to the densest region of that stream but to lie at the very heart of it, suggesting it may be the dwarf's nucleus.

Astronomers have linked three other globular clusters to that tidal stream: Terzan 7, Terzan 8, and Arp 2. One other, Palomar 12, appears to have belonged to the stream but was tidally stripped from it.

Although M54 probably formed at the same time as the dwarf galaxy, only to be accreted into the Milky Way, some astronomers now believe that many globulars are the products of galactic cannibalization. Indeed, in the hierarchical scenario for galaxy formation, dwarf galaxies are the basic building blocks out of which larger galaxies are assembled. And while no one knows for sure whether galaxies form by merging over time or in one

multiple (4–5) star-formation bursts over the entire life of the satellite, including the time since the Sagittarius dwarf spheroidal galaxy began disrupting.

A more recent study by M. Bellazzini (INAF-Osservatorio Astronomico di Bologna, Italy) and colleagues, published in 2008 in *Astronomical Journal* (vol. 136, p. 1147), has also shown that the nucleus of the Sagittarius dwarf spheroidal galaxy may have formed independently of M54, which probably plunged to its present position because of significant decay of the original orbit caused by dynamical friction, the authors say. Try imagining this the next time you peer at seemingly insignificant M54!

In 2009, R. Ibata (Observatoire Astronomique, Université de Strasbourg, France) and colleagues reported the detection of a black hole with a mass of about 9,400 solar masses in the center of M54. In 2011, Joan M. Wrobel (National Radio Astronomy Observatory, Virginia) and her colleagues analyzed new Chandra and recent Hubble Space Telescope data that ruled out the x-ray counterpart proposed by Ibata et al. If an intermediate black hole exists in M54, the Wrobel team concluded, then it is similar to the one in M15.

To find M54, use the wide-field chart at the back of the book to locate Zeta (ζ) Sagittarii. Center that star in your telescope and then use the chart here to sweep to the globular 1 1/2° west-southwest of Zeta Sagittarii, at the base of the teapot asterism. Sighting M54 from a dark sky should be no problem. Through 7 × 35 binoculars, it appears as a bright, starlike object. But being so distant, the cluster does not easily reveal its secrets to telescopic viewers. Through the 4-inch at 23×, the globular is a softly glowing fuzzy star, sort of what Omega Centauri looks like to the naked eye. (Shapley considered M54 perfectly round.) At 72×, the

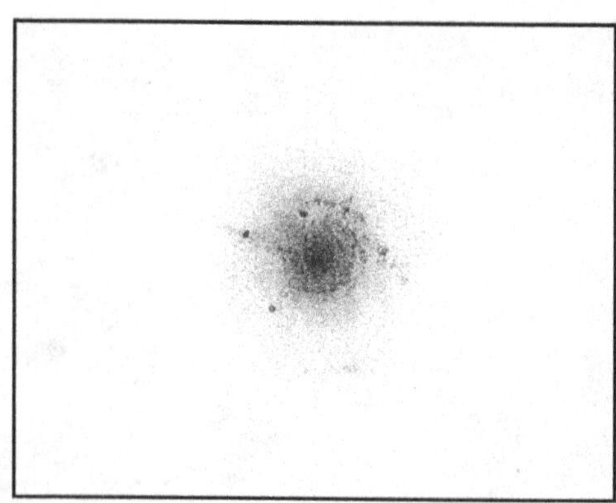

globular remains largely unresolved, though some outliers in the halo pop into and out of view – a perception shared by Skiff using a 10-inch telescope.

This view confirms John Herschel's impression that the brightest stars in the cluster hover around 14th magnitude, though photometry of the cluster shows the brightest members are actually about 15th magnitude. Regardless, M54 displays a bright, starlike core surrounded by a tight, yellowish inner halo that gradually diffuses outward. With persistent averted vision, the cluster looks more like a planetary nebula with a brilliant central star (something Glyn Jones noted too) than a globular. Helping to reinforce this impression is that the outer halo appears distinctly blue, glacial blue, as do some planetaries.

The view changes once again at 130×. Now M54 begins to look like a face-on spiral galaxy just beginning to show resolution. Namely, two arms of very faint stars in the cluster's northern envelope appear to swirl counterclockwise around the nucleus. Clear gaps separate these features from one another and from the core.

Through her 12-inch Schmidt-Cassegrain telescope, South African Magda Streicher resolved the cluster into "very faint stars with a diffuse envelope (218×). A good example of an average globular in all its facets. Soft and hazy with a relatively wide centre that slowly brightens with a thin ... soft peripheral halo."

I did note an orange star superposed on one of these arms. Can you determine which one?

M55

NGC 6809
Type: Globular Cluster
Con: Sagittarius

RA: $19^h40.0^m$
Dec: -30°58′
Mag: 6.3
Diam: 19′
Dist: ~17,600 light-years
Disc: Nicolas-Louis de Lacaille, 1752

MESSIER: [Observed July 24, 1778] Nebula that appears as a whitish patch, about 6 minutes across. Its light is evenly distributed and has not been found to contain any star. Its position was determined relative to ζ Sagittarii by using an intermediate, seventh-magnitude star. This nebula was discovered by M. l'abbé de la Caille, *Mémoires de l'Académie 1755*, page 194. M. Messier searched for it in vain on 29 July 1764, as reported in his paper.

NGC: Globular, pretty bright, large, round, very rich, very gradually brighter in the middle, stars from 12th to 15th magnitude.

ONE OF THE MOST SOUTHERLY GLOBULARS observed by Messier, M55 was included in a study of globular cluster motion that revealed that these stellar agglomerations are moving at high speeds (greater than 100 miles per second) in elliptical orbits through the distant halo of our Galaxy – like long-period comets in orbit around the Sun. At a distance of only 17,600 light-years, however, M55 is relatively nearby (consider that M54 is 86,400 light-years distant). You can see it with the naked eye on the outskirts of the Milky Way band about 8° east-southeast of Zeta (ζ) Sagittarii.

The cluster lies a bit closer to the Galactic center (13,000 light-years) and has an overall spectral type of F4. On average, each of its members has about 1/37 as much metal per unit hydrogen as the Sun, and we see it approaching us at 108 miles per second. The cluster spans nearly 100 light-years in physical extent.

Using a large set of CCD images collected with the du Pont telescope between 1997 and 2008, K. Zloczewski (Nicolaus Copernicus Astronomical Center, Warsaw, Poland) and colleagues derived an absolute proper motion of the cluster of about 3.31 milliarcseconds per year relative to background galaxies. They reported their findings in the *Monthly Notices of the Royal Astronomical Society* (vol. 414,

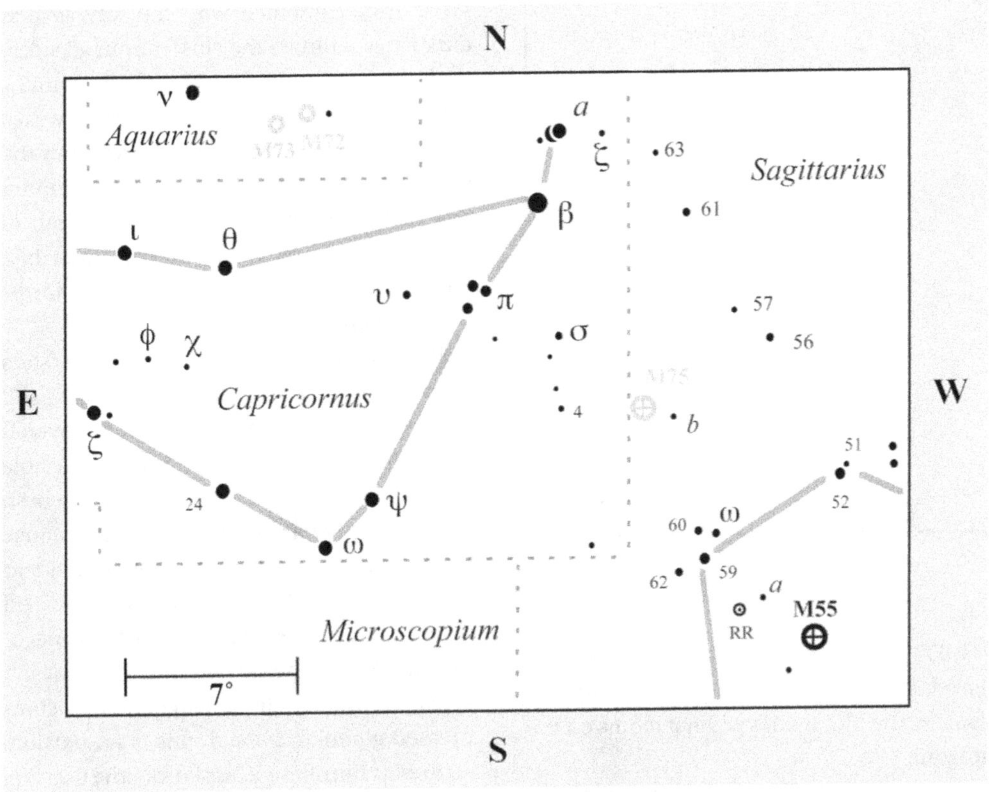

p. 3711, 2011). They also found 43 variable stars sharing the cluster's proper motion, so they are probable members. This sample was dominated not only by pulsating blue straggler stars but also five eclipsing binaries, three of which are main-sequence objects. The survey also identified several candidate blue, yellow, and red straggler stars belonging to the cluster. Finally, the researchers detected 15 likely members of the Sagittarius dwarf spheroidal galaxy (Sgr dSph; see M54) located behind M55. In 2000, the IRAS satellite traced dust aligned with stellar overdensities toward the Galactic center that could be associated with the cluster, but more recent studies have not demonstrated confidence that these features are beyond the cluster's tidal radius, so the cluster may not have a true tidal tail.

In 7 × 35 binoculars, M55 looks like a hairy star. With the 4-inch at 23×, the cluster starts to splinter across its large and loose surface – a refreshing sight after looking at the smaller, more difficult globulars M54, M69, and M70 nearby. In fact, M55's disk looks huge when compared with the disks of those globulars, even at high power.

One night, I had the pleasure of watching an Earth-orbiting satellite sail across M55's misty bay of light. Such transits are becoming increasingly common to deep-sky observers. Such an event puts the scale of the universe, or our little corner of it anyway, in perspective. Here was an artificial craft about the size of a Volkswagen van and a few hundred miles away drifting in front of a cluster of stars a mind-boggling 93 light-years wide and

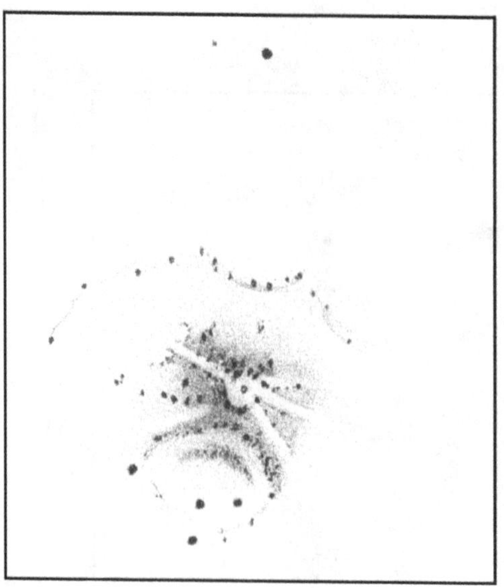

~17,600 light-years distant. In my mind's eye, the cluster swelled to magnificent proportions, as the Moon does when it looms on the horizon.

At 72×, a multitude of bright stars are scattered across a homogeneous background of fainter stars. Many of the cluster's members are nicely resolved into stellar arcs, somewhat reminiscent of the parallel streams in M25. Patterns visible in the background haze are out of sync with those formed by the brighter stars. This, combined with the way several dark lanes infiltrate the cluster from all sides, leads to a most impressive, albeit complicated, view. Ripples of stars and dark waves seem to be flowing into the cluster from the southeast, forming a cove, or what Luginbuhl and Skiff referred to as a "large bite" out of the cluster's outer halo. Another such bite appears to have been taken out of the northwestern part of the outer halo.

The central, southern curve of stars becomes very pronounced at 130×, but I still prefer the moderate power for an overall view. That's when a bright star in a dark hole at the center of the cluster shows up the best. A strong, dark bar extends an equal distance southwest and northeast of that central star and dark hole. Also, dark forks branch off from the southwest axis, forming the negative image of a "peace" symbol. One doesn't expect to find a hole in the center of a compressed globular. Indeed, this is an illusion. As Robert Burnham Jr. explains, "The unusual 'openness' of this cluster ... is due to the fact that only a relatively small percent of the members exceed a brightness of 13th–14th magnitude, and the cluster does not begin to 'fill in' until one reaches about 17th magnitude, where a vast swarm of stars quite suddenly appears."

M56

NGC 6779
Type: Globular Cluster
Con: Lyra

RA: $19^h16.6^m$
Dec: +30°11′
Mag: 8.3
Diam: 7′
Dist: ~30,600 light-years
Disc: Charles Messier, 1779

MESSIER: [Observed January 23, 1779] Nebula without a star, which is very faint. M. Messier discovered it on precisely the same day as he discovered the comet of 1779, on 19 January. On the 23rd, he determined its position by comparing it with Flamsteed 2 Cygni. It is close to the Milky Way and near it is a tenth-magnitude star. M. Messier plotted it on the chart for the comet of 1779.

NGC: Globular cluster, bright, large, irregularly round, gradually very much compressed in the middle, well resolved, stars of 11th to 14th magnitude.

WITH ITS SOFTLY GLOWING CORE, DELICATE spherical halo, and lack of resolution at low power, M56 has always been my favorite non-comet. I still remember many a warm summer night when I would intentionally sweep my telescope back and forth, up and down, between Beta (β) Cygni and Gamma (γ) Lyrae, until I chanced upon M56 about halfway between them. A chill would race up my spine whenever its comet-like form entered the eyepiece. In those magical moments, I was Messier, experiencing the thrill of discovery.

Reality would nevertheless soon set in. But what a perfect snowball M56 is, 7′ across and shining at 8th magnitude, amid the blizzard of the Milky Way. Actually the globular looks more like a dirty snowball because of its grayish pallor. I find the color startling, considering that so many other Messier globulars display delicate pastel hues. Interstellar dust dims the cluster by about 0.3 magnitude.

M56 is a relatively rich but metal-poor globular cluster in the halo of our Galaxy – nearly the same distance from us as it is from the Galactic center (~30,000 light-years). It is receding from us at about 85 miles per second. With 1/95 of it composed of metals, it is not only one of the most metal-poor globulars known but also one of the oldest (~13 billion years) in the Galactic halo.

M56

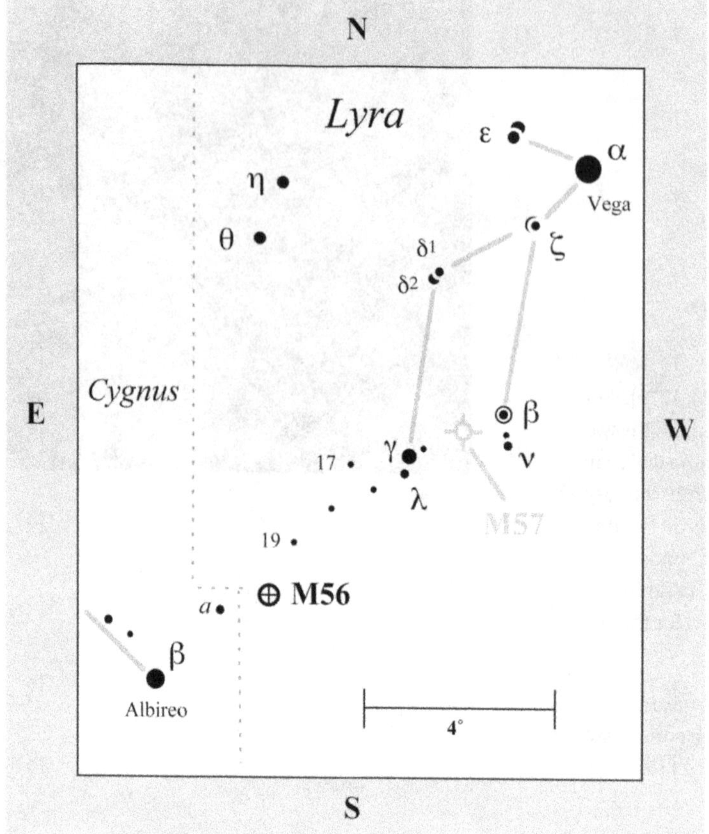

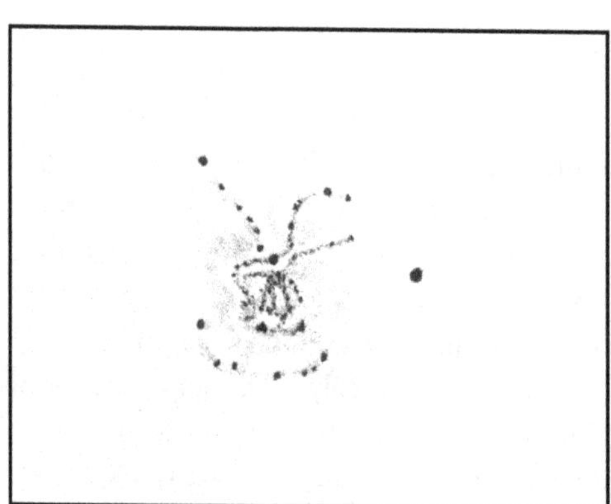

In the low-power view, a faint stream of stars flows from the southeast and drains into M56's foggy pool. The tiny, round glow (which is really ~62.5 light-years across) is punctuated by a 10th-magnitude star to its west. A magnitude 5.8, orange M2 star about 1/2° to the northeast adds some color. There's also a fine 8th-magnitude double about the same distance to the southwest.

Moderate power starts to resolve the cluster, which immediately appears elongated north to south. In fact, the inner core is highly asymmetrical. With a thin fan of material to the south and nothing like it immediately opposite to the north, the innermost region of M56 is quite unique, looking like a light bulb or an exclamation mark. My pencil sketch shows the location of a rather strong, 10th-magnitude central condensation, the point of the exclamation mark.

Interestingly, John Mallas noted that, unlike most other globulars, M56 does not have a bright core. And his drawing of the cluster shows a substantially large elliptical vacancy in the center. Glyn Jones reported that M56 has "no very marked central condensation," as did Burnham (almost verbatim). Skiff speaks of an irregularly round and broad concentration to the center, which corroborates what Smyth drew in his *Cycle of Celestial Objects*.

High power did reveal a strong, dark lane separating my 11th-magnitude "central" star and the southern fan of mottled starlight. I believe this dark feature almost certainly is what Mallas observed. Still, someone should keep an eye out for a possible variable star lurking at the heart of M56. Meanwhile, don't miss out on enjoying the vast web of darkness surrounding M56 at 23×. It causes many neighboring regions to look as if the stars have been deposited there by the ebb and flow of some unseen galactic surf.

M57

Ring Nebula
NGC 6720
Type: Planetary Nebula
Con: Lyra

RA: $18^h53.6^m$
Dec: +33°02′
Mag: 8.8
Dim: 65″
Dist: ~2,000 light-years
Disc: Antoine Darquier de Pellepoix, 1779

MESSIER: [Observed January 31, 1779] Patch of light between the stars γ and β Lyrae, discovered while observing the comet of 1779, which passed very close. It seemed that this patch of light, which has rounded borders, must be composed of very faint stars. It has not, however, been possible to see them, even with the best telescopes, but the suspicion remains that there are some. M. Messier plotted this patch of light on the chart for the comet of 1779. M. Darquier, at Toulouse, discovered this nebula when observing the same comet, and he reported "Nebula between γ and β Lyrae, it is extremely faint, but perfectly outlined. It is as large as Jupiter and resembles a fading planet."

NGC: A magnificent object, annular nebula, bright, pretty large, considerably extended (in Lyra).

> Among the curiosities of the heavens should be placed a nebula that has a regular, concentric, dark spot in the middle. – William Herschel

When a star with a mass similar to that of our Sun nears the end of its life, it blows off a shell of gas that, from our perspective, appears like a ring centered around the dying star. M57, the Ring Nebula, represents the remains of one such disgorging episode about 7,000 years ago. The second planetary nebula discovered, it has worked its way ever since into the hearts of virtually all telescopic observers. Rightly so, because few planetary nebulae appear so bright and distinctive through small apertures.

When French astronomer Antoine Darquier (1718–1802) discovered it through his 2 1/2-inch-long-focus refractor, he described it as a "very dull nebula, but perfectly outlined; as large as Jupiter and looks like a fading planet." (Perhaps M57 should be recognized as the original "Ghost of Jupiter" nebula, a more modern moniker that describes the planetary NGC 3242 in Hydra.) As you can see in the quotation opening this essay, William Herschel first spied the nebula's "hole."

M57 is a challenging binocular object, well placed in the northern sky about 6 1/2° southeast of brilliant Vega (Alpha [α] Lyrae) and nearly halfway between the eclipsing binary star Beta (β) Lyrae (whose brightness fluctuates between magnitude 3.3 and 4.3 every 12.9 days) and 3rd-magnitude Gamma (γ) Lyrae. Telescopically, M57's tiny 9th-magnitude annulus of gray smoke floats against a rich Milky Way field crisscrossed with dark streamers, some of which appear to be as gray and smoky as the Ring Nebula itself. John Herschel first saw the "interior filled with a feeble but very evident nebulous light: like gauze stretched over a hoop." And German astronomer Friedrich von Hahn (1742–1805) added the central star in 1800.

The two-dimensional "ring" we see through our telescopes is actually a torus (doughnut shaped) viewed looking down the hole. This is unlike the planetary M27, which is seen side-on. But, as C. R. O'Dell (Vanderbilt University) and colleagues explain in a 2007 paper in *Astronomical Journal* (vol. 134, p. 1679), the ring's three-dimensional structure is a triaxial ellipsoid (radii of 0.3, 0.4, and 0.6 light-years) seen nearly pole-on, with each arcsecond of the nebula expanding ballistically at about 0.65 kilometers per second.

The researchers also note that the equator of the Ring Nebula is optically thick and much denser than the optically thin poles. The "ring" we see is the nebula's inner halo surrounding a post-asymptotic central star that burns with a temperature of about 120,000 K. The inner halo, the authors say, represents the pole-on projection of the AGB (asymptotic giant branch) wind at high latitudes (circumpolar) directly ionized by the central star. The ring is surrounded by an outer, fainter, and circular halo, which they say is the "projection of the recombining AGB wind at mean to low latitudes, shadowed by the main nebula."

Given the central star's temperature and nebular age of 7,000 years, O'Dell et al. propose that the central star is approaching the white dwarf cooling sequence. In a 2010 paper in *Astronomy and Astrophysics* (vol. 518, p. 137), however, P. A. M. van Hoof (Royal Observatory of Belgium) and colleagues say that the star is on the cooling track, causing the outer parts of the nebula to recombine. The accompanying Hubble Space Telescope image reveals the doomed central star surrounded by ionized gases at the edge of the nebula, which contains elongated dark clumps and knots of material.

The van Hoof team found that H_2 (deuterium) resides in those high-density knots, which, they say, were formed after the recombination of the gas started when the central star entered the cooling track. "Hydrodynamical instabilities due to the unusually low temperature of the recombining gas are proposed as a mechanism for forming the knots," they explain. "H_2 formation in the knots is expected to be substantial after the central star underwent a strong drop in luminosity about one to two thousand years ago, and may still be ongoing at this moment, depending on the density of the knots and the properties of the grains in the knots."

To find the Ring, use the wide-field chart at the back of the book to locate Beta (β) and Gamma (γ) Lyrae. Then use the chart here to find it shining roughly midway between them. From very dark sites, it is visible in 10 × 50 binoculars, but you have to know exactly where to look with averted vision. Look for a small "star" that appears ever so slightly swollen.

One night, when I viewed M57 nearly overhead through the 4-inch at 72×, I found the sight almost shocking. Not only was the "hole"

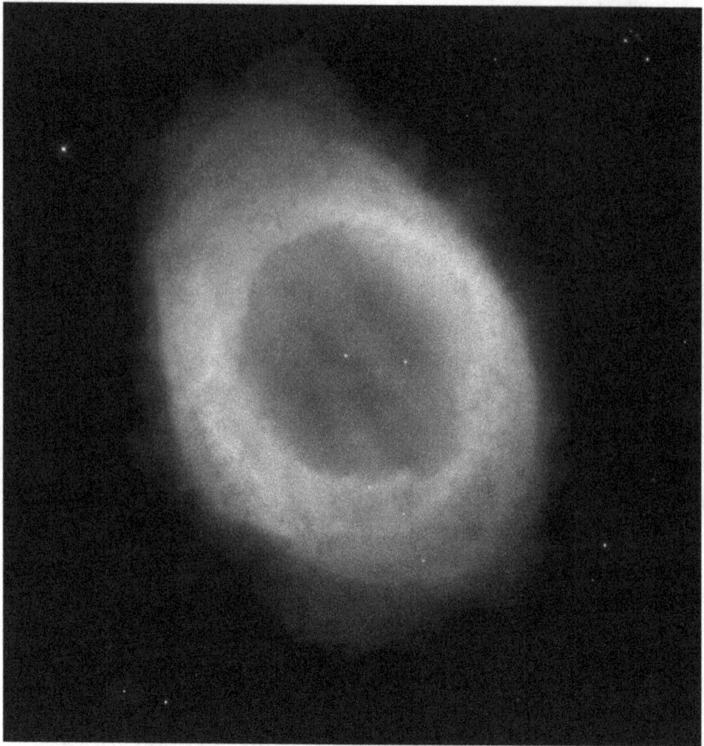

rim. A 14th-magnitude star lies just inside the Ring's southwest rim and is commonly mistaken for the central star. In most star catalogues, the blue dwarf central star is listed as 15th magnitude, but its brightness is the subject of considerable debate. About 80 years of photometric data reported in the *Catalogue of the Central Stars of True and Possible Planetary Nebulae* (Acker and Gliezas, Observatory of Strasbourg, 1982) show the central star varying from magnitude 13 to magnitude 15. Skiff notes that several recent photoelectric measurements give values of magnitude 15.0 to 15.2. Many veteran skywatchers have estimated it to be as bright as 14th magnitude (putting it within range of a good 4-inch telescope under a dark sky). Yet Burnham notes that the star was fainter than 16th magnitude when he looked at it in 1959 through the 40-inch reflector at Lowell Observatory. Although the star appears to be variable, this may only partly explain the disparate magnitude estimates.

clearly visible but the entire ring appeared to scintillate with stellar jewels – foreground stars superposed on the gaseous loop. As my eye raced to see one star in the nebula, my averted vision captured another. As soon as that one was spied, my averted vision jumped to the next one. In this way, the stars seemed to revolve around the ring like the flashing of light bulbs around a Las Vegas billboard. Some nineteenth-century observers alluded to seeing this effect. For example, astronomer Angelo Secchi believed he resolved the ring into minute stars "glittering like stardust," and Webb saw M57's light as "unsteady ... probably an illusion." (See the essay on M82 for a similar phenomenon.)

The brightest star in the gaseous ring shines at 13th magnitude and is located just 1′ east of center, 20″ beyond the extreme outer

One summer evening in 1988, I was visiting comet discoverer Michael Rudenko at the 18-inch refractor at Amherst College in Massachusetts. Some observers had turned the telescope to the Ring Nebula and were complaining that they could not see the central star. When I looked through the eyepiece, I noticed that the apparent size of the Ring

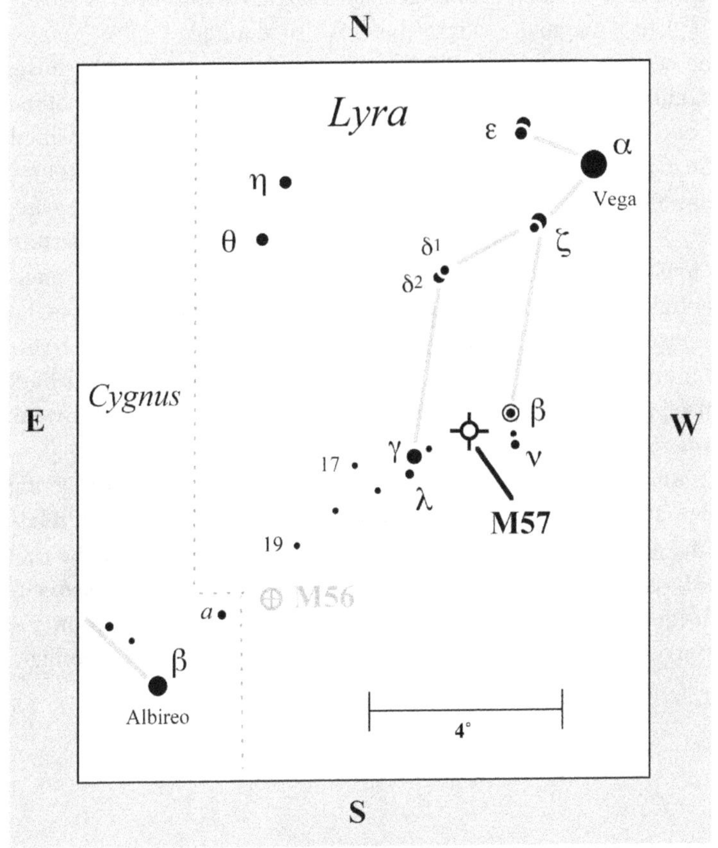

provided about 1,000× magnification. And – voilà! – the central star burst into view, suggesting that the apparent brightness of the star *is* affected by the surrounding nebulosity.

I also looked at the Ring through two 40-inch reflectors, one atop Pic du Midi in the French Pyrenees, the other at Lick Observatory on Mount Hamilton in California. At Pic du Midi, William Sheehan and I centered M57 in the finder scope but were then surprised that we could not see it in the main telescope; all we saw was a pair of stars. It turned out that at 1,200× the gaseous ring was outside our tiny field of view; we were seeing only the central star and its similarly bright neighbor to the northwest, which allowed us to conclude that the central star shines at around magnitude 14.5. Interestingly, though I saw the central star through the 40-inch reflector at Lick Observatory, I did so only with great difficulty – the view was at low power, and I was not able to change the eyepiece. Again, the contrast between the star and its background was diminished. Now, consider that on September 2, 1981, Brent Archinal made a detailed drawing of M57 based on his observations with Lowell Observatory's 72-inch reflector on Anderson Mesa in Arizona, which shows that the central star's neighbor was noticeably fainter than the central star (or the

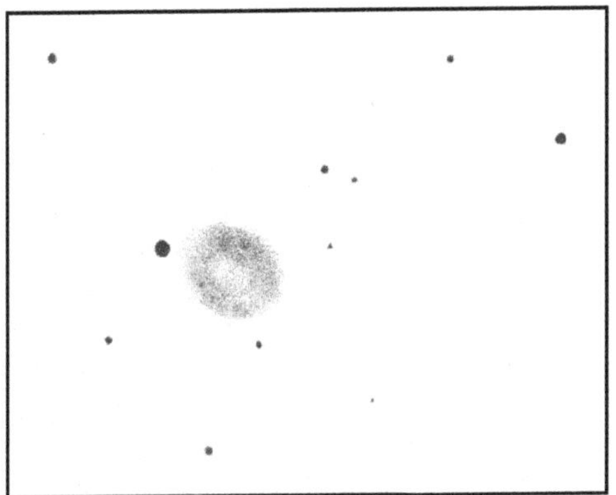

was so small that the "hole" actually appeared bright, diminishing the contrast between the star and its background. I removed that eyepiece and replaced it with another one that

central star was noticeably brighter than the neighboring star), perhaps by 1 to 2 magnitudes! He concludes that the central star is between 15th and 16th magnitude. "When I've had problems making it out," he reports, "I can tell that it's usually due to poor seeing or poor-quality large-aperture telescopes or eyepieces."

The late Walter Scott Houston argued that the ability to see the central star is also enhanced by one's color sensitivity: those with blue-sensitive eyes are more apt to see it. He confirmed this for himself after having a cataract operation, where his yellowed lens was replaced with a fresh, transparent one, thus allowing more ultraviolet radiation into the eye. I have seen the central star with the 9- and 15-inch refractors at Harvard College Observatory at high magnifications, and I'm convinced that a skilled observer could see it in even smaller telescopes - ones that can tolerate high magnifications well. A steady atmosphere is also required.

High power really is necessary because, as John Herschel noticed, the interior of the ring is "filled with a feeble but very evident light." Lord Rosse, through his great reflector, saw striations. The trick to perceiving Herschel's feeble light with a small aperture is to use 23× and 72× and compare the intensity of the interior of the ring with that of the sky just outside the ring. Some observers stop "looking" after they see the Ring's dark hole, missing the opportunity to see the weak interior glow.

Finally, I always like to end an observing session by returning to low power and "relaxing." When I do this with M57, I see the ring as an inflated inner tube afloat in a semicircular black pond. Now concentrate on the dark "water." Do you see glints of moonlight reflecting off the wave crests?

M58

NGC 4579
Type: Mixed Spiral Galaxy (SAB(rs)b)
Con: Virgo

RA: $12^h37.7^m$
Dec: +11°49′
Mag: 9.6
SB: 13.0
Dim: 5.9′ × 4.7′
Dist: ~63 million light-years
Disc: Charles Messier, 1779

MESSIER: [Observed April 15, 1779] Very faint nebula discovered in Virgo, almost on the same parallel as third-magnitude ε. The slightest illumination of the micrometer crosshairs causes it to disappear. M. Messier plotted it on the chart for the comet of 1779, which will be found in the Academy volume for the same year.

NGC: Bright, large, irregularly round, very much brighter in the middle, mottled.

MODESTLY BRIGHT AND UNASSUMING, M58 is, nonetheless, among the most prominent Messier galaxies in the Virgo cluster of galaxies. Spanning 108,000 light-years, this mixed spiral galaxy has about the same mass as our Milky Way Galaxy, or about half that of the Andromeda Galaxy (M31). We see it receding from us at about 940 miles per second.

In large telescopic images, M58 displays a small, bright, diffuse nucleus in a smooth lens with multiple dust lanes. A larger disk with many thin, dusty lanes surrounds this region. Among that dust lie two massive, tightly wound spiral arms dappled with H II regions. The arms fly off the ends of a bar that stretches through the inner lens. The bar is offset by about 30° north of the major axis. The more prominent arm emanates from the bar's southwestern end and wraps almost completely around the galaxy. The northeast arm can be traced for about 200°.

M58 also exhibits an asymmetric nuclear spiral structure called the "loop," with a radius of about 5,000 light-years. The galaxy has long been classified as a LINER (low-ionization nuclear emission region) with an active galactic nucleus (AGN). Marcella Contini (Tel Aviv University, Israel) and colleagues modeled M58's AGN and found that its emission is relatively low compared with other similar Seyfert galaxies, reporting their results in a 2004 paper in the *Monthly Notices of the Royal Astronomical Society* (vol. 354, p. 675). While the main emission may be from a young starburst region, a small contribution appears to emanate from an older one with an

M58

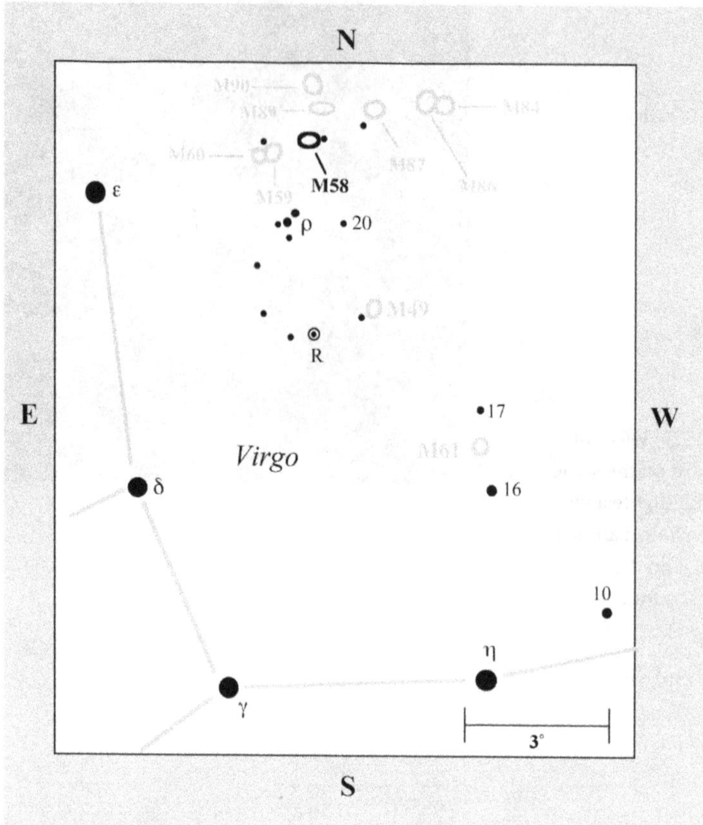

and make that large sweep to 20 Virginis. Once you locate the star, switch to the chart here. You will find M58 stationed 8′ east of an 8th-magnitude star, about 2° northeast of 20 Virginis. You could also make a smaller hop about 40′ to a pair of magnitude 7.5 stars, separated by about 10′ and oriented northwest to southeast. M58 is just 1 1/4° farther to the northeast.

If you live under dark skies, try locating M58 first with binoculars; look for a mere glint of diffuse light. The galaxy shares the same low-power telescope field with the bright ellipticals M59 and M60; it is also north of an intriguing double galaxy, NGC 4567 and NGC 4568, also known as the Siamese Twins. At 23×, M58 displays only a starlike nucleus surrounded by a uniform halo that gradually fades into the background. But 72× immediately brings out a mottled halo and a patchy inner glow, and 130× starts to reveal some of the galaxy's subtle structure.

When looking at galaxies with high power, I usually alternate between using hyperventilation and relaxation techniques to visually chisel out detail. For M58, the first features to materialize out of the galactic haze are knots on the east and west ends of the nuclear region, followed by the dim extensions that outline the galaxy's gently curving arms. Only the northeastern part of the bar

age of 4.5 million years, probably located in the external nuclear region. Radio emissions reveal a jet flowing out of an accretion disk within 0.3 light-years from the active center. These jets have collided with clouds, creating detectable shock waves. "All types of emissions observed at [a] different radius from the center can be reconciled with the presence of the jet," the Contini team said.

To find M58, use the wide-field chart at the back of this book to locate 5th-magnitude Rho (ρ) Virginis. The star is easy to identify in binoculars, since it is the central bright star in a 30′-wide upside-down Y of stars (oriented north to south). Now look for 6th-magnitude 20 Virginis 2° due west; use binoculars first, then center Rho Virginis in your telescope

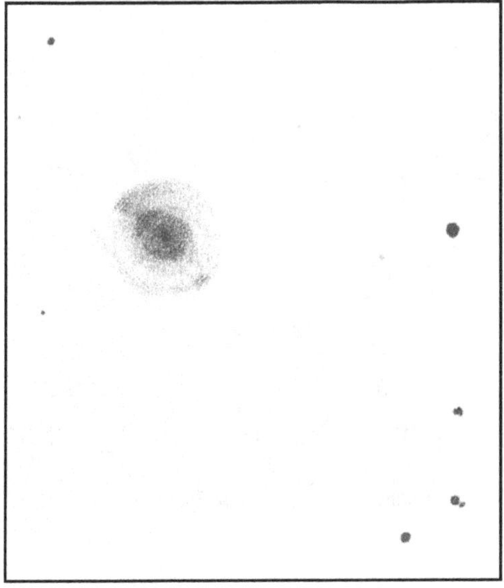

connecting the outer knots to the nucleus is definitely visible through the 4-inch; the bar to the southwest was merely a suggestion, probably because our eyes like to create patterns. Kenneth Glyn Jones, however, reported clearly detecting the bar from England in an 8-inch telescope!

Concentrate on the innermost nuclear region. Here I do see a bar connected to faint arcs. The details in this inner region wonderfully replicate those in the outer halo, only they're just slightly skewed.

M58 has hosted two known supernovae: Kaora Ikeya found SN 1988A, a Type II event, shining at magnitude 13.5 just 40″ south of the galaxy's center. The next year, G. N. Kimeridze found a 12th-magnitude Type I supernova (SN 1989M) 33″ north and 40″ west of the galaxy's center.

M59

NGC 4621
Type: Elliptical Galaxy (E5)
Con: Virgo

RA: $12^h42.0^m$
Dec: +11°39′
Mag: 9.6
SB: 12.5
Dim: 5.4′ × 3.7′
Dist: ~51 million light-years
Disc: Johann Gottfried Kohler, 1779

MESSIER: [Observed April 15, 1779] Nebula in Virgo and close to the preceding one [M58], on the parallel of ε, which was used to determine its position. It is of similar luminosity to the previous nebula, being just as faint. M. Messier plotted it on the chart for the comet of 1779.

NGC: Bright, pretty large, little extended, very suddenly very much brighter in the middle, two stars [eastward].

M60

NGC 4649
Type: Elliptical Galaxy (E2)
Con: Virgo

RA: $12^h43.7^m$
Dec: +11°33′
Mag: 8.8
SB: 12.8
Dim: 7.1′ × 6.1′
Dist: ~55 million light-years
Disc: Johann Gottfried Kohler, 1779

MESSIER: [Observed April 15, 1779] Nebula in Virgo, slightly more conspicuous than the previous two [M58 and M59], again on the same parallel as ε, which was used to determine its position. M. Messier plotted it on the chart for the comet of 1779. He discovered these three nebulae when observing the comet, which passed very close to them. The latter passed so close on 13 and 14 April that both were in the same field of the telescope, but he was unable to see the nebula. It was not until the 15th, when searching for the comet, that he perceived the nebula. These three nebulae do not appear to contain any stars.

NGC: Very bright, pretty large, round, east of a double nebula.

THE TWO ELLIPTICAL GALAXIES M59 AND M60 fit neatly in the same field of view with M58 in 7 × 35 binoculars. M60 is clearly the brightest of this trio in Virgo, shining nearly a full magnitude brighter than both of its neighbors in the sky: M5 lies nearly 5 million light-years more distant than M60, and M58 is nearly twice that far.

On April 11, 1779, Johann Gottfried Koehler (1745–1801) discovered M60 from his private observatory in Dresden just days before Charles Messier independently found it; both observers found the objects while following the comet of that year. Koehler described them as "two small nebulae" through his 3-foot telescope.

Let's first look at M59. This sizable elliptical galaxy spans 80,000 light-years of space, and we see it receding from us at 255 miles per second in the eastern outskirts of the Virgo Cluster. It is a large and typical diffuse elliptical system that contains a system of nearly 2,000 globular clusters and shows a clear, edge-on, thin, faint stellar disk, extending at least 10″ from the center. The galaxy's tips are not pointed as one would expect in a

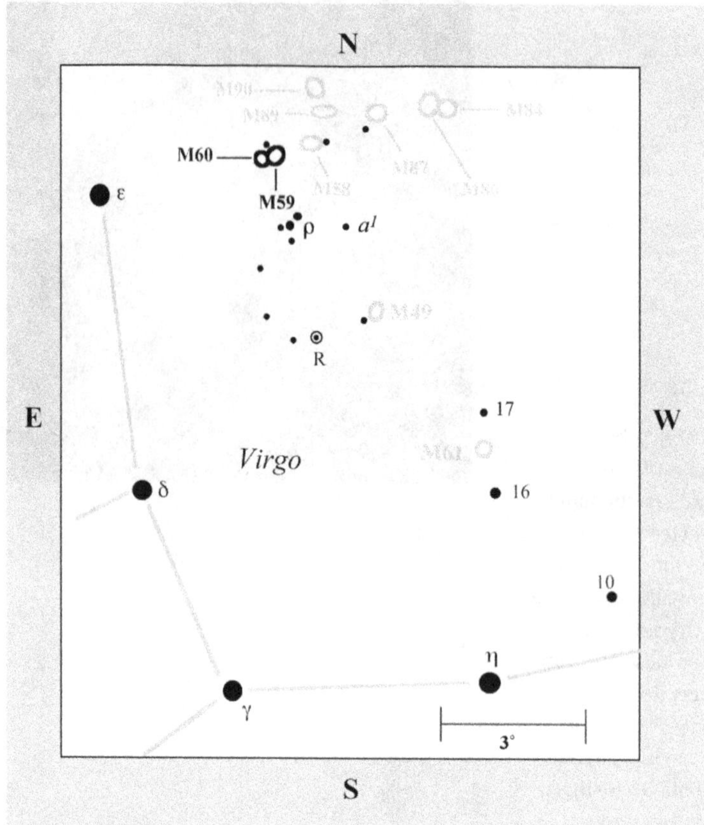

M60, among the largest ellipticals, measures 115,000 light-years across and is some 200 times more massive than the elliptical galaxy M32, the more condensed companion of the Andromeda Galaxy. It forms a close pairing with NGC 4647 – a spiral with feathery structure and heavy dust lanes – just 2.5′ to the northwest, though no tidal disruptions to either galaxy have been detected. This suggests that the two are at different distances along a line of sight, and perhaps separated by nearly 40,000 light-years. We see M60 receding from us at 695 miles per second.

lenticular galaxy, so it has been classified as a very typical E5 elliptical system.

Despite its visual impression of a dense core, M59 doesn't appear to have a distinct nucleus; there is a strong indication that the disk's inner 200 light-years harbors a counter-rotating core. In a 2008 paper in *Astrophysical Journal* (vol. 675, p. 1041), J. M. Wrobel of the National Radio Astronomy Observatory and colleagues described a supermassive black hole with a mass greater than 100 million solar masses at the core of M59. Their analysis of archival Chandra and recent Very Large Array data also show that M59 contains a low-luminosity active galactic nucleus energized by the black hole and dominated by a radio outflow rather than an accretion flow.

In a 2010 paper in *Astrophysical Journal* (vol. 711, p. 484), Juntai Shen of Shanghai Astronomical Observatory and Karl Gebhard of the University of Texas at Austin used Hubble Space Telescope data on M60's stellar and globular cluster populations to determine a total mass of 4.5 billion solar masses (about two times larger than previously published values) and the presence of a dark matter halo.

M60 contains 5,100 globular cluster systems. Studies of them that analyze the radial velocities of planetary nebulae in the galaxy have revealed a dark matter halo and a total dust mass of about 400 billion solar masses. In a 2012 paper in *Astrophysical Journal* (vol. 757, p. 121), Michael Loewenstein and his colleagues presented a spectroscopic x-ray

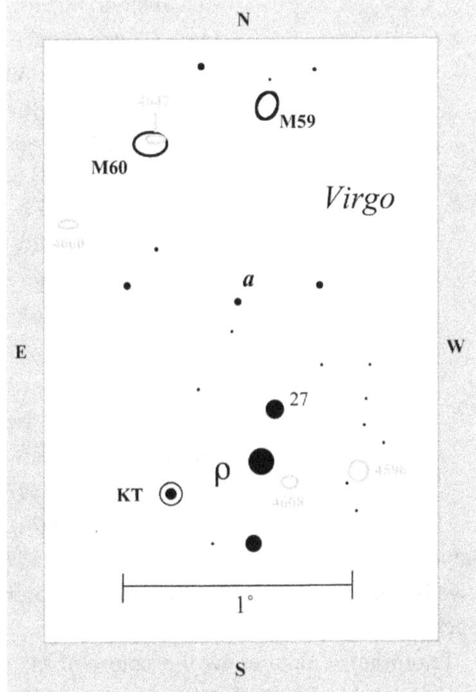

analysis of M60 with data from Suzaku and XMM-Newton Observatory exposures of the Virgo Cluster. The study shows a hot interstellar medium with a complex history that reflects a continual evolution caused by the heating and circulation of accreted gas from cannibalized members of the cluster.

To find M59 and M60, use the chart at the back of the book to locate 3rd-magnitude Epsilon (ε) Virginis. Now use the charts here for M59 and M60 to locate Rho (ρ) Virginis about 5° to the west-southwest. Rho Virginis is easy to identify because it is the brightest star at the center of a roughly 30′-wide upside-down Y-shaped asterism with three stars of magnitude 6.5 to 7, one of which is 27 Virginis to the north-northwest. M59 is about 1 1/2° north of Rho Virginis, or about 40′ north-northwest of magnitude 9.5 Star *a*; M60 is 25′ to the east-southeast.

When I was younger and saw old photos of M59 and M60, I could not help but think of an elliptical galaxy as little more than a uniformly bright and structureless glow, so I was always surprised whenever I looked at an elliptical galaxy through a small telescope and actually saw some detail.

For example, I was once surprised to see that M59 contains a sharp, starlike nucleus, because I hardly ever saw one in a long-exposure photograph. (At the time, I wasn't aware that John Herschel had noted that M59 was "very suddenly much brighter in the middle," and that Kenneth Glyn Jones had also detected the galaxy's starlike core.)

When I first saw M59 through the 4-inch Tele Vue refractor, it not only had a starlike nucleus but also looked mottled, though I could not tell whether this was because of faint stars superposed on it, from actual dust lanes and patches, or from physiological effects. The brightness across the disk is definitely asymmetrical, with the northwest quadrant decidedly brighter than the southeast one. Most puzzling is what appears to be a faint, needle-like bar running through its major axis – a feature I have seen associated with other ellipticals. It is probably just an illusion. Modern CCD images do show, however, a pinching in the center, as well as a tiny star along the major axis, so it may be a combination of both effects.

M59 has sported one known supernova, SN 1939B, which appeared 53″ south of the galaxy's center, toward the edge of the galaxy. The progenitor star of this Type I event was likely old and evolved.

Like M59, M60 also has a starlike nucleus. Admittedly, when I first saw M60's crisp central star and then thought of its fuzzy photographic image, how could I not suspect a supernova? But that isn't the case. As Christian Luginbuhl and Brian Skiff note,

M60

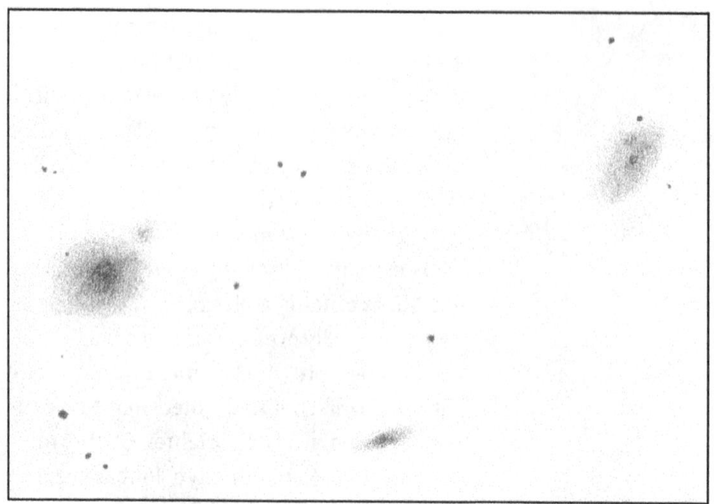

M60 is an evenly fading circular glow with a faint central "pip."

M60 has a tiny (2.7′) magnitude 11.3 companion, NGC 4647, 2.5′ to the northwest, whose seemingly transparent glow is a pleasing sight. This galaxy duet reminds me of fuzzy Alcor and Mizar, the famous naked-eye optical double star in the handle of the Big Dipper. Unlike M60, though, NGC 4647 is a face-on spiral comprising 10 billion stars. William Herschel discovered the small companion (H III-44), and Admiral Smyth noted it was "extremely faint," though that may be because of his long-focus instrument. Although the nineteenth-century observer Rev. Thomas W. Webb saw M60's companion galaxy, he also said he couldn't detect it in a 3.7-inch telescope.

But consider the following experience. Midway between M59 and M60, and to the south, is the 12th-magnitude elliptical galaxy NGC 4638. I find that by staring directly at NGC 4638 at 130×, M60's tiny companion burns into view in my peripheral vision! But, when I look directly at M60, the companion disappears. Thus I refer to NGC 4647 as the "disappearing galaxy." With averted vision, NGC 4647 really swells and displays definite inner and outer sections. It also looks gray.

Photometric data show the colors of M59 and M60 to be almost identical, but to me M59 looks bluer than M60, which has a deeper yellow tint. In fact, if you use high power and relax, M60 appears to have an intense yellow core that seems to be surrounded by countless glistening beads that form a diamond-shaped structure around it.

On January 28, 2004, the Lick Observatory Supernova Search discovered a magnitude 18.8 Type I supernova (SN 2004W) 52″ west and 79″ south of M60's center.

M61

Swelling Spiral
NGC 4303
Type: Mixed Spiral Galaxy (SAB(rs)bc)
Con: Virgo

RA: $12^h21.9^m$
Dec: +04°28′
Mag: 9.7
Dim: 6.0′ × 5.9′
SB: 13.4
Dist: ~53.5 million light-years
Disc: Barnabus Oriani, 1779

MESSIER: [Observed May 11, 1779] A nebula that is very faint and difficult to see. M. Messier took this nebula for the comet of 1779 on 5, 6, and 11 May. On the 11th he realized that it was not the comet, but a nebula that happens to lie on its path and at the same point in the sky.

NGC: Very bright, very large, very suddenly brighter in the middle to a starlike center, binuclear.

M61 IS ONE OF THE MOST PLEASING OPEN-faced spirals in the Messier catalogue for small-telescope users. This may be the result of its small apparent size (6.0′ × 5.9′), which condenses the galaxy's light, making features more apparent. You'll find it only about 5° north and slightly east of Eta (η) Virginis, roughly halfway between the 6th-magnitude stars 16 and 17 Virginis. Although I could detect its feeble 10th-magnitude light with 7 × 35 binoculars, it is essentially at the binocular limit, and I imagine it would be difficult to glimpse in skies that were not very clear and dark.

On May 5, 1779, Italian astronomer Barnabus Oriani discovered M61 while following the comet of that year. He described it as "very pale, looking exactly like the comet." His words proved prophetic. Messier had also encountered M61 that same night but had mistaken it for the comet; he repeated his error the following night, and again on the 11th, when he noticed that the comet had not moved against the stars. The object he had been following was "but a nebula that happens to lie on [the comet's] path and at the same point in the sky."

The nineteenth-century British astronomer Admiral William Henry Smyth later commented that M61 is "an outlier of a vast mass of discrete but neighbouring nebulae, the spherical forms of which are indicative of compression." M61 is indeed one of the more prominent members of the Virgo cluster of galaxies – even though it lies about 8° south of its main concentration and recedes from us at

M61

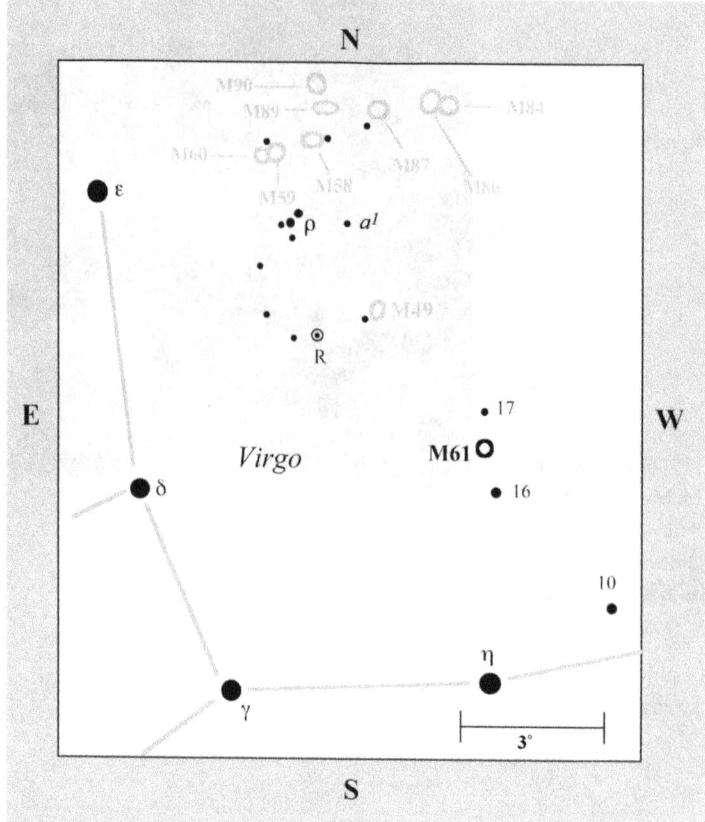

~970 miles per second (a velocity larger than that of most galaxies in the cluster).

M61 has a true diameter of about 95,000 light-years, making it comparable in size to the Milky Way. It is in fact one of the largest spirals in the Virgo Cluster. A magnificently detailed system with crooked arms, this face-on galaxy displays an extremely bright nucleus within a broad, diffuse bar. Many dust lanes thread through its round central bulge, forming a spiral structure near the ends of the bar. These "dust spirals" wind outward through the inner disk to form a pseudo-ring. There the spiral structure greets the galaxy's two main luminous arms, one of which bends into two straight sections that meet at a sharp angle. A multitude of fainter arms branch off the main arms. All are studded with star-forming H II regions, but an especially large concentration appears near the northern end of the bar.

M61 has other structures that resemble those in grand-design barred spirals, including a central nuclear ring. The ring consists of very young stars, which do not show up in the near infrared. H II regions are also prominent in the nucleus and circumnuclear region. A pair of dust lanes connect the bar to the nuclear ring in the south and north, and the nuclear spectrum reveals a faint Seyfert nucleus hidden by H II regions.

To locate the galaxy, use the chart at the back of the book, or the wide-field chart here, to locate 16 and 17 Virginis, which lie about 5° north-northeast of Eta Virginis. You can then use the detailed finder chart here to pinpoint the galaxy.

The NGC description uncharacteristically misled me into thinking M61 would appear brighter and larger than it really is. At 23× through the 4-inch, the galaxy is so small and faint that I almost swept over it, perhaps because it is tucked away in the crook of a nice wishbone asterism of 9th-magnitude stars. Take the time to concentrate on M61's light, which with averted vision appears to gradually dilate into an appreciable disk. Don't be surprised if your eye also begins to notice other ghostly glows hiding among a

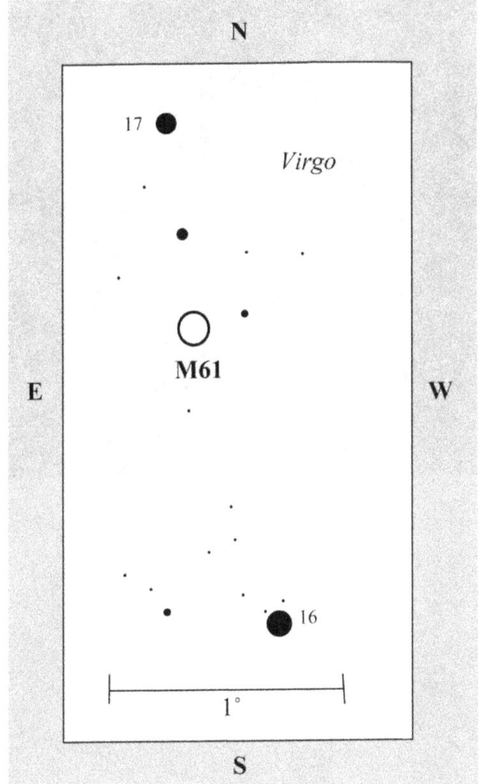

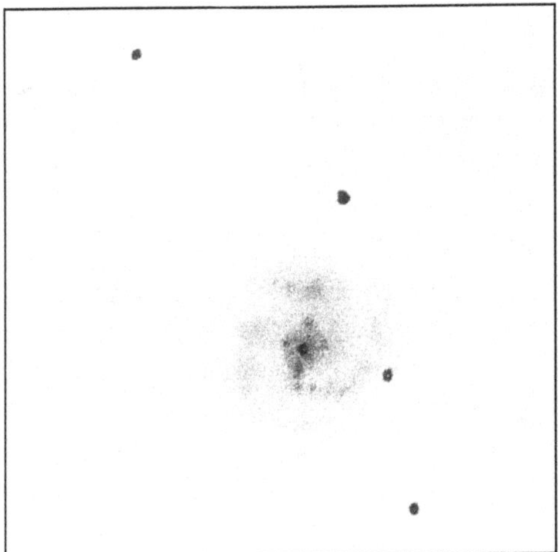

forest of similarly bright stars populating the region. These are the dim and distant shapes of at least a dozen nearby NGC galaxies!

M61 itself reveals a brightening at the center that pops into view only with averted vision. This nucleus is centered in a fairly condensed halo of light that is wrapped in an even fainter shell. The fading into view (or out of view, depending on how you stare at it) of this faint outer envelope is the reason the galaxy appears to swell (and shrink), which is why I refer to M61 as the Swelling Spiral.

Moderate power does not add much to the view, though it does start to reveal star-forming regions, which appear as knots in the outer halo. But because this galaxy is so tight, high power is required to resolve it into its separate parts. And what a very interesting view it is at 130×! A starlike nucleus is surrounded by a mottled, diamond-shaped inner core. With averted vision, the northernmost knot shows an extension that trails off irregularly to the east in a wavelike fashion, like a gracefully thin wake of smoke.

Closer to the nucleus and to the south is another star-studded region that trickles off in lumps to the west before it curves sharply to the north. A faint but definite arm can be glimpsed to the east, and it boxes in the nuclear region. Overall, at high power, the galaxy looks like a square with slightly rounded edges. Through his 4-inch refractor, John Mallas also observed three luminous patches outside the nucleus, which match the positions of M61's spiral arms seen in photographs.

Here's a challenge. Try to detect the three stars running along the western edge of the galaxy. The southernmost star is the most obvious at 14th magnitude. The northernmost one shines closer to 15th magnitude. The one in the middle is roughly magnitude 14.5.

M61 is one of the most abundant extragalactic supernova producers known. As of this writing, we know of five: SN 1926A, SN 1961I, SN 1964F, SN 1999gn, and SN 2006ov. So keep an eye (or CCD) out for the next!

M62

Flickering Globular
NGC 6266
Type: Globular Cluster
Con: Ophiuchus

RA: $17^h01.2^m$
Dec: –30°07′
Mag: 6.4; 6.7 (O'Meara)
Diam: 14′
Dist: ~22,200 light-years
Disc: Charles Messier, 1771

MESSIER: [Observed June 4, 1779] A very fine nebula, discovered in Scorpius; it resembles a small comet. The center is bright and is surrounded by faint luminosity. Its position was determined relative to the star τ Scorpii. M. Messier had seen this nebula before on 7 June 1771, but was only able to determine the approximate position. Observed again 21 March 1781.

NGC: Remarkable globular cluster, very bright, large, gradually much brighter in the middle, well resolved, stars of 14th to 16th magnitude.

ODD AND MYSTIFYING ARE THE WORDS THAT seem to best describe M62 in Ophiuchus. A metal-rich globular cluster 22,200 light-years from the Sun, it is singled out in the literature mostly for its strong asymmetry. At 5,500 light-years distant, M62 is the Messier globular closest to the nucleus of the Milky Way Galaxy, so the strong gravitational tug on the globular might account for its peculiar visage. Specifically, the western half of the cluster is dramatically brighter than the eastern half.

Ninety light-years wide, the cluster contains many variable stars, including 200 RR Lyraes. M62 also has the fourth largest – next to Terzan 5, 47 Tuc, and M15 – population of millisecond pulsars (MSPs) among its type. All are in binary systems. Indeed, in a 2006 paper in *Astronomical Journal* (vol. 131, p. 2551), G. Beccari (Osservatorio Astronomico di Bologna, Italy) and colleagues describe how they used multiband high-resolution Hubble Space Telescope and wide-field ground-based images of M62 to find that it has experienced (or is experiencing) a phase of very high binary star production while the cluster has not yet undergone the collapse of the core.

Interestingly, M62 is the only Galactic globular cluster known to possess radio emission coincident with the cluster's core that does not appear to be of stellar or binary origin. M62 has long been thought the most

suitable cluster for hosting an intermediate-mass black hole (IMBH) at its core. But studies attempting to confirm this possibility have been inconclusive. In a 2012 paper in *Astrophysical Journal* (vol. 745, p. 175), Bernard J. McNamara of New Mexico State University and his colleagues reported that Hubble Space Telescope observations show that M62's properties do not suggest the existence of a centrally located IMBH, though a black hole with a rough upper-limit mass of less than a few thousand solar masses cannot be excluded. Their best-fitting model of NGC 6266 without an IMBH yields a cluster mass of about 822,000 solar masses.

McNamara and his colleagues also used the Hubble Space Telescope to study the proper motion of M62's stars, and its RR Lyrae star population, which yielded a distance to the cluster of about 23,000 light-years and an age of 11.4 billion years.

To find this enigmatic cluster, one of the brightest globulars in Ophiuchus, use the star chart in the back of the book to locate 2nd-magnitude Epsilon (ε) Scorpii in the Scorpion's hindquarters and then switch to the chart here. M62 lies roughly midway between Epsilon Scorpii (an orange K2 star at the beginning of the Scorpion's tail) and 5th-magnitude 36 Ophiuchi (an orange dwarf 500 light-years from us). It is also only about 3 1/2° south of globular star cluster M19 in Ophiuchus. Can you see the globular with your naked eye?

Through the 4-inch at 72×, the cluster's western half stands out boldly against the stars of the Milky Way, while the cluster's weaker, eastern stars quickly fade as you move away from the center. At the eastern edge, several strings of stars dangle like loose threads in a tattered cloth. Looking at it differently, that shredded eastern section makes M62 appear like a still photograph of a balloon just beginning to burst, with the "air" rushing out to the east-southeast.

But I find M62 mystifying for reasons other than its asymmetry, which I have seen in other globulars (to a lesser degree), such as M56 in Lyra. What fascinates me is that the cluster's core appears to flicker, like a dying flame. The effect is hypnotic. One night, while watching that flame dance at 72×, I quickly broke the spell and changed to an eyepiece yielding 130×. Still, my gaze was magically drawn toward the center! The core retained its ruddy cast and continued to pulsate. Seeking an explanation for this intriguing effect, I soon figured out that it was not real – the core isn't actually pulsating in brightness – but an optical illusion.

The western section of the tight inner core has three distinct waves of starlight that seem to ripple out from the ruddy nucleus. Given the wealth of detail in such a small area, my eye cannot help but leap outward from wave crest to wave crest and then return to the nucleus, where the process repeats. As that is happening with my eyes, my mind is trying to fathom another illusion: faint rings of starlight appear to be rippling out from the core

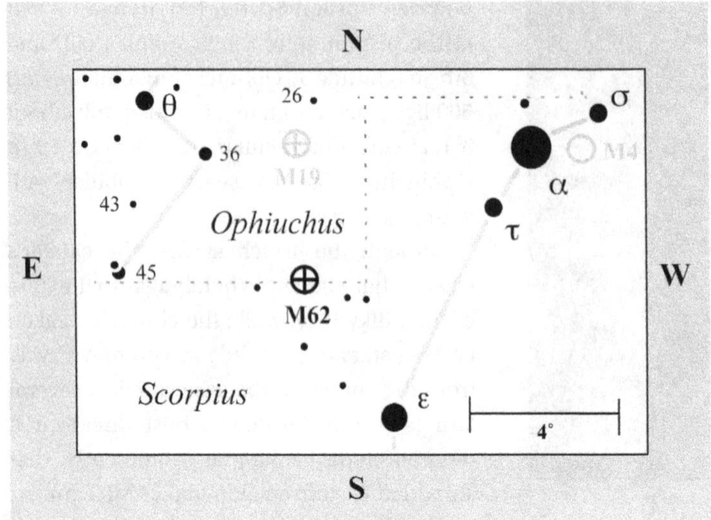

and toward the viewer! These changing optical moiré effects seem to set the cluster's core into animation.

Equally intriguing is that as the nucleus "flickers," its color changes like that of a flame in a breeze; one moment it is smoky yellow, the next it's dusky red. This, I believe, results from the eye shifting rapidly back and forth between the reddish center and the yellowish inner halo. The outer envelope also wavers with a pale bluish light. These effects appear to be not atmospheric (the cluster is some 40° high from my observing site) but physiological (differential refraction can also cause color changes). I saw a similar visual effect one day when staring at a neighbor's needlepoint of a cat.

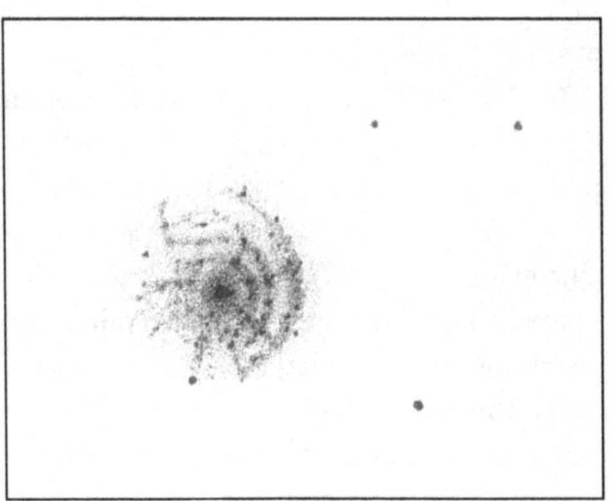

M63

Sunflower Galaxy
NGC 5055
Type: Spiral Galaxy (SAbc)
Con: Canes Venatici

RA: $13^h15.8^m$
Dec: +42°02′
Mag: 8.6
Dim: 13.5′ × 8.3′
SB: 13.6
Dist: ~27 million light-years
Disc: Pierre Méchain, 1779

MESSIER: [Observed June 14, 1779] Nebula discovered by M. Méchain in Canes Venatici. M. Messier searched for it; it is faint and is approximately as bright as the nebula described here under number 59. It does not contain any stars and the slightest illumination of the micrometer's crosshairs causes it to disappear. Close to it there is an eighth-magnitude star, which crosses the meridian crosshair before the nebula. M. Messier plotted the position on the chart for the comet of 1779.

NGC: Very bright, large, pretty much extended in about position angle 120°, very suddenly much brighter in the middle to a bright nucleus.

IN IMAGES, M63 LOOKS LIKE A SPIRAL GALAXY that has lost control of its gravity, and we are catching a rare sight of its arms being tossed into space. This spiral, which is 10 billion times brighter than the Sun, is a prime example of a type of galaxy that displays a lack of cohesion between its inner and outer arms. The inner region of M63's 100,000-light-year-wide disk (about the size of the Milky Way) is ringed by strong spiral structure, while the plentiful outer arms appear loose, patchy, and haphazard. Not surprisingly, M63 is nicknamed the Sunflower Galaxy because of its resemblance to that towering plant, whose dense, seedy head is ringed by an abundance of bright, overlapping petals.

When Admiral Smyth observed the "nebula" in the nineteenth century, he spied a milky-white tint, a bright core, and a stellar nucleus. Later Lord Rosse recognized its spiral structure, making it one of 14 "spiral nebulae" discovered by 1850. Actually, Smyth's and Rosse's descriptions together describe the basic visual structure of this galaxy in modern-day instruments.

What they couldn't see is that M63's partially resolved, inner filamentary arms can be traced as dusty fragments almost to the center

M63

significantly richer than the bulge population, inferring that the disk may have undergone relatively recent, strong starbursts that significantly increased that population, although ongoing starbursts are also observed in the nuclear region.

At radio wavelengths, M63's gaseous disk extends out to about 13,000 light-years and shows a pronounced warp that starts at the end of the bright optical disk.

of the galaxy, where a very small, very bright active galactic nucleus (AGN) shines within a lenticular core. Nor could they have known that the AGN may include a compact star cluster surrounded by five luminous young star clusters with ages ranging from one million to tens of millions of years.

In images, the surface brightness of M63's multiple-armed pattern decreases abruptly at a radius of around 10,000 light-years from the center. The outer, feathery arm fragments have a lower surface brightness and wind outward to surround the extended disk. The arms contain many small, generally unresolved H II (star-forming) regions. These can be seen as glittering pinpoints of light (like lint on a sweater) with NASA's Spitzer telescope, the most sensitive infrared space observatory ever launched (see the accompanying photo). The "arms" you see are actually dust lanes spiraling in all the way down to a dust ring around the densest region of stars at the galaxy's center. These dusty patches are where new stars are being born. In 2007, Chandra x-ray data showed that the galaxy's disk population of binary stars is

This very extended warp has large-scale symmetry, which along with the rotation period of its outer parts (~1.5 billion years) suggests a long-lived phenomenon. Many galaxies have warps (including our Milky Way), and while their cause remains somewhat mysterious, it most likely results from interactions between galaxies. Indeed, M63 interacts gravitationally with M51 (the Whirlpool Galaxy) and several smaller extragalactic systems.

Otherwise, M63 shows remarkable overall regularity and symmetry. The galaxy's inner region appears to be dominated by the stellar disk, while invisible dark matter exerts its influence on the outer region, causing its stars to rotate about the galaxy's center at a speed so high that they should fly off into space.

To find this special-case Messier galaxy, use low power and sweep about 1 1/2° north-northwest of 20 Canum Venaticorum, a magnitude 4.5 star about 4 1/2° northeast of Alpha^{1,2} ($\alpha^{1,2}$) Canum Venaticorum, a fine double composed of a blue, magnitude 2.9 primary and a white, magnitude 5.5 secondary separated by 19″.

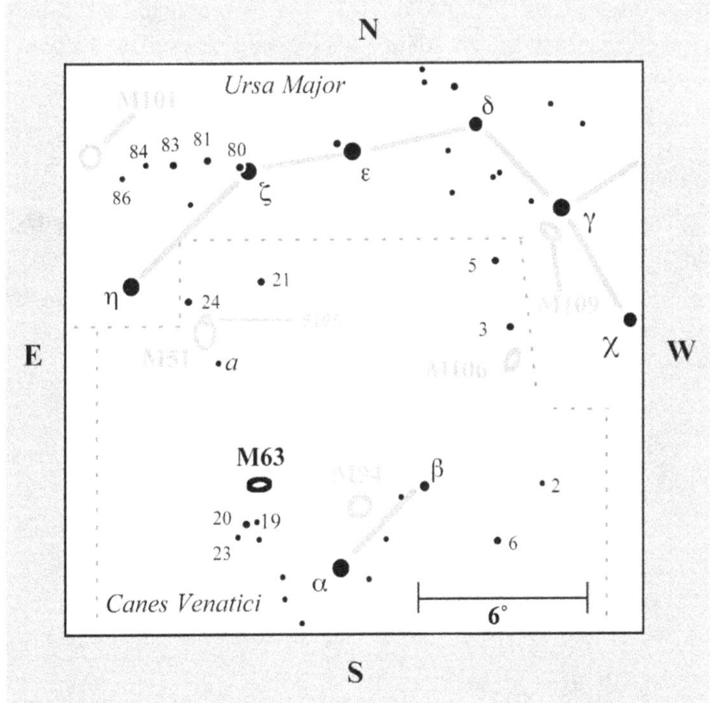

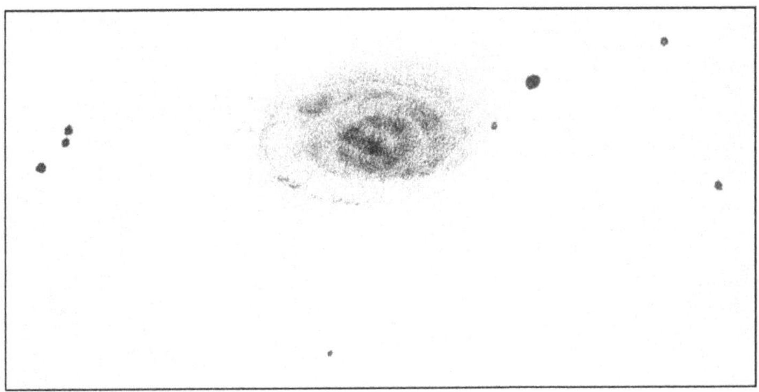

The challenge is to use moderate and high powers to make sense of the arms. This will probably require several nights at the telescope making drawings. The problem I encountered when starting my drawings was that I could not instantly determine which way the spiral arms wind around the core – clockwise or counterclockwise – because all the details are so delicate.

In fact, M63 is one of the most finely detailed galaxies in the Messier catalogue in small telescopes. You will need patience. I start by focusing in on the innermost glow, where the spiral pattern is more strongly suggested. It almost helps to blur your vision or defocus the image in the telescope ever so slightly, so that the patches blend into arms. Archinal notes that this resolution problem is similar to that observed with M51, where early observers noticed only a ring, not spiral arms.

At first glance, the galaxy might appear shy, as if it is trying to hide behind the fiery blaze of a 9th-magnitude star just 4′ to the west. Take advantage of this illusion to "see" this galaxy beyond the stars of our own system.

Through a telescope at 23×, its soft glow reveals hints of a mottled structure in the pale outer disk, while the core is tack sharp.

With 130×, I noticed a strange alignment of patches near the nucleus, which I call the "crooked cross." First, I saw four prominent condensations in a dappled ring surrounding the starlike nucleus. When I concentrated on the core, suddenly a bar seemed to cross the nucleus from southwest to northeast. This bar appears skewed about 30° from the galaxy's major axis and lines up with two of the bright knots just described. The two other bright condensations, however, do not quite line up to form a straight cross. No matter, it is all an illusion, but try and see it, because it is a good example of how Percival Lowell was able to "see" canals on Mars – his eye and brain played a game of connect the dots.

On May 25, 1971, G. Jolly of Corralitos Observatory discovered a Type I supernova (SN 1971I) 2″ west and 147″ south of the galaxy's center that may have reached peak magnitude on May 17 at magnitude 11.3.

M64

Black Eye Galaxy, Evil Eye Galaxy, Sleeping Beauty Galaxy

NGC 4826

Type: Spiral Galaxy ((R)SA(rs)ab)

Con: Coma Berenices

RA: $12^h56.7^m$

Dec: +21°41′

Mag: 8.5

Dim: 9.2′ × 4.6′

SB: 13.3

Dist: 17 million light-years

Disc: Edward Pigott, 1779

MESSIER: [Observed March 1, 1780] Nebula discovered in Coma Berenices, which is slightly less apparent than the one that is below the hair [M53]. M. Messier plotted the position on the chart for the comet of 1779. Observed again 17 March 1781.

NGC: Remarkable, very bright, very large, very much extended roughly toward position angle 120° bright in the middle with a small, bright nucleus.

IN PHOTOGRAPHS, M64, THE FAMOUS BLACK Eye Galaxy in Coma Berenices, is a very distinctive spiral system. Its smooth, silken arms wrap gracefully around a porcelain core, whose northern rim is lined with dust. The galaxy resembles a closed human eye with a "shiner." The dark dust cloud (called the Evil Eye) looks as thick and dirty as tilled soil. But, in his classic book *Galaxies*, astronomer Timothy Ferris notes that a jar of its material would be difficult to distinguish from a perfect vacuum. Yet M64's black cloud is so expansive – some 45,000 light-years in diameter – that it contains enough atoms to loam the gardens of billions of planets. Appreciate the soil you walk on, Ferris reminds us, because every atom in it once belonged to an interstellar dust cloud like the one we see so strikingly in M64.

Most references (including the first edition of this book) credit Johann Elert Bode with the discovery of M64 on April 4, 1779. But credit should actually go to Edward Pigott (1753–1825), who had found it 12 days earlier from Glamorganshire, England. Messier was unaware of Pigott's discovery, as it was not published until 1781. It remained unrecognized until 1907, when French astronomer Guillaume Bigourdan apparently linked Pigott's position of the "nebula" to M64.

M64

Studies of the properties of M64's counterrotaing disks have led astronomers to believe that we're witnessing the effects not of a galaxy collision but an accretion event of a gas-rich dwarf object that caused significant turbulence in M64's bulge – a situation ideal for stimulating the observed bursts of star-formation activity. In fact, the bulge has several regions of ongoing star-formation activity, which has most likely contributed to the bulge's blue tint as opposed to the red found at the galaxy's core. The accretion event could have happened more than one billion years ago. The dwarf has since been almost completely consumed, but signs of the collision persist in the backward motion of gas at the outer edge of M64.

M64 is a nearby early-type spiral galaxy. It appears at a moderately high inclination and hosts a prominent dust lane asymmetrically placed across the galaxy's prominent bulge. But M64 burst into prominence in 1994 when Vera Rubin of the Carnegie Institution of Washington found that while in M64's inner disk (which contains the Evil Eye) the gas and stars rotate in concert, the galaxy shows two counterrotating gas disks (ionized gas and neutral hydrogen) in the outer region. In other words, the gas stars in M64's inner region are rotating clockwise, while the gaseous outer regions are rotating counterclockwise.

In the Hubble Space Telescope image shown here, we see extraordinary detail in the Evil Eye and its surroundings. Active formation of new stars is occurring in the shear region, where the oppositely rotating gases collide, are compressed, and contract. Particularly noticeable in the image are hot young stars that have just formed, along with clouds of glowing hydrogen gas that fluoresce when exposed to ultraviolet light from newly formed stars.

To find M64, use the chart in the back of this book and the one here to locate 35 Comae

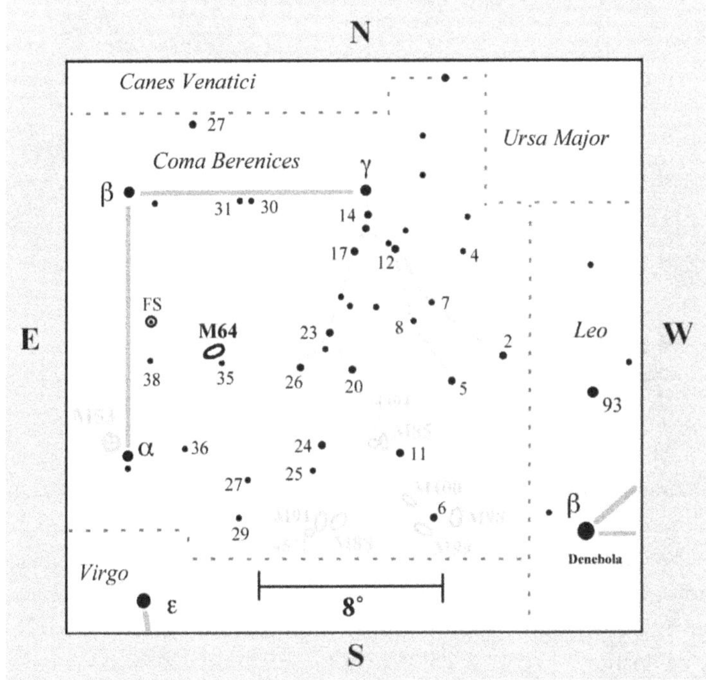

telescope is needed to see it with certainty, whereas John Mallas claimed to have spied it in a 2.4-inch glass. The first time I looked at M64 with the 4-inch Genesis, I resolved the black eye without knowing it. That is, I noticed the feature only upon studying my drawing: First I noted an area in the drawing lacking detail and then I realized that the shape of this void matched the location of the black eye. Therefore, I recorded it indirectly. The feature certainly is not visually obvious in small telescopes, so don't let the photograph fool you. Use high power to search for the black eye, because the apparent size of the galaxy, and especially its inner core, is small. You need magnification to get in there and separate the inner arms. The galaxy's outer details, though, have too low contrast for high power and show up best at 72×.

I was surprised to read Luginbuhl and Skiff's comment on the galaxy's "non-stellar nucleus." My impression was quite different: I perceived a very bright stellar nucleus. Supporting Skiff's view is a drawing by Mallas in *The Messier Album* that shows nothing even resembling a core. Contrary to that, however, the NGC description states that M64 has a "small, bright nucleus." Glyn Jones sided with Luginbuhl and Skiff, saying, "the central nucleus is small but decidedly not starlike." John Herschel said, "I am much mistaken if the nucleus be not a double star." This quotation

Berenices. The galaxy lies about 1° northeast of that star. It shines at a modest magnitude 8.5 and spans only 10′ of sky, so it appears as a little puff of light through 10 × 50 binoculars under a dark sky.

Through a 4-inch telescope, this modest spiral appears as elegant as it does in photographs, though not as detailed. Its brilliant nucleus lies within an extremely smooth oval disk with a milky texture. The disk also has a hint of blue coloration. I found the view at 72× very confusing: I was expecting to see a large, uniform glow, but so much subtle detail suddenly materialized that I thought I must have chanced upon the wrong galaxy.

The debate over the visibility of the dark cloud forming the "black eye" is nearly as intense as the debate about the visibility of the central star in M57. Here are the extremes: Kenneth Glyn Jones felt that an 8-inch

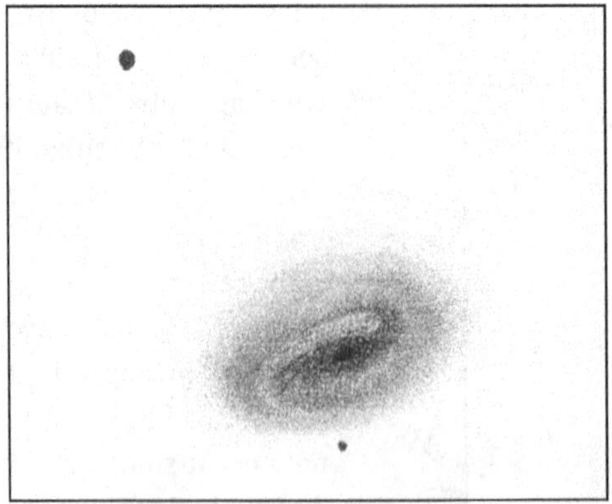

originates from Herschel's 1833 log notation describing M64 as "vsmbm," meaning "very suddenly much brighter in the middle." In another observation, Herschel says it is "very suddenly very much brighter in the middle, almost to a star, but magnifying destroys this effect." Thus, the appearance of M64's core depends on the magnification used.

By the way, in 1994, Vera Rubin also suggested that M64 be called the Sleeping Beauty Galaxy instead of the Black Eye or Evil Eye, simply because when "viewed at the telescope, [M64] is an unusually beautiful galaxy." So see if visitors to your telescope appreciate the subtle beauty of the galaxy when they hear that name.

M65

NGC 3623
Type: Mixed Spiral Galaxy (SAB(rs)a)
Con: Leo

RA: $11^h18.9^m$
Dec: +13°05′
Mag: 9.3; 8.8 (O'Meara)
SB: 12.4
Dim: 8.7′ × 2.2′
Dist: 41 million light-years
Disc: Pierre Méchain, 1780

MESSIER: [Observed March 1, 1780] Nebula discovered in Leo. It is very faint and does not contain any stars.

NGC: Bright, very large, much extended roughly toward position angle 165°, gradually brightening in the middle to a bright nucleus.

M66

NGC 3627
Type: Spiral Galaxy (SAB(s)b)
Con: Leo

RA: $11^h20.2^m$
Dec: +12°59′
Mag: 8.9
SB: 12.5
Dim: 8.2′ × 3.9′
Dist: 33 million light-years
Disc: Pierre Méchain, 1780

MESSIER: [Observed March 1, 1780] Nebula discovered in Leo; very faint and very close to the preceding one [M65]. They both appear in the same telescopic field. The comet observed in 1773 and 1774 passed between these two nebulae on 1 and 2 November 1773. Doubtless M. Messier did not see it then because of the comet's light.

NGC: Bright, very large; much extended roughly toward position angle 150°, much brighter in the middle; two stars to the northwest.

LEO'S M65 AND M66 PROBABLY RANK second only to M81 and M82 as sought-after galaxy pairs. Located about halfway between Theta (θ) and Iota (ι) Leonis and separated by only 20′, M65 and M66 are both visible in 10 × 50 binoculars under a dark sky. Look for two slightly oval glows, with M66 appearing a little more prominent, since M65 hugs a 7th-magnitude star immediately to its north.

These spiral wonders belong to the Leo spur of galaxies – an outlying branch of galaxies some 20 million light-years distant between us and the great Virgo cluster of galaxies, and we see both highly inclined, revealing substantial amounts of dust. The inclination of M65 is nearly optimum to show the galaxy's dust silhouette, which traces a well-defined spiral pattern (with two main spiral arms) that surrounds a diffuse nucleus and bar embedded in a tiny nuclear bulge. The spiral pattern forms a pseudo-ring around the nuclear ring. On the near side, M65's dust pattern crosses the disk and projects into the center as a series of spiral fragments.

M66 is the larger of the two, spanning 80,000 light-years in true physical extent. The dust pattern in M66 is particularly heavy, especially in the lane going from the southeast to the northeast quadrant, along the inner edge of the most luminous spiral arm. Otherwise, M66 has a small and very bright nucleus in a complex bar and lens with many dark lanes. Like M65, it has two main partially resolved arms. But as the Hubble Space

Telescope image here clearly shows, these arms appear asymmetric and seem to "climb above" the galaxy's main disk. The Hubble image also shows that the nucleus is apparently displaced.

But the principal feature of M66's spiral pattern is its prodigious dust, which, when combined with the galaxy's high inclination, makes it difficult to see many features. Many H II regions exist throughout the disk, often on the outside of the principal dust lanes that define the spiral pattern. X-ray emission from M66 has revealed the presence of a source most likely comprised of several starburst regions of active star formation.

M65, M66, and nearby NGC 3628 form what's known as the Leo Triplet (see image). Yet, despite the proximity and prominence of M65 and M66, they do not appear to be interacting, as M65's morphology shows no sign of a tidal disturbance. However, M66 does appear to be involved in a graceful gravitational dance with NGC 3628, around which M66 orbits. Astronomers have now found evidence

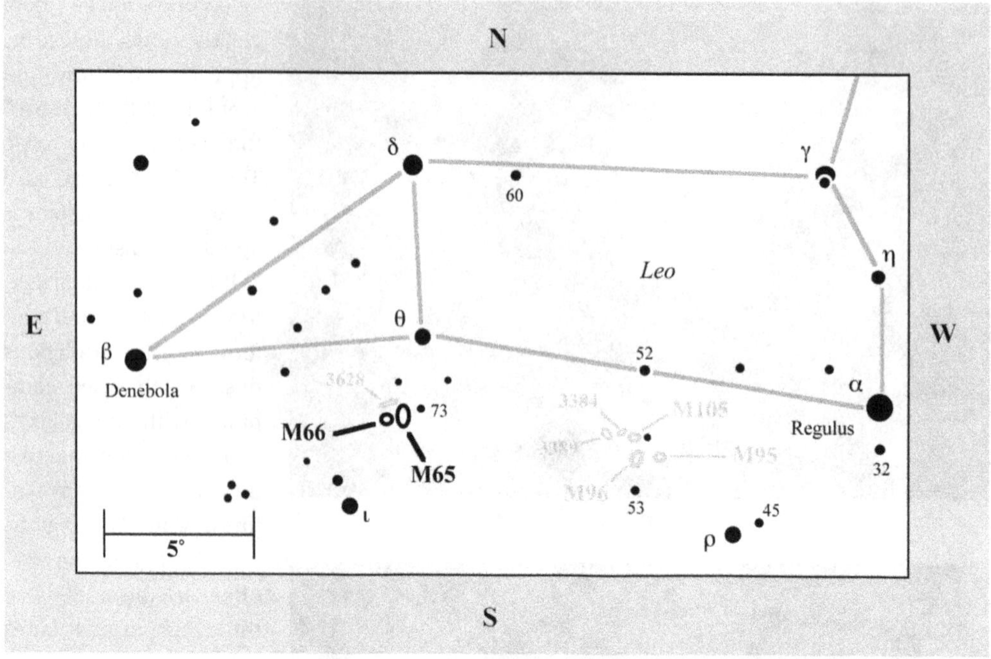

for the interaction in a faint optical plume extending about 45′ east of M66 that was found to be coincident with a neutral-hydrogen filament. It's believed that gravitational forces alone create the plume.

In a 2012 paper in *Astronomy and Astrophysics* (vol. 544, p. 113), M. Wezgowiec (Astronomisches Institut der Ruhr-Universität Bochum, Germany) and colleagues detail how they analyzed soft x-ray emission from M66 to study the distribution of the hot gas and its temperature in different areas of the galaxy as well as two ultraluminous point sources in the eastern part of the disk, which are probably tightly linked to the evolution of the galaxy.

The researchers found an increase in the temperature of the hot gas in the area of a radio ridge in the western arm of the galaxy, which, they believe, most likely formed from ram pressure as the galaxy moved through the gases populating the space between the galaxies in the Triplet. They also found distortions in the eastern part of the disk in polarized radio maps, an asymmetry in the galaxy's bar, and distortions of the eastern arm, which, they say, "propose a recent collision of [M66] with a dwarf companion galaxy."

To find this beautiful triplet, use the chart at the back of the book or the one here to locate magnitude 3.5 Theta Leonis. Center it in your telescope at low power. Then make a careful 2° sweep south-southeast to 6th-magnitude 73 Leonis, which you can confirm in your binoculars. You'll find M65 only about 45′ to the east-southeast. M66 will be 20′ farther to the east, and NGC 3628 will lie about 35′ north of M66.

Through the 4-inch under a dark sky, all three galaxies vie for attention in the same low-power field. At 23×, M65 appears immediately oval and quite large. To me, it looks a half-magnitude

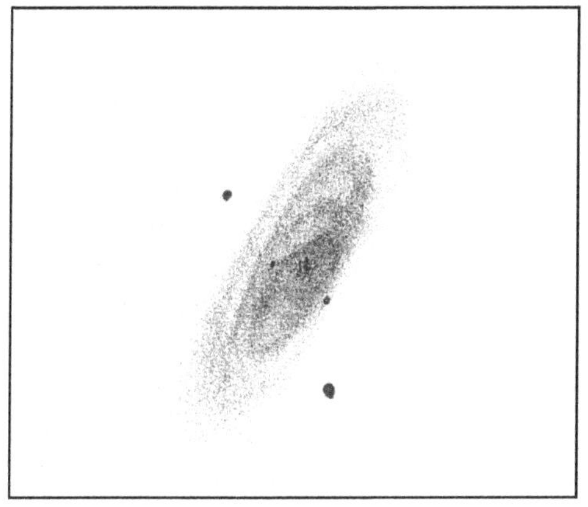

brighter than M66, but some sources claim just the opposite. For example, Kenneth Glyn Jones asserted that M66 is actually the brighter of the two but that M65 tends to appear more conspicuous because of its streaky outline. At a glance, it might appear that the discrepant brightness estimates are related to the surface brightnesses of the galaxies. Surface brightness is, in a very general sense, akin to dividing the object's magnitude by its area.

The surface brightnesses of M65 and M66 are roughly the same, magnitudes 12.4 and 12.5, respectively. This means that each arcminute of the galaxy shines roughly with the brightness of a magnitude 12.5 star. So the difference in surface brightness between M65 and M66, a mere 0.1 magnitude per square arcminute, is not sufficient to account for the brightness discrepancy. Without question, in photographs, the tightly wound mass of M66, with its burning core and clumps of star-forming regions, looks brighter than the more normal-looking spiral M65. But photographic images do not reflect what is seen visually. I also made my magnitude estimate with 7 × 35 binoculars, so as not to be fooled by the objects' telescopic appearance.

An optical phenomenon can make M65 appear larger than it really is at low power. I found that at first glance the galaxy looks very long and extended – the tips of the major axis seem to trail faintly off to unknown lengths. But when I concentrate on the more brightly illuminated area, the galaxy appears quite stubby. This illusion of infinite length is caused by the dust lane that sharpens the galaxy's eastern limb, which the eye likes to extend.

M65 is a joy to see but an awfully difficult galaxy to observe. The details within its bright, nearly edge-on disk vary only slightly from the background brightness, so they're hard to pick out from the "noise." But they can be recorded with accuracy once you familiarize yourself with the galaxy after a few observing sessions. Just focus on a different part of the galaxy each night.

Use 72× to survey the diamond-shaped nuclear region, which is partly an illusion caused by two faint stars nearby – a 12th-magnitude one southwest of the core and a 13th-magnitude one to the northeast. Aside from the dust lane, the most prominent feature lies northwest of the nucleus, where the galaxy's arms can first be seen as faint patches, then in three distinct spirals. The details in the southwest section are far more difficult and less defined. Look for some knots just outside the nuclear region connected by a looping arc.

Visually, the details of M66 are even more subtle than those of M65. Like the Black Eye Galaxy (M64), M66 appears very soft and graceful; it's nothing like the strong and dynamic image you see in photographs. The galaxy's bright, starlike nucleus is its most noticeable feature. Use low power to compare the cores of M65, M66, and NGC 3628. M65 has a somewhat stellar nucleus, and NGC 3628 reveals absolutely nothing! Now look for a bright knot immediately northwest of M66's nucleus. A dark patch lies east of the knot, followed by a mere stump of a spiral arm. The region surrounding the nucleus is oval

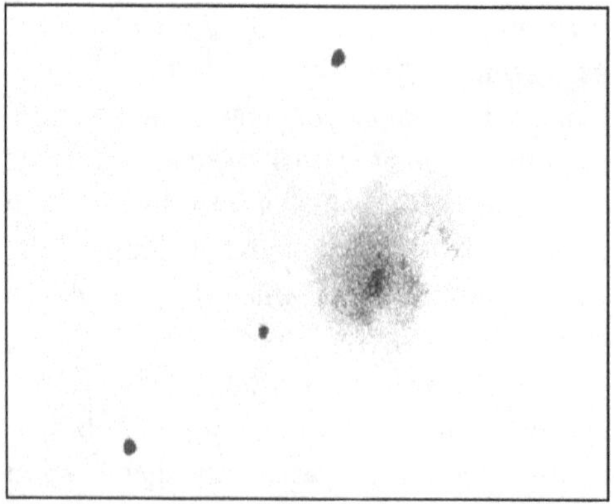

shaped and oriented southeast to northwest. See if your eye catches a streak of light running off to the south. The orientation of this streak can help you determine the galaxy's spiral pattern. The rest of the details are a chaotic mix of faint streaks of dark and light. But don't let this confusion deter you from drawing the detail, because, if you look at the photograph, the galaxy really is a dizzying world of curdled starlight.

By the way, I call NGC 3628 the "Vanishing Nebula" because, with each increase of power, the galaxy blends more and more into the background of deep space until it all but vanishes. This happens because the thick dust lane, which runs across the galaxy's entire length, overpowers the feeble light outlining it. So, as you increase power, you magnify the dark lane, while spreading the faint light across a larger area of sky. At magnitude 9.5, NGC 3628 shines almost as brightly as M65 and M66, but at magnitude 13.7 it has a much lower surface brightness.

M66 has had four known supernova events since 1973: SN 1973R, SN 1989B, SN 1997bs, and SN 2009hd. At the time of discovery, they appeared at magnitudes 14.5, 13, 17.0, and 15.8, respectively.

M67

King Cobra
NGC 2682
Type: Open Cluster
Con: Cancer

RA: $08^h51.4^m$
Dec: +11°49′
Mag: 6.9; 6.0 (O'Meara)
Diam: 25′
Dist: 2,600 light-years
Disc: Johann Gottfried Kohler, between 1772 and 1779

MESSIER: [Observed April 6, 1780] Cluster of faint stars with nebulosity, below the southern claw of Cancer. Its position was determined from the star α.

NGC: Remarkable cluster, very bright, very large, extremely rich, little compressed, stars from 10th to 15th magnitude.

M67 IS A FASCINATING OPEN CLUSTER 2° WEST of 4th-magnitude Alpha (α) Cancri, though it is often neglected for Cancer's brighter, showier Beehive Cluster (M44), 9° to the north. By way of comparison, M67 measures 25′ across and shines at magnitude 6.0 (by my estimate), whereas M44 measures 70′ across and beams more brightly at magnitude 3.1. Try to detect these clusters simultaneously with the unaided eye because they can give you a sense of intergalactic depth: M44 is 577 light-years distant, while M67 is 2,600 light-years distant, or 4 1/2 times farther away! The diameter of M67 is 19 light-years – nearly twice that of M44.

At 5 billion years old, M67 is one of the most ancient open clusters known; the only others that compare are NGC 188 in Cepheus (at 5 billion years) and NGC 6791 in Lyra (at 7 billion years). M67 contains some 500 stars within its roughly 20 light-year diameter. In a 2012 paper (http://arxiv.org/PS_cache/arxiv/pdf/1201/1201.0987v1.pdf), Barbara Pichardo of Universidad Nacional Autonoma de Mexico and her colleagues tell how they used the most recent proper-motion determination of M67 in orbital computations to explore the possibility that the Sun once belonged to this cluster.

"One of the peculiar and perhaps most intriguing aspects of M67 is its chemical composition similar to that of the Sun," they explain. "All recent high quality works, based on observations of both evolved and Solar type stars, indicate an impressive similarity between the M67 chemical composition and the Solar one. The similarity in composition and age is so close that several authors have

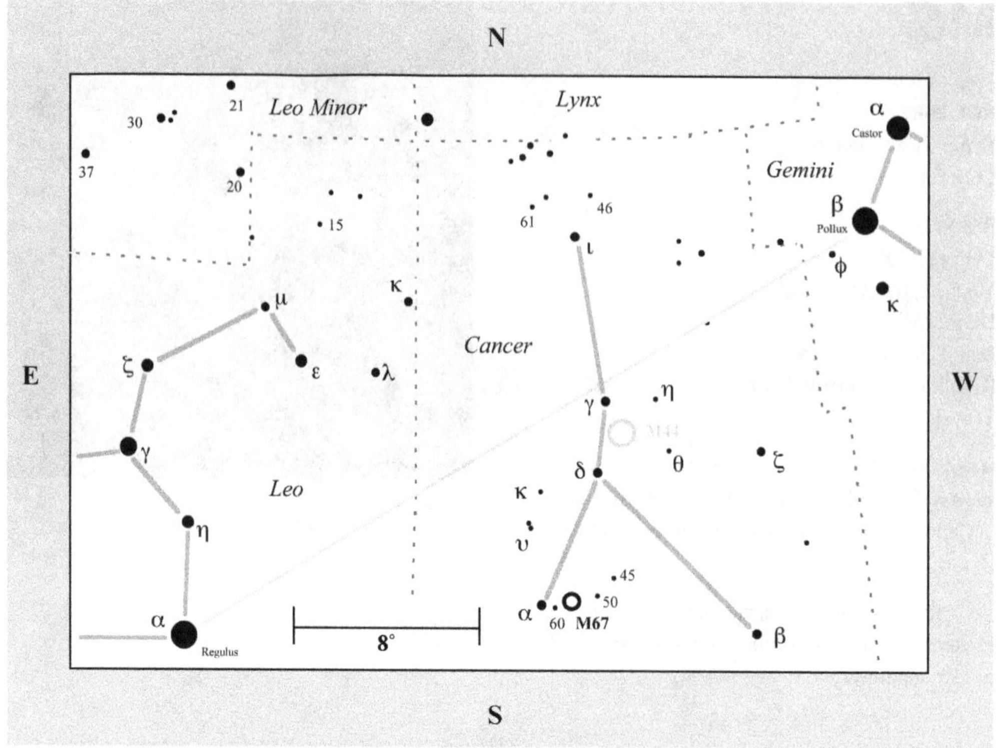

indicated in M67 stars the best-ever Solar analogues so far discovered."

One possibility for the similarities is that the Sun had a close encounter with cluster members near the tidal radius and could have been ejected from the system in a three-body encounter to become a renegade star. If this had occurred, the Sun's high escape velocity would have destroyed any initial circumstellar disk around our star or have dispersed its already formed planets.

"Now, an interesting question is: if we come from a stellar cluster, is part of this cluster still there? Starting from this idea and considering the very similar metallicities, ages, and distance from the Galactic center of the Sun and M67, it becomes tempting to place the Sun origin within this open cluster," Pichardo and her team say.

The research team ran 350,000 simultaneously computed pairs of orbits for the Sun and M67, looking for close encounters in the past. The analysis of these close encounters showed that the corresponding relative velocity between the Sun and M67 (>12.5 miles per second) is too high.

"Such a three-body encounter within M67, with the Sun being one of the three bodies," they say, "and giving this ejection velocity to the Sun, would destroy an initial circumstellar disk around the Sun, or disperse its already formed planets. Thus, the Sun was not ejected by a three-body encounter in M67. Also, by analyzing a possible encounter of M67 with a giant molecular cloud, we find a very low probability, much lower than 10^{-7}, that the Sun was ejected from M67 by such an encounter. The high values of the relative

to the south-southeast. Far afield, weak arms of stars fly off to the north, but are these associated with M67 or are they just chance alignments?

With a little imagination, the cluster looks as if it is dangling from a rack of three relatively bright field stars, the brightest member of which shines at 8th magnitude with incandescent yellow light. A dark rift runs east of that star, where it is rimmed to the north by a faint string of stars. More than the Beehive, M67 looks like a swarm of insects that seem particularly attracted to the light of the yellow star. The faintest stars cannot be resolved in a 4-inch, but many huddle together in a "nebulous" mass. As a whole, M67's stars form a distinct reverse S-shaped pattern, whose swollen (hazy) midsection reminds me of a king cobra that has just swallowed a meal. A pointed arrow of stars in the southern stellar clump forms the snake's head, and the faint stream of stars to the north marks its tail.

I hope this cluster finds its way into your mind's eye. Smyth called it a "Phrygian cap" – a brimless, conical cap with a limp top, worn by the ancient Greeks and by the French revolutionaries as a Liberty Cap. Kenneth Glyn Jones said, "The stars [of M67] form a pleasing pattern but Smyth's description of it being like a Phrygian cap – does not exactly leap to the eye." In case my snake asterism doesn't work for you either, try Camille Flammarion's "sheaf of corn," or Christian Luginbuhl and Brian Skiff's "fiber-optic tree." Better yet, try creating your own visual metaphor!

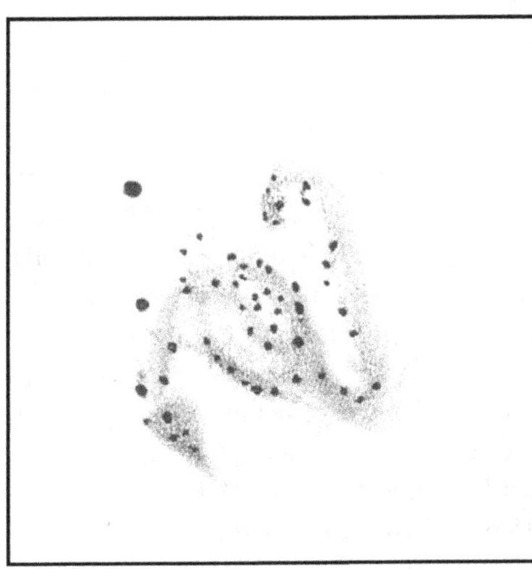

velocity also exclude the possibility that the Sun and M67 were born in the same molecular cloud."

At 23×, the cluster first appears as a loose and uniform sprinkling of bright stars across a carpet of fainter stars. But the view quickly changes into one of a slightly oval sphere of stars separated from another stellar clump

M68

NGC 4590
Type: Globular Cluster
Con: Hydra

RA: $12^h39.5^m$
Dec: –26°45′
Mag: 7.3; 7.6 (O'Meara)
Diam: 11′
SB: 12.5
Dist: 33,600 light-years
Disc: Charles Messier, 1780

MESSIER: [Observed April 9, 1780] Nebula without stars below Corvus and Hydra. It is very faint, very difficult to detect with refractors. Close to it is a sixth-magnitude star.

NGC: Globular cluster of stars, large, extremely rich, very compressed, irregularly round, well resolved, stars of magnitude 12.

IF YOU HAVE A GOOD SOUTHERN HORIZON AND are under dark skies, M68 is a marvelously challenging naked-eye globular. I estimated the cluster's magnitude to be 7.6, which is 0.3 magnitude fainter than the published photometric value. The cluster, another Méchain discovery, is located in Hydra just 30′ northeast of a magnitude 5.5 star and about 4 1/2° south-southeast of magnitude 2.6 Beta (β) Corvi.

In fact, if you stare at that magnitude 5.5 star and concentrate on M68's position, the cluster might just pop out at you. With a dedicated effort, I could just make it out with the unaided eye and using averted vision, so it joins M79 as being another of the fainter globulars visible without optical aid. Needless to say, M68 is a cinch in binoculars, unresolved yet obviously not a star.

In the first edition of this book, I had credited Pierre Méchain with the discovery of M68. But, as detailed by the Students for the Exploration and Development of Space (SEDS) Web site (http://messier.seds.org/m/m068.html), Charles Messier is the rightful discoverer. The error had been passed on by Admiral Smyth – who, in his *Cycle of Celestial Objects*, described it as a "round large nebula ... discovered by Méchain in 1780" – and picked up and copied by Kenneth Glyn Jones in his 1968 book *Messier's Nebulae and Star Clusters.*

William Herschel first resolved it "into a rich cluster of small stars so compressed that most of the components are blended together." And his son John found it irregularly round, which is a perfect description given the range of visual impressions (from "round" to "oval") that have been reported over the years by various skilled observers.

Today we know that the cluster spans roughly 100 light-years – making it about one-

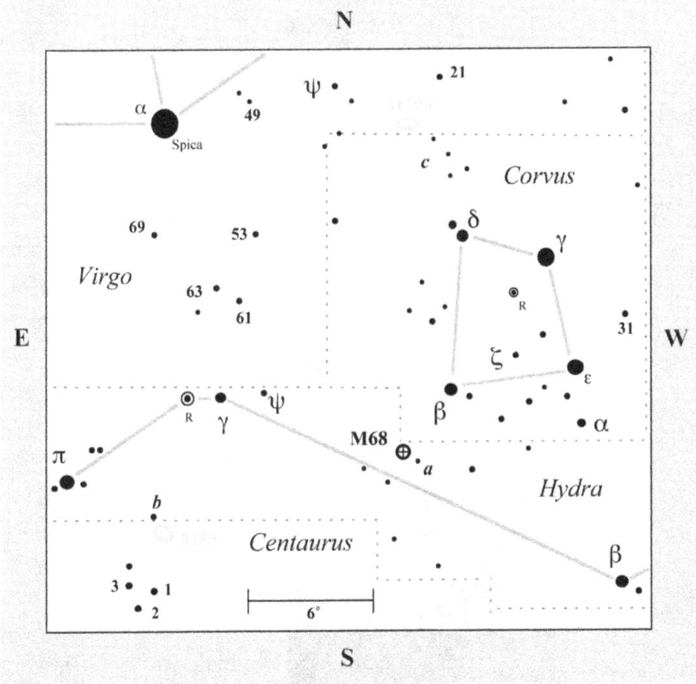

tenth the size of the Milky Way – and contains hundreds of thousands of, and perhaps even a million, stars. The cluster is metal-poor, with each of its stars, on average, having about 1/170 as much metal per unit hydrogen as the Sun. The stellar clock marking the age of a globular cluster is indicated by the luminosity of its main-sequence turnoff. Recently, astronomers have used high-quality color-magnitude diagrams and other reliable data to deduce an age for M68.

Astronomers measure the ages of globular clusters by investigating the chemical elements seen in their spectra. Since globular clusters are so old, they generally contain fewer heavy elements than stars like the Sun. Stars gradually create these elements through nuclear fusion, so the older they are, the fewer metals they have as fusion dies out. Using these interpretive methods, astronomers estimate that M68 is about 10 billion years old, making it 3 to 4 billion years younger than determined from earlier studies.

In a 2011 paper in the *Bulletin of the American Astronomical Society* (vol. 153, p. 13), Sloane K. Simmons (University of Texas) and colleagues explain their chemical composition study of 25 stars in M68. The study includes 11 red giant branch (RGB) stars, 9 red horizontal branch (RHB) stars, and 5 blue horizontal branch (BHB) stars, spanning an effective temperature range of approximately 5,000 K. The researchers found a generally consistent metallicity across all the evolutionary groups, with an iron-to-hydrogen ratio of 1/251, thus confirming the cluster's elderly status in the universe.

Interestingly, however, like M15 in Pegasus, M68 shows elevated abundances of silicon compared with other globular clusters. But M68 singles itself out by having a relative underabundance of titanium as well. In a 2005 paper in *Astronomical Journal* (vol. 129, p. 251), Jae-Woo Lee (Sejong University, Seoul, Korea) and colleagues interpret this result as "implying that the chemical enrichment seen in M68 may have arisen from contributions from supernovae with somewhat more massive progenitors than those that contribute to abundances normally seen in other globular clusters."

In August 2012, the Space Telescope Science Institute released the amazing image of M68 seen here, taken with the

M68

NASA/ESA Hubble Space Telescope. The view reveals an incredibly crowded "stellar encampment," whose components are under a gravitational embrace. The image, which combines visible and infrared light, has a field of view of approximately 3.4 × 3.4 arcminutes.

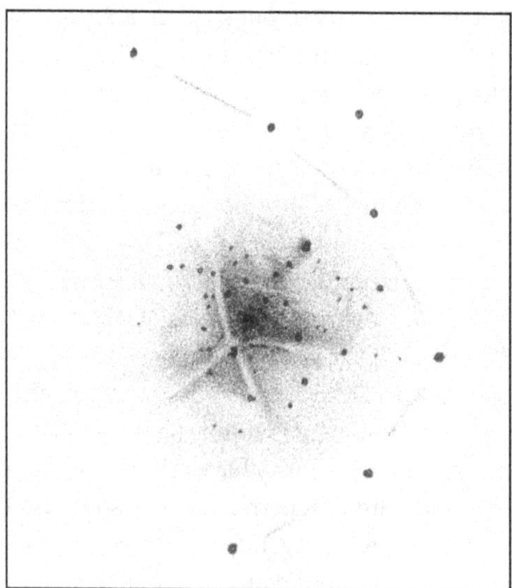

Through the 4-inch at 23×, M68 is very compact but highly mottled, hinting at how nice it will appear at higher powers. Of this globular's telescopic view, Admiral Smyth wrote, "It is very pale but so mottled that a patient scrutiny leads to the inference that it has assumed a spherical figure in obedience to attractive forces." Take a moment to look around the cluster with east up. Do you see how it sits in a V-shaped asterism, a bucket? Now look to the east; a chain of stars leading from the magnitude 5.5 star seems to be attached to this bucket, drawing it up from a celestial well.

And the cluster certainly is beautiful at medium power. Dozens of "bright" stars (the brightest being about magnitude 12.5) burst forth from a seething glow of faint stars that seem ready to boil out of the cluster's core. Indeed, the nuclear region displays a cauldron of bright stars that at 130× looks highly fractured. The brightest section is skewed to the northwest, where the stars concentrate in a wedge-shaped pattern. I've spent hours studying that mysterious center, trying to fathom its complexity. The wait was worth it. Look at the drawing and notice the four darkest lanes (white in this negative rendering), which form a windmill-like formation over the southeastern half.

Other dark patches are clearly created by gaps between strings and arcs of bright stars, especially to the south. The "windmill" is but one of many patterns of dark lanes in this region. What you see really depends on how you look and what's on your mind. While your imagination is still engaged, look for a striking detached portion in the northern halo, where you will find a dark "footprint."

M69

M69

NGC 6637
Type: Globular Cluster
Con: Sagittarius

RA: $18^h31.4^m$
Dec: −32°21′
Mag: 7.7; 7.4 (O'Meara)
Diam: 10′
SB: 12.7
Dist: ~28,700 light-years
Disc: Nicolas-Louis de Lacaille, 1752

MESSIER: [Observed August 31, 1780] Nebula without a star in Sagittarius, below the left arm and close to the bow. Nearby there is a ninth-magnitude star. The luminosity is very faint and can be seen only under good conditions, and the slightest illumination of the micrometer's crosshairs causes it to disappear. Its position was determined from ε Sagittarii. This nebula was observed by M. de la Caille, and given in his catalogue. It resembles the nucleus of a small comet.

NGC: Globular cluster, bright, large, round, well resolved, stars of 14th to 16th magnitude.

MESSIER'S SIXTY-NINTH CATALOGUE ENTRY IS one of several tiny objects in "globular cluster alley" – a 10°-long strip of sky along the bottom of the Sagittarius teapot between Epsilon (ε) and Zeta (ζ) Sagittarii that contains three Messier globulars and one NGC globular. M69, a 7th-magnitude globular, is the one closest to Epsilon Sagittarii, about 2 1/2° to the northeast.

M69 is a sizable cluster some 80,000 light-years in true physical extent. It lies only about 5,500 light-years from the Galactic center and is considered part of the Galactic bulge. Studies of M69, then, may be of importance in that the evolution of the bulge is not well understood. Two competing models exist for its formation: (1) the hierarchical scenario, where the bulge is built up through mergers over time; and (2) disk instability. But neither suffices, owing to the fact that they cannot explain why our bulge is dominated by old, metal-rich stars.

Like the bulge, M69 is old and metal-rich – having stars with one-quarter as much metal per unit hydrogen on average as the Sun. Hubble Space Telescope studies have provided an age estimate for the cluster of 14 billion years, about the same as for 47 Tucanae. The close similarity of the ages of M69 and 47 Tucanae (a disk population globular cluster) indicates that the age-metallicity relations of

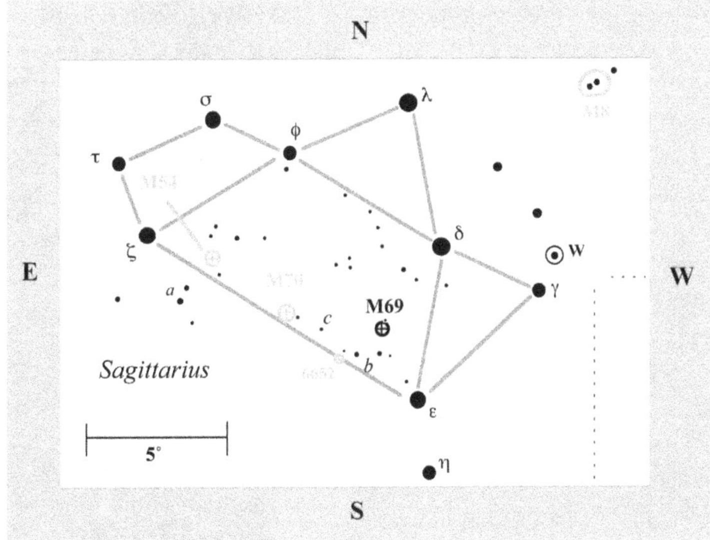

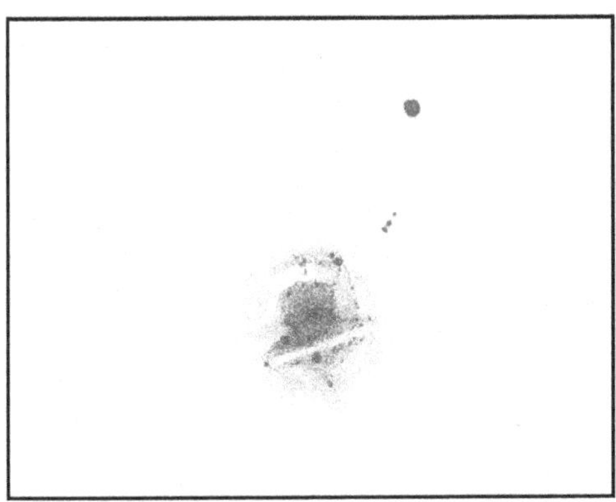

these populations intersect, suggesting a common origin.

The color–magnitude diagrams show a predominantly red-clump horizontal branch morphology with hints of a blue horizontal branch extension. Recent surveys, however, reveal significant contamination by field stars. In a study of variable stars performed by M. E. Escobar (Pontifícia Universidad Católica, Santiago, Chile) and colleagues at Las Campanas Observatory, however, the researchers found 61 candidate variable stars, 54 of which were new discoveries, including 8 RR Lyrae stars, 10 eclipsing binaries, and 15 long-period variables.

To find M69, use the chart here to locate Epsilon Sagittarii. Then look about 2° northeast for a pair of 6th-magnitude stars (*b*), oriented east to west and separated by about 30′. M69 lies only about 40′ north of the westernmost star in Pair *b*.

At 23×, it is a uniformly bright fuzzball next to which shimmers an 8th-magnitude topaz star. When I relaxed my gaze, several patches of light in the area pulled my eye away from M69. These patches turned out to be sections of Milky Way sliced into little shards by dark, nebulous veins. The most prominent section lies to the southeast, where it forms a bank of patchy light through which a black river flows. All this leads to the diminutive 9th-magnitude globular NGC 6652, which lies about 8′ south of a magnitude 6.9 aqua-tinged star. After looking at these colorful stars, I suddenly realized how sickly M69's pallor is. In fact, all the clusters in globular alley remind me of ancient artifacts that have faded and weathered with time, parts of them disintegrating into loose particles. This area is charged with a fair amount of intergalactic dust, which dims the light emanating from these distant objects.

Although moderate power will resolve a few stellar members, the brightest of which shine around magnitude 13.7, the cluster is much more satisfying at 130×. Still, only large telescopes will achieve full resolution at around 16th magnitude. Immediately the core appears asymmetrical to the north – an effect caused largely by a dark rift slicing through the southern half. With a longer gaze, the entire cluster fragments into patches. Here we have a patchy cluster mimicking its Milky Way surroundings! In fact, M69 contains too many faint patches of darkness for me to render, though not much detail reveals itself to the east. Can you see the dark lagoon north of the core?

M70

NGC 6681
Type: Globular Cluster
Con: Sagittarius

RA: $18^h43.2^m$
Dec: -32°18′
Mag: 7.8
Diam: 8′
SB: 14.5
Dist: ~29,300 light-years
Disc: Charles Messier, 1780

MESSIER: [Observed August 31, 1780] Nebula without a star close to the previous one [M69], and on the same parallel. Nearby there is a ninth-magnitude star and four faint, telescopic stars almost in a straight line, very close to one another, and which lie above the nebula, as seen in an inverting telescope. The nebula's position was determined from the same star, ε Sagittarii.

NGC: Globular cluster, bright, pretty large, round, gradually brighter in the middle, stars from 14th to 17th magnitude.

ALTHOUGH M70 LIES AT ABOUT THE SAME distance as M69 and is slightly smaller (70,000 light-years across in true physical extent), it is approaching us five times as fast – 136 miles per second versus 25 miles per second. And, like M69, it is a bulge globular, lying only about 7,000 light-years from the Galactic center. Yet, unlike metal-poor M69, each of M70's stars has, on average, 1/41 as much metal per unit hydrogen as the Sun.

The Hubble Space Telescope image of M70 shown here captures in glorious detail the cluster's densely populated postcollapse core. Globular clusters can collapse when more stars than normal squeeze into the core's already tight quarters. This happens as stars interact with each other over time in their orbits, sending lighter stars speeding toward the cluster's edges while forcing heavier stars to slow and congregate in orbits toward the cluster's center. So over time the core tends to compress inward (collapse) while its halo expands outward. What we see in the end is an expansive globular with a small and intensely dense and bright core. Only about 20 percent of the 150 or so known globular cluster stars have experienced a core collapse. In the case of M70, what's surprising is that, given the cluster's proximity to the Galactic center and the intense gravitational forces on it, why didn't more of its stars pull away from the cluster's core rather than descend deeper into it?

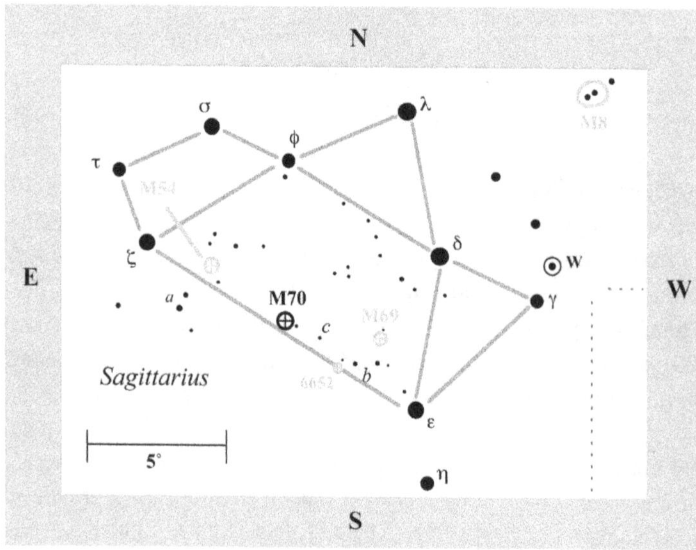

The HST also detected a hot stellar population of 122 stars at far-ultraviolet wavelengths. The cluster's far-UV–visible color-magnitude diagram shows a well-defined horizontal branch with no evidence for hot, more evolved descendants (*Astrophysical Journal*, vol. 435, p. L55). The core also harbors two blue stragglers. Another HST ultraviolet study, by Jennifer L. Connelly of the American Museum of Natural History and her colleagues, published in a 2006 paper in the *Bulletin of the American Astronomical Society* (vol. 38, p. 937), produced the deepest near- and far-ultraviolet survey of a globular star cluster to date.

To find M70, use the chart here to locate Epsilon Sagittarii. Then look about 2° northeast for a pair of 6th-magnitude stars (*b*) oriented east to west and separated by about 30′. M70 lies about 2° northeast of the easternmost star in Pair *b* and 1° from 8th-magnitude Star *c*. You could also use your binoculars to find the

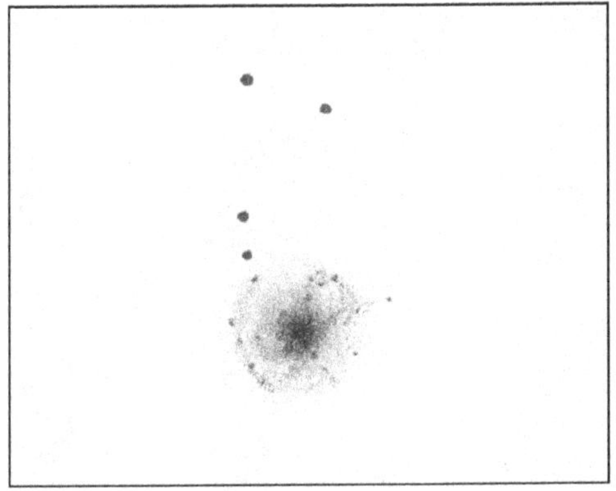

globular's tight, hazy glow about 2° east and slightly north of M69.

Telescopically at low power, the cluster looks like the sizzling end of a fuse of stars to the north. Kenneth Glyn Jones offered a similar metaphor: "As if attached to the cluster, there is a little slightly curved 'tail' of small stars, shooting off like sparks to the [north-northeast]. These may be the stars mentioned by Messier." Camille Flammarion spoke of the cluster being decorated with a pretty double star to the northeast, and this star pair is a welcome sight even at high power.

If I didn't know better, I probably would have mistaken M70 for a spiral galaxy. Three distinct "arms" curl out of a very tight, fuzzy core. Most noticeable are the northern and southern arms, both of which are riddled with starlight. The brightest stars in M70 shine around 13th magnitude, which is excellent for a globular, but to achieve decent resolution, you must be able to see to magnitude 15.6.

The southern string ends abruptly at a single bright star. Can you resolve the cluster's outer halo? I cannot; it remains an elusive haze with teasingly knotty sections. Use high power and you might have some success. I noted that the core remained unresolved in the 4-inch. It is much more compressed than M69's core and, as Brian Skiff observed, it is slightly elongated. Try comparing M69 and M70 on the same night.

Once, when I was looking at M70 with low power, I picked up with averted vision what appeared to be a very faint nucleus to a comet nearly 1 1/4° to the south-southeast. I checked the object with high power, and it remained an unresolved fuzzy star. When the object did not move after an hour, I suspected it was a planetary nebula, which it turned out to be. It is IC 4776, at magnitude 10.4.

Also be prepared to scout out NGC 6652, another globular cluster just a little more than 20′ east of the easternmost star in Pair *b*. But it shines at magnitude 8.5. Fortunately, it's also only 6′ across, so its surface brightness is a reasonable 12.4 – about the same as that for M70. What's more, it's 0.7 magnitude brighter than M72 in Aquarius and roughly the same size, though its surface brightness is less (13.3). NGC 6522, with an age of 12 billion years, lies ~32,600 light-years from the Sun and 8,800 light-years from the Galactic center, in the inner halo.

M71

NGC 6838
Type: Globular Cluster
Con: Sagitta

RA: $19^h53.8^m$
Dec: +18°47′
Mag: 8.4; 8.0 (O'Meara)
Diam: 7′
Dist: ~13,000 light-years
Disc: Probably Philippe Loys de Chéseaux, 1746; Johann Gottfried Kohler observed it around 1775

MESSIER: [Observed October 4, 1780] Nebula discovered by M. Méchain on 28 June 1780, between the stars γ and δ Sagittae. The following 4 October, M. Messier searched for it. The light is very faint and it contains no stars. The slightest illumination causes it to disappear. It lies about 4 degrees below the nebula that M. Messier discovered in Vulpecula; see number 27. It is plotted on the chart for the comet of 1779.

NGC: Cluster, very large, very rich, pretty much compressed, stars from 11th to 16th magnitude.

M71 WAS ONCE AN OBJECT OF CONTROVERSY among professional astronomers. Some argued that it is a loose globular, while others claimed it's an extremely dense open cluster. There is no doubt today, however, that M71 is a globular. In fact, it is a very near one, being only 13,000 light-years distant, which is why it is so easily resolved and appears to lack the dense center typical of more distant globulars. In the spectacular HST image here, we peer deep into the core of M71, demonstrating well the extremely loose nature of the stars.

This great ball of ancient stars, some 27 light-years across, presently resides about 21,800 light-years from the Galactic center. We see it receding from us in its orbit around the Galaxy at about 14 miles per second.

Raminder S. Samra (University of British Columbia, Vancouver) and colleagues used Gemini North in Hawaii, together with the NIRI-ALTAIR adaptive optics imager in the near infrared, to explore the core of M71. They reported their results in a 2012 paper in *Astrophysical Journal* (vol. 751, p. L12). They obtained proper motions for 217 stars and found that the stars in the core disperse by ~179 microarcseconds per year without any structural ordering or radial-velocity

one certain and seven candidate cataclysmic variables.

M71 is easily spotted midway between the bright stars Gamma (γ) and Delta (δ) Sagittae and a mere 20′ northeast of 9 Sagittae, all in the shaft of the famous Sagitta, the Celestial Arrow. Anyone viewing this cluster through a telescope will immediately understand that former classification controversy. Here is a sizable (7′) yet compact glow of moderately bright to faint stars (typical of globulars), whose loose center resolves well with medium magnification (typical of open clusters).

variation (the rate of change in distance toward or away from the viewer over the course of time) with respect to the cluster center. They also set the upper limit on any central black hole to be ~150 solar masses at the 90 percent confidence level.

In a 2012 book titled *Star Clusters in the Era of Large Surveys*, Alessandra Di Cecco (European Southern Observatory) and colleagues used photometric data collected at the Canada-France-Hawaii Telescope and the Hubble Space Telescope (HST) on M71 to find an age of about 11 billion years, which, they say, "agrees well with the bulk of the more metal poor globular clusters." Data from the HST also helped astronomers uncover optical counterparts to 25 out of 29 Chandra x-ray sources, including

In a low-power field of view, M71 occupies the center of an oval area outlined by four distinctive Y-shaped asterisms, all oriented in different directions. M71 itself is rather Y-shaped, especially if you include the two 12th-magnitude stars just outside its round and moderately condensed haze. Webb called it an "interesting specimen of the process of stellar evolution" – an interpretation seemingly shared by astronomer Isaac Roberts, who, after imaging the cluster in 1890, commented that the curves in its crowded star regions were suggestive of having been produced by the effects of spiral movements.

Through the 4-inch Genesis, no central condensation is apparent, even with high power.

M71

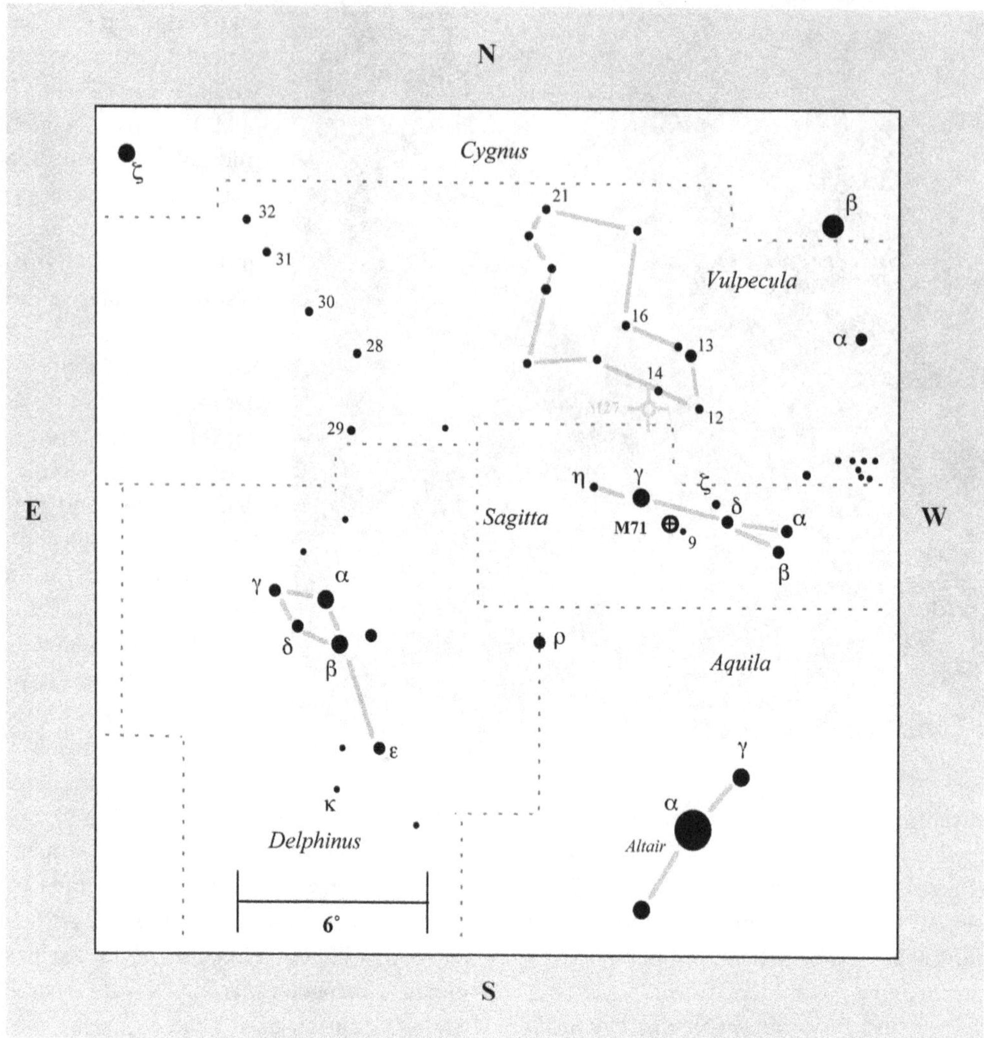

The cluster does sport an arrow-shaped concentration to the southwest. Dark lanes cut through that region in a gridlike fashion. Three of the darkest lanes run southeast to northwest, alternating with waves of starlight flowing from the densest wedge of light to the southwest. Now, relax your mind. Do you see these stars calving away from that wedge like heavy snow down an angled roof (a chilly thought for warm summer nights, when M71 is best placed in the evening sky)?

Surprisingly, John Mallas claimed he could not resolve any stars within the cluster in his 4-inch refractor. Brian Skiff believes Mallas may have assumed that the faint stars he resolved near the cluster were field stars, which is

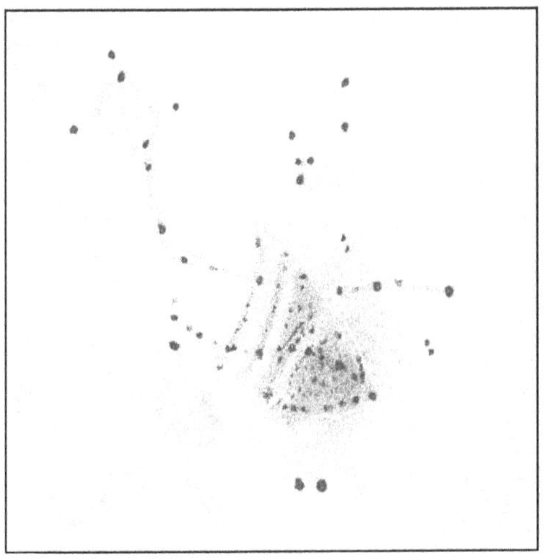

usually the case, but here they're actually cluster members. Otherwise, Mallas's observation remains a mystery since, a century before him, Webb saw the cluster "yield to a cloud of faint stars" with high power in a 3.7-inch - presumably a telescope of inferior quality.

M72

NGC 6981
Type: Globular Cluster
Con: Aquarius

RA: 20^{h}53.5^m
Dec: -12°32′
Mag: 9.2
Diam: 7′
Dist: ~55,400 light-years
Disc: Pierre Méchain, 1780

MESSIER: [Observed October 4, 1780] Nebula seen by M. Méchain on the night of 29–30 August 1780, above the neck of Capricornus. M. Messier searched for it the following 4 and 5 October. Its light is faint like the previous one [M71]. There is a faint telescopic star nearby. Its position was determined relative to fifth-magnitude ν Aquarii.

NGC: Globular, pretty bright and large, round, much compressed in the middle, well resolved.

ABOUT 4° (NEARLY TWO FINGER WIDTHS AT arm's length) southeast of magnitude 3.8 Epsilon (ε) Aquarii, or about 10° (a fist width) due east of the naked-eye double star Alpha (α) Capricorni, you come to the faintest Messier globular, M72. At magnitude 9.2 and just 6′ in apparent diameter, this globular is easy to pass over. Look for a 9th-magnitude "double star" separated by 5′ – the eastern component is, in fact, a star; the western component is M72. Once found, use moderate magnification to enlarge the cluster's disk.

One of the reasons M72 appears so feeble is its distance; the cluster lies about 55,400 light-years from the Sun and 42,000 light-years from the Galactic center. Yet, in true physical extent, it spreads across about 110 light-years of space. And, like M71, M72 has been classified as a very open globular, with an age of ~12.7 billion years. Each of its stars, on average, has 1/26 as much metal per unit hydrogen as the Sun. The cluster is rich in variable stars, but some suspected variables have turned out to be nonvariables. Indeed, in a 2011 paper in *Monthly Notices of the Royal Astronomical Society* (vol. 413, p. 1275), D. M. Bramich (European Southern Observatory) and colleagues explain how their photometry of M72 revealed that "20 suspected variables in the literature are actually non-variable." They did, however, confirm the variable nature of 29 other variables. They also detected 11 new RR Lyrae stars and three new SX Phoenicis stars, bringing the total confirmed variable star count in M72 to 43.

The cluster's brightest members are near the limit of resolution in small apertures. Some 14th-magnitude stars can be detected

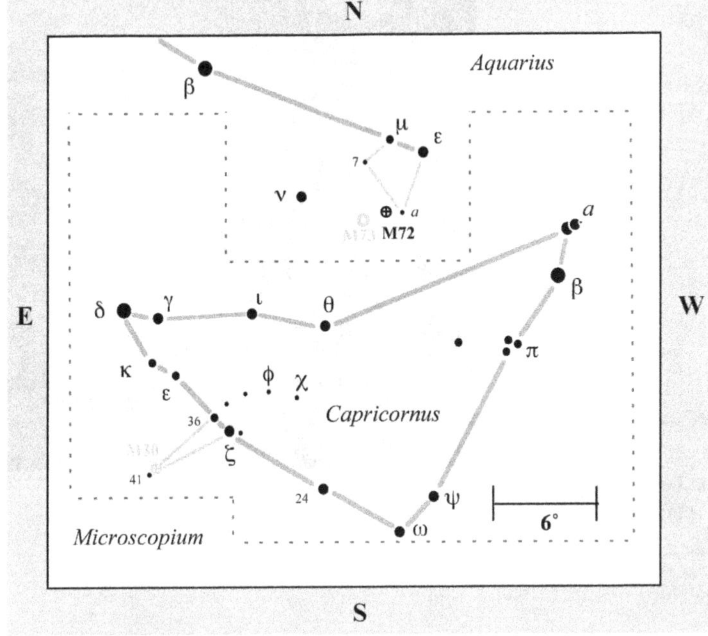

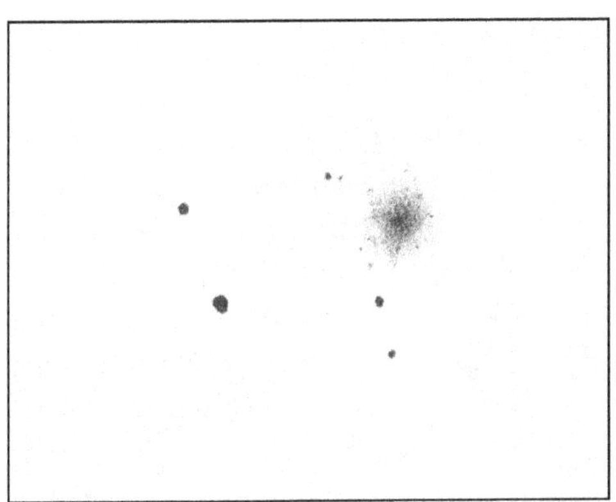

in the outer envelope at 72×, but only with difficulty: look for a pair of stars to the south and another, closer pair to the northeast. Heinrich d'Arrest partially resolved the cluster at 9× and found it well resolved at 123×, though his observation might have been made with an 11-inch refractor.

Although M72 is still difficult at 130× in the 4-inch, I can resolve a fair number of members and trace some faint "arms." But this is not at all a simple task! By resolving the stars, I mean being able to see insignificantly faint speckles pop in and out of view in an otherwise foggy moor of starlight. Photometry of this globular reveals that its brightest star shines at magnitude 14.2. The core, which is apparently loose when viewed through large instruments, looks slightly diamond shaped and impenetrable through the 4-inch.

It's not that M72 is a bad cluster; it is just better suited for large telescopes and high magnifications. Full resolution occurs around 17th magnitude.

M73

NGC 6994
Type: Asterism?
Con: Aquarius

RA: $20^h58.9^m$
Dec: –12°38′
Mag: 9.7
Diam: 1.4′
Dist: Unknown
Disc: Charles Messier, 1780

MESSIER: [Observed October 4 and 5, 1780] Cluster of three or four faint stars, which, at first glance, resembles a nebula, and which does contain some nebulosity. This cluster lies at the same declination as the previous one [M72]. Its position was determined using the same star, ν Aquarii.

NGC: Cluster, extremely poor, very little compressed, no nebulosity.

"A TRIO OF 10 MAG. STARS IN A POOR FIELD" is how Admiral William Henry Smyth simply described this curious M object in his *Cycle of Celestial Objects*. A Y-shaped asterism conveniently situated 1 1/2° east of M72, M73 is actually a grouping of four stars – specifically General Star Catalogue (GSC) numbers 05778-00802, -00492, -00509, and -00594 – with no apparent connection except that they lie in the same line of sight.

This is no longer suspect. Brian Skiff at Lowell Observatory, in a private communication to Steven Hynes and Brent Archinal that appears in their book *Star Clusters*, stated that the proper motions and colors of the stars all argue "(if not conclusively) for pretty much what we've expected all along, which is that the stars are unrelated in space." Hynes and Archinal go on to note that Skiff's conclusion is fully supported by high-resolution spectra of the six brightest stars in the area, and Tycho-2 catalogue proper motions show that "these stars definitely do not form a physical system."

No matter, I enjoy the shape and texture created by these four stars because, if nothing else, they incite mental flights of fancy. Once, I brought a tape of Gustav Holst's *The Planets* with me into the field and played the track "Venus" as I stared into the eyepiece. The Y-shaped asterism, which is aligned east to west, suddenly transformed into a Flash Gordon–style rocket ship sailing through interstellar space. Because the stars gradually fade to the west, the ship is seen obliquely from behind. The two bright end stars of the Y are the burning rockets. It's a stretch, granted, but for this object in particular, a little flash is just what's needed!

I do not experience the same thrill of discovery that Messier must have felt when he

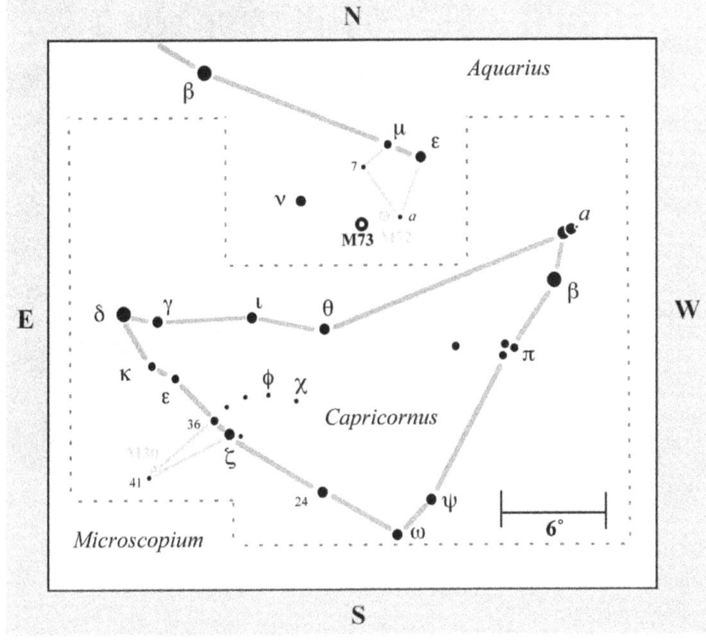

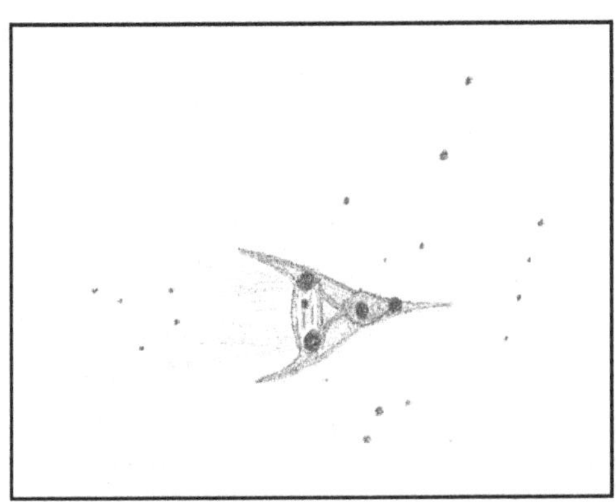

encountered these stars, because, even at 23× in the 4-inch, M73 does not look fuzzy – in the sense of being comet-like, or nebular, as Messier described it. Obviously, the optics of Messier's telescope were not of the quality of optics in modern telescopes, and that is probably one reason why this grouping looked fuzzy to him. Another reason could have been poor atmospheric conditions. So I went in search of another, more comet-like asterism suitable for the 4-inch. And I found one!

Try sweeping your telescope at low power about 1° to the southwest of M73. Do you encounter a small haze? It is yet another asterism of four stars in the shape of a Y, only this one is a mirror image of M73. These stars are closer together than those in M73, giving them a nebular appearance at 23×. Archinal independently chanced upon these stars as well, and noted their similarity to the M73 grouping. This also corroborates Vehrenberg's comment that asterisms of this nature are not uncommon!

For another challenge, sweep your telescope 2° northeast of M73 and look for a slightly swollen 8th-magnitude "star." This is the Saturn Nebula, a pale green planetary nebula. At 440×, a magnitude 12.8 star blazes at the center of the nebula's 23″ disk. Use high magnification to see that central star.

I once had an opportunity to view this nebula through the great 60-inch reflector atop Mount Wilson in California. On that fine, moonless night in August 1986, the planetary displayed two sharply defined green oval disks surrounding the central star, which at 400× burned brightly. On closer inspection, each ring had a series of condensations highlighting it and two long rays with tight knots stretched beyond the ansae of the rings. The nebula bore a close resemblance to the planet Saturn seen edgewise, with bright moons adorning the tips of the hairline rings.

M74

The Phantom
NGC 628
Type: Spiral Galaxy (SA(s)c)
Con: Pisces

RA: $01^h36.7^m$
Dec: +15°47′
Mag: 9.4; 8.5 (O'Meara)
Dim: 11.0′ × 11.0′
SB: 14.4
Dist: ~30 million light-years
Disc: Pierre Méchain, 1780

MESSIER: [Observed October 18, 1780] Nebula without a star, close to star η in the ribbon of Pisces. Seen by M. Méchain at the end of September 1780, and about which he reported "This nebula does not contain any stars. It is quite broad, very dim, and extremely difficult to observe; it may be distinguished more accurately during fine frosts." M. Messier searched for it and found it to be as M. Méchain described. It was directly compared with the star η Piscium.

NGC: Globular cluster, faint, very large, round, pretty suddenly much brighter in the middle, some stars seen.

BLAZING WITH THE LIGHT OF 40 BILLION suns, flinging spiral arms across 96,000 light-years of space, M74 is a prima donna among open-faced spiral galaxies. At least, that is how it comes across in long-exposure photographs taken through large telescopes. In small telescopes, it is more like a phantom, which is the nickname I have given it. No object in the Messier catalogue has proven more troublesome, more elusive, or more provocative to amateur astronomers than this giant spiral. The problem is that the galaxy's large apparent size (11′) and very low surface brightness (14.4) require a very dark sky for it to be seen well, if at all!

Méchain was right on the mark when he said, "This nebula ... is quite broad, very dim, and extremely difficult to observe; it may be distinguished more accurately during fine frosts." The fine frosts he refers to, I am sure, are those incredibly transparent evenings following the passage of a cold front, when the night sky is free from moisture and atmospheric contaminants, and the stars can be seen with crystal clarity away from city lights. On these nights, you have the best chance to see dim and diffuse objects.

I searched for M74 on many nights with Harvard Observatory's 9-inch refractor

dramatic are the dust lanes, especially along the most luminous arms. If you could see the galaxy in "negative" form, its spiral structure would appear equally impressive. Actually, there is no evidence for dust structure within ~65 light-years of the nucleus. Had this galaxy been twice as distant or farther away, M74 might have been classified as a loosely wound or chaotic spiral. Unseen to the eye is a larger gas halo that extends a whopping 10′ beyond the optical disk.

In images with large telescopes, such as the Hubble Space Telescope image shown here, the galaxy starts to resolve around a blue magnitude of 20, which is in the range of many amateur scopes equipped with sensitive CCD cameras. Easier targets for CCD imagers are the many complex H II regions, which display several nuclei and halo diameters up to 5″. These "beads on a string" are embedded in diffuse emission and trace the spiral pattern.

In ultraviolet images, the brightest knots (regions of star formation) show an excess of emission, which tends to lie on the inside edges of the spiral arms. This suggests that the disk of M74 rotates through the spiral pattern from inside to outside, such that the youngest and most massive stars are located at the inner edge of the spiral pattern. The entire disk has undergone active star formation within the past ~500 million years, but the inner regions

without success. The fact is that so large an aperture under less than perfect skies doomed my quest before it even started. It's best to use a small-aperture instrument, low power, and a wide field of view on the finest of nights.

But even armchair astronomers would enjoy the grace and beauty of M74. This stunning spiral (inclined only 5° from face-on) resembles M101 in Ursa Major, but its arms appear more regular and symmetrical, and without as many branches of smaller arm segments, though they do appear quite dramatically along the galaxy's two main arms. The grand-design spiral pattern winds outward from a very small but bright nucleus. Thick dust lanes join them out to a quarter of a rotation before they begin to thin. Actually, the nucleus appears no different from the multitude of dense knots in the galaxy. Equally

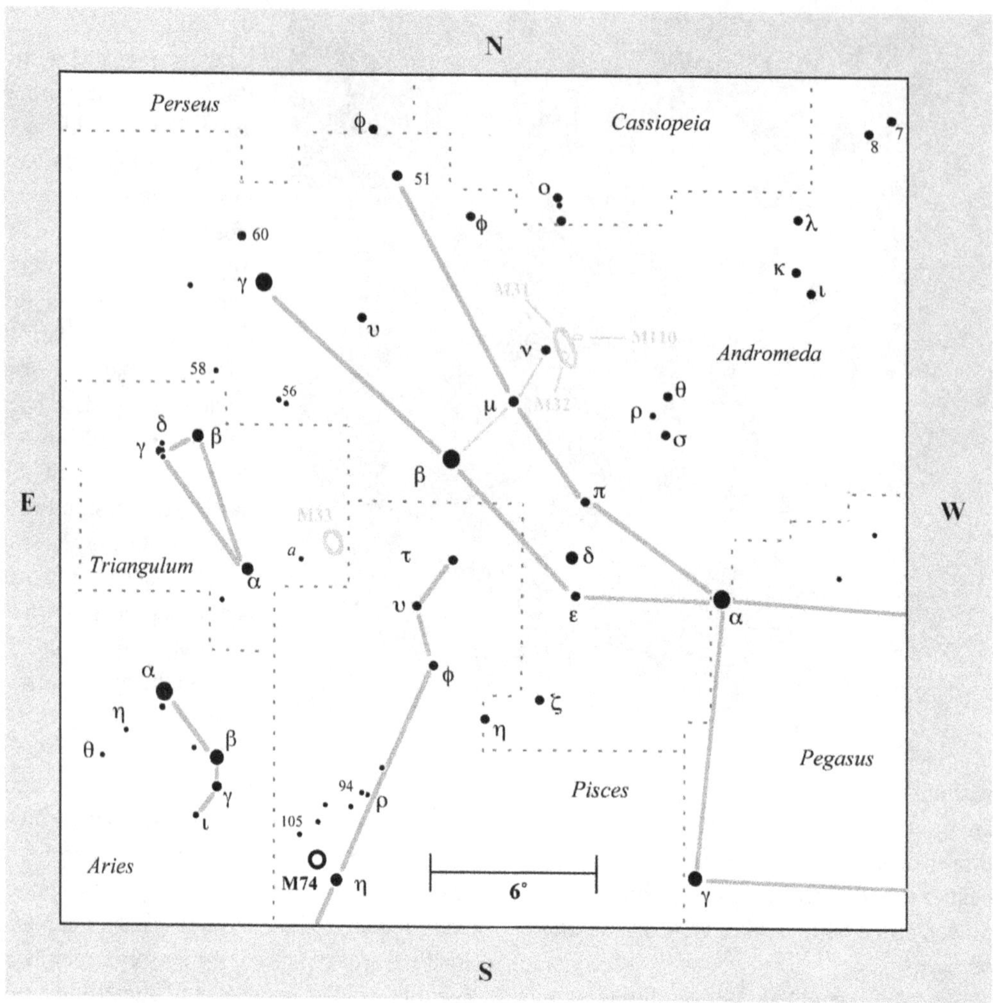

have experienced more rapidly declining star formation than the outer regions. "What is surprising is not that the star formation history changes with radius but that it exhibits such a remarkably smooth and organized pattern of change," say Pamela M. Marcum (California Institute of Technology) and her colleagues (*Astrophysical Journal Supplement*, vol. 132, p. 129). "Such patterns are important clues to the global mechanisms by which disk galaxies regulate star formation."

Locating the field is easy. Use the chart at the back of the book to locate magnitude 3.6 Eta (η) Piscium. Then use the chart here to find the galaxy about 1 1/4° east-northeast of that star. From dark skies, 23× shows M74 as an obvious but pale disk of uniform light. The longer you look, the more detail you should see. Watch how the core slowly materializes into a compact orb punctuated by a pinpoint of light. With averted vision, a diffuse outer skirt also takes shape.

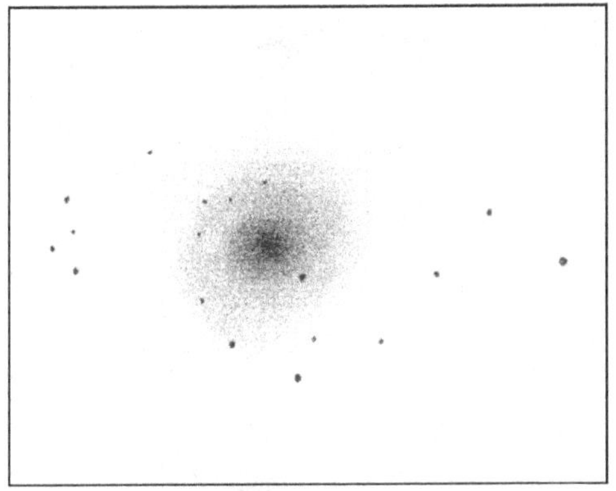

It might appear that M74 cannot stand anything but low power, but give it a try. If you can make out the galaxy's inner core, then that is the time to change to higher power. The core itself is mottled with the light of many dim stars, which emerge only with high powers. It is this chance alignment of field stars projected against the nucleus that obviously caused John Herschel to mistake M74 for a globular cluster, which in turn led Dreyer to list it as a globular in the *New General Catalogue* (see the NGC description earlier).

M74 is also a very difficult galaxy to sketch. Its spiral patterns are like spirits that weave in and out of view. So fleeting are these moments that I needed hours (and nights) of observing to confirm and reconfirm what I thought I was seeing. It is ironic that once I identified the field around M74 in the telescope, I used 7 × 35 binoculars to spot the galaxy. Imagine, here is the object I couldn't see with a 9-inch refractor from Cambridge, Massachusetts, yet here in my sparsely populated corner of Hawaii, it is visible in 7 × 35s. This is an excellent example of the degrading effect of light pollution.

M75

NGC 6864
Type: Globular Cluster
Con: Sagittarius

RA: $20^h06.1^m$
Dec: –21°55′
Mag: 8.6
Diam: 7′
Dist: 68,200 light-years
Disc: Pierre Méchain, 1780

MESSIER: [Observed October 18, 1780] Nebula without a star between Sagittarius and the head of Capricornus. Seen by M. Méchain on 27 and 28 August 1780. M. Messier searched for it the following 5 October, and on the 18th compared it with sixth-magnitude Flamsteed 4 Capricorni. It seemed to M. Messier that it consists only of very faint stars, but contains some nebulosity. M. Méchain described it as a nebula without stars. M. Messier saw it on 5 October, but the moon was on the horizon, and it was not until the 18th of the same month that he was able to make out its form and determine its position.

NGC: Globular cluster, bright, pretty large, round, very much brighter in the middle to a brighter nucleus, partially resolved.

MESSIER'S SEVENTY-FIFTH DEEP-SKY CURIOSITY is also one of the more challenging to find, because it lies in a southern region of sky devoid of bright guidepost stars. Once thought to be a fugitive globular fleeing the bounds of our Galaxy, M75 is, on the contrary, a healthy member of the Milky Way's halo community. In fact, its distance (68,000 light-years) pales in comparison with that of the truly faraway globulars, such as NGC 2419 in Lynx, which is an incredible 380,000 light-years distant! Despite its great distance, M75's highly compact 7′ disk shines at a respectable magnitude 8.6. That's because, among other things, it is a sizable system, spanning a whopping 140 light-years of space.

The globular lies beyond the Galactic center, some 48,000 light-years from it. The cluster appears to be moving away from us at about 120 miles per second, in a retrograde orbit. Some astronomers have speculated that such systems formed in large satellite galaxies that subsequently merged with the Galaxy

M75

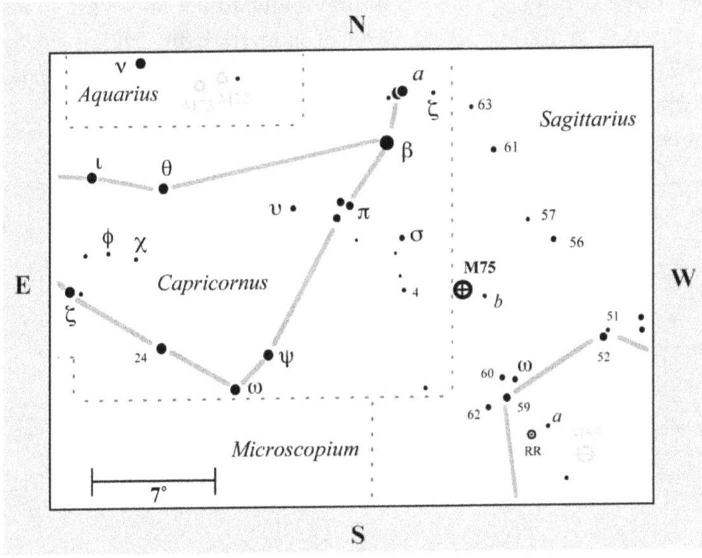

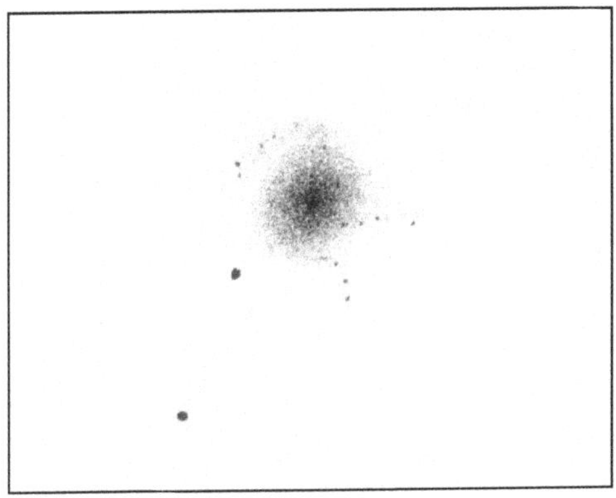

to become the Galactic halo. These systems have an extraordinary chemical homogeneity, with identical calcium abundances. NGC 6864 does exhibit a tidal tail, which may indicate a merger event.

In images, M75 has a relatively high degree of central concentration, and Shapley placed it in his Class I. It has an integrated spectral type of F9, and each of its stars, on average, has about 1/19 as much metal per unit hydrogen as the Sun. It also contains about a dozen variable stars.

For amateurs, this isolated, southern cluster resides 8° southwest of 3rd-magnitude Beta (β) Capricorni. If your telescope has setting circles, you can start at Pi (π) Sagittarii in the spoon asterism northeast of the teapot and then move the telescope 1° south in declination and about 1 hour (14°) east in right ascension, but it is more fun just to sweep. You'll want to point your telescope about midway between Pi Capricorni and 52 Sagittarii. You will find the cluster about 1 1/2° northeast of 6th-magnitude Star *b*, nestled in a faint asterism that bears a striking resemblance to the constellation Scorpius – but it is a backwards Scorpion.

At 23×, I swept over the cluster a few times before sighting its slightly swollen starlike disk. At moderate power, the cluster looks larger but remains largely unresolved, though a few outlying 14th-magnitude field stars do start to materialize. The brightest stars in the cluster are dimmer than magnitude 14.6, and full resolution won't come unless you're using a large telescope that can reach to magnitude 17.5! Messier claimed to have resolved this cluster, but obviously he was seeing stellar patches rather than individual starlight. Aside from the task being impossible through the telescopes he used, consider also that he failed to resolve stars in the brighter giants, like M13 and M5!

At 72×, the globular shows three tiers of brightness: a starlike nucleus burns at the heart of a dense vaporous shell, which is surrounded by a round outer envelope. High power brings out at least three "arms." The most prominent one, to the northeast, looks like a band of faint starlight drifting slowly away from the cluster, like renegades fleeing from the fugitive pack.

M76

Little Dumbbell Nebula
NGC 650 and NGC 651
Type: Planetary Nebula
Con: Perseus

RA: 01^{h}42.6^m
Dec: +51°35′
Mag: 10.4
Diam: 163″ × 107″
Dist: 3,900 light-years
Disc: Pierre Méchain, 1780

MESSIER: [Observed October 21, 1780] Nebula near the right foot of Andromeda, seen by M. Méchain on 5 September 1780, and which he described thusly: "This nebula does not contain any stars; it is small and faint." The following 21 October, M. Messier searched for it with his achromatic telescope and it seemed to him to consist of very faint stars with some nebulosity, and the slightest illumination of the micrometer crosshairs caused them to disappear. The position was determined from fourth-magnitude φ Andromedae.

NGC 650: Very bright, western part of double nebula.

NGC 651: Very bright, eastern part of double nebula.

SMALLER, FAINTER, AND LESS POPULAR THAN its cousin the Dumbbell Nebula (M27) in Vulpecula, the Little Dumbbell in Perseus is nonetheless a very dramatic planetary. Sometimes, unfortunately, the popularity of a Messier object seems to be based more on how bright or large it appears in the night sky than on how much detail it reveals through the telescope. But in the case of M76, its beauty lies not in its visual punch but in its wealth of subtle detail, which lures small-telescope users into a web of visual suggestions. I'd place it among the most surprising and mysterious objects in the Messier catalogue for viewing with backyard telescopes.

Certainly astronomers at the turn of the twentieth century must have been awed by its image in long-exposure photographs. Instead of a ghostly ring or disk of light encircling a central star – as is generally characteristic of planetaries – here was a 16th-magnitude central star in a rectangular bar of light made up of two prominent nebulous patches flanked by an irregular array of luminous knots and nebulous arcs.

When William Herschel swept it up in 1787, he saw not one but two nebulae in contact, so he decided to give each nebula a separate number in his catalogue, with NGC 651 being the northeastern component. In 1891,

M76

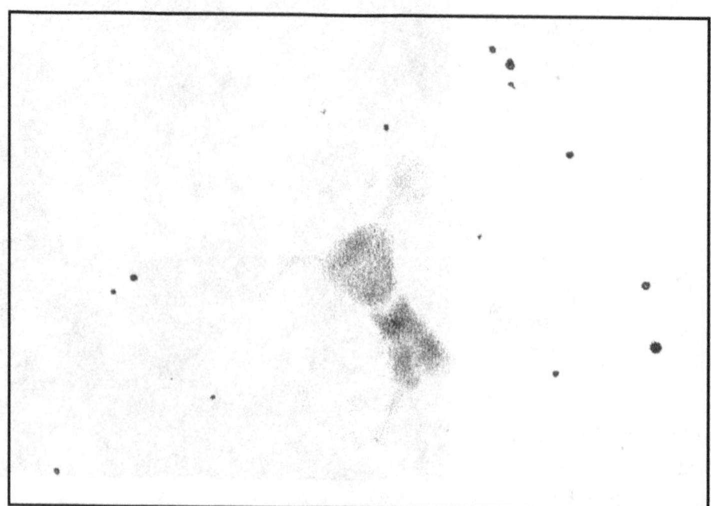

component of the pair appears bluer and brighter, suggesting it's the parent central star. In 1995, Howard Bond of the Space Telescope Science Institute discovered that the fainter secondary was itself a double. His results appeared in the *Bulletin of the American Astronomical Society* (vol. 27, p. 1401).

But in a 1998 paper in *Astronomy and Astrophysics* (vol. 335, p. 277), J. Koorneef and S. R. Pottasch of Groningen University in The Netherlands used the HST to study the central star and its two companions, finding a probable distance to the companions of between 16,000 and 20,000 light-years. "Using the spectral and photometric characteristics of the central star," they say, "it appears unlikely that the central star is physically related to the two visible companions."

Isaac Roberts suggested that M76's unusual character is the result of our seeing its broad ring of material edgewise. The object's complex appearance might, however, derive more from the way gases of varying densities near the central star illuminate and obscure the star's light. But today we know that Herschel's two nebulae are part of one ring-type planetary nebula whose central, barrell-shaped torus is seen nearly edge-on. The butterfly wings form the sides of expanding bubbles of gas ejected from the polar regions.

The distance to M76 is poorly known, though Hubble Space Telescope observations indicate that it is likely about 3,900 light-years. M76 is about five times more distant than M27, but they are about the same actual size (about 3 light-years across), so M76 only appears smaller. The highly obscured central star of M76 shows quite a high excitation, indicating a high temperature. In 1973, Kyle Cudworth of Lick Observatory discovered that the central star is a binary with a separation of about 1.4″. His results were reported in the *Publications of the Astronomical Society of the Pacific* (vol. 85, p. 401). The northeastern

Spitzer infrared observations of the central torus indicate that indeed it is a dusty barrel seen nearly edge-on, with two distinct emission structures. In a 2006 paper in *Astrophysical Journal* (vol. 650, p. 228), Toshiya Ueta of NASA Ames Research Center explains these two features as (1) a highly ionized region behind the ionization front and (2) low-temperature dust in the remnant asymptotic giant branch wind shell. Apparently, mass loss at the end of the central star's ASB stage ensued only in the equatorial direction, not in the polar direction.

Some catalogues severely underestimate the brightness of the Little Dumbbell, listing it as 12th magnitude, which is much too faint.

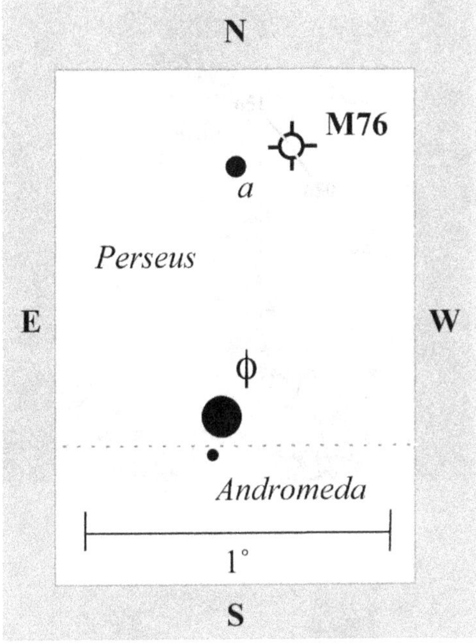

Indeed, as proof, M76 is at the limit of detectability in 7 × 35 binoculars, which is about 10th magnitude. At 23× in the 4-inch, its small, opalescent disk reposes in a rich field of faint stars, many of which are pairs, less than 1° north-northwest of yellowish Phi (ϕ) Persei. If you want to hop to it, move 50′ due north of Phi Persei to magnitude 6.5 Star *a*. M76 is about 12′ west-northwest of Star *a*. To avoid passing over it with low power, sweep slowly and use averted vision. With peripheral vision, the planetary seems to swell. This is not an illusion. Your eyes are picking up the faint light from the nebulous loops that connect to the four corners of the brighter rectangular bar.

At 72×, I found that the Little Dumbbell resembles the Crab Nebula (M1), probably because both objects display bright patches of nebulosity separated by a dark lane and bordered by misty frills. And these details are immediately apparent in M76. Despite its irregular appearance in photographs, the planetary looks symmetrical through small-aperture telescopes; there is an hourglass-shaped inner nebula, oriented northeast to southwest, with two semicircular arcs, like butterfly wings: one to the southeast, the other to the northwest – much like M27!

Interestingly, when I look at M76 with moderate power, the southeastern loop seems more pronounced, but at high power the northwestern one does. Although I can see the arcing extensions attached to each corner of the rectangle, the brightest sections (knots) within each loop are clearly separated from the hourglass. The hourglass itself is made up of bright knots, which could be faint stars superposed on the nebula. Lord Rosse also detected "subordinate nodules and streamers," which led him incorrectly, though understandably, to deduce that this was a spiral nebula. There appears to be a hint of brightening near the center of the rectangle, but do not confuse this isolated glow with the 16th-magnitude central star. As the drawing shows, M76 has a heart of darkness (light in the negative rendering), and the central star hides in the deep recesses of that bleak cavity.

M77

NGC 1068
Type: Spiral Galaxy (R)SA(rs)b
Con: Cetus

RA: $02^h42.7^m$
Dec: –00°01′
Mag: 8.9
SB: 13.2
Dim: 8.2′ × 7.3′
Dist: ~44 million light-years
Disc: Pierre Méchain, 1780

MESSIER: [Observed December 17, 1780] Cluster of faint stars, which contains nebulosity, in Cetus, and at the same parallel as the star δ, which is reported to be of third magnitude, but which M. Messier estimates to be only of fifth magnitude. M. Méchain saw this cluster on 29 October 1780 as a nebula.

NGC: Very bright, pretty large, irregularly round, suddenly brighter in the middle, some stars seen near the nucleus.

M77 IS THE PROTOTYPE OF A PECULIAR class of extragalactic objects known as Seyfert galaxies. These systems have very active nuclei, which are potent emitters of radio-wavelength energy and whose spectra show strong emission lines – characteristics also displayed by the distant quasistellar objects, or quasars. But quasars are infant systems billions of light-years away, whereas energetic M77 is a mere 44 million light-years distant. Apparently, gas clouds (some with a mass 10 million times the Sun's) are blasting away from the nucleus of M77 with velocities up to 360 miles per second and enough energy to power several million supernova explosions! M77 is the closest you will come to seeing a quasar in action.

M77 is as large as the Milky Way, measuring about 100,000 light-years across. It shows a significant discontinuity between the surface brightnesses of its inner and outer regions. The inner region consists of tightly wrapped, knotty spirals near the nucleus, with two main dust arms. The outer region consists of faint spirals that form an outer ring, while an intermediate range of arms lies in between that are actually prominent H II (star-forming) regions at the interface. Since the numerous short dust lanes in the outer spiral pattern do not have the coherence of the loosely wound nuclear spirals, M77 has been classified as a chaotic spiral (see the Hubble Space Telescope image).

At infrared wavelengths, a beautiful elliptical, starburst ring surrounds the nucleus

the two look like a fine double star. Look for a faint halo of light surrounding M77, whose contribution raises the galaxy's brightness to 9th magnitude. You will need moderate or high magnification to see the galaxy's peculiar arms, which long-exposure photographs reveal as tight and knotty close to the brilliant nucleus but fainter and less tightly wound farther out. Burnham notes that the knots of the inner arms can be glimpsed through a 4-inch telescope on the finest nights, and he is absolutely right! In fact, these knots are what caused earlier astronomers like the Herschels and Smyth to perceive M77 as a resolved star cluster. Lord Rosse noted it as a blue spiral. Can you detect any color?

After observing the galaxy on three different nights with the 4-inch at high power, I traced its inner spiral patterns with a high degree of confidence and accuracy. A tiny but strong bar slices across the nucleus in an east–west direction and quickly curves away into tight but graceful arms. In 1993, infrared images of M77 made at Kitt Peak National Observatory revealed a nebulous bar extending through the nucleus. The question is whether this infrared bar matches the bar seen visually.

about 6,500 light-years out. This has been deemed one of the best examples of a circumnuclear starburst coexisting with high-excitation gas ionized by the galaxy's active nucleus.

And, indeed, the nucleus of M77 is bright. You can see it shining like a 10th-magnitude star about 1° southeast of 4th-magnitude Delta (δ) Ceti, just west of a real 10th-magnitude star, so at low power, and at first glance,

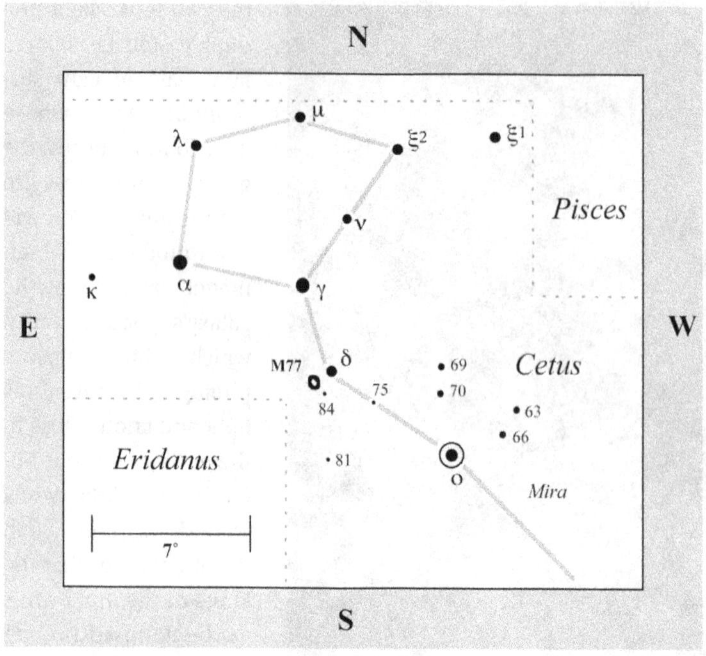

patches, so you have to find a way to piece the chaos together into a coherent picture. The trick is to use as much power as possible, to hyperventilate to heighten visual acuity, and to play with the focus and use averted vision until you can cause the features to blend. Can you resolve the faint knot to the east of the nucleus, in the innermost spiral?

Returning to Delta Ceti, look about 6° southwest, where you'll find the fascinating variable star Omicron (o) Ceti, also known as Mira. David Fabricius (1564–1617) is credited with the discovery of Mira in August 1596, though Korean and Chinese astronomers may have noticed this "guest star" several months before him. This winking star brightens into and fades out of naked-eye visibility like clockwork every 333 days. It fluctuates in brightness from a maximum of magnitude 3.4 (usually) to a minimum of 9.3. However, it has peaked as bright as magnitude 0.9. William Herschel made that historic observation in November 1779, when he found that Mira "excelled Alpha Arietis so far as almost to rival Aldebaran; and continued in that state a full month." The star varies in brightness because it has entered the pulsating, red giant stage of its life cycle. Mira has roughly the Sun's mass, so by studying its behavior, we can learn more about how our own star will act once it enters old age.

Most perplexing, at first, is the galaxy's northern arm, which appears to branch from a knot close to the nucleus. You really have to study this area long and hard before you can begin to distinguish the individual arms from the nucleus. The difficulty is that these arms, like those of M66, are broken into knots and

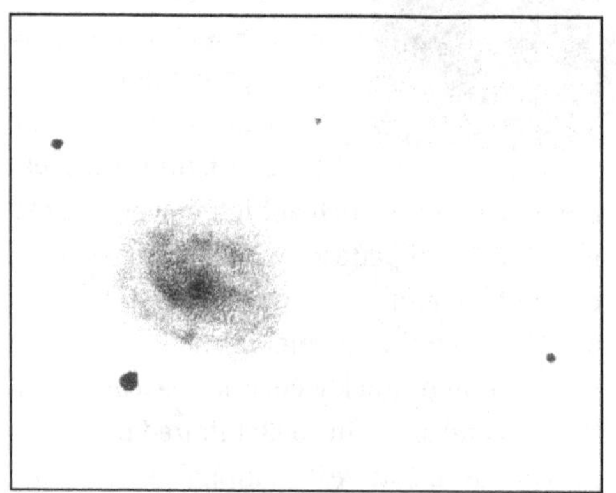

M78

NGC 2068
Type: Diffuse Nebula
Con: Orion

RA: $05^h46.7^m$
Dec: -00°03′
Mag: 8.0 (O'Meara)
Dim: 8′ × 6′
Dist: 1,630 light-years
Disc: Pierre Méchain, 1780

MESSIER: [Observed December 17, 1780] Cluster of stars, with a lot of nebulosity in Orion, and on the same parallel as δ in the belt, which was used to determine its position. The cluster crosses the meridian 3°41′ after the star and is 27′7″ higher in the sky. M. Méchain saw this cluster at the beginning of 1780, when he reported it as follows: "On the left-hand side of Orion, 2 to 3 minutes in diameter. Two fairly bright nuclei are visible, surrounded by nebulosity."

NGC: Bright, large wisp, gradually much brighter to a nucleus, three stars involved, mottled.

BEFORE BEGINNING THIS BOOK, I HAD looked at M78 only once. Some 20 years ago, Peter Collins, a visual discoverer of four novae, had turned the Harvard College Observatory's 9-inch Clark refractor to the nebula, and I took a peek. Once home, I looked it up in *Burnham's Celestial Handbook*, which supported my view of very little detail. Now, having spent many evenings with it using a much smaller telescope, I realize my misfortune in neglecting this wonderfully mysterious object (more on the mysteries later).

Located about 2 1/2° northeast of Zeta (ζ) Orionis, M78 is far enough away from the nebular madness surrounding Orion's belt and sword to be easily overlooked. In fact, M78 is the brightest of three specter-like glows in the region. Its companion nebulae, NGC 2067 and NGC 2071, will be described subsequently. All three shine by reflected light from hot, young B-type stars and are part of a greater complex of nebulosity sweeping through much of Orion. Most sources don't publish a magnitude for M78, even though it is visible with 7 × 35 binoculars. The reason is that there are very few good measurements of nebula magnitudes. My magnitude estimate of 8.0 was made with binoculars. If you have good to excellent eyesight, challenge yourself by trying to detect M78 without aid.

A close pair of roughly 10th-magnitude stars burn at the heart of M78. Even at low power, they look like bloodless eyes peering back at you through a frosty window.

M78

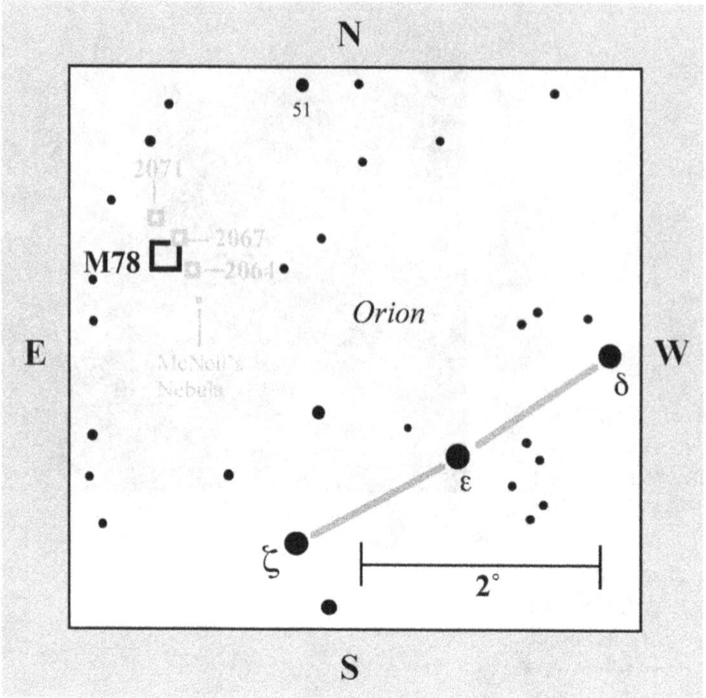

With high magnification, focus your gaze immediately west of the two 10th-magnitude nuclei. At first, you will see what appears to be a star, then two stars. But are these really stars? Perhaps they are knots of nebulosity. There is yet another very faint star or knot immediately to the southwest of the southern component. No image I have seen shows the inner details of M78, so it has been difficult to confirm these sightings. Perhaps someone with a CCD camera and an image-processing program with unsharp masking can bring out these details.

Smyth and Webb described M78 as a wispy nebula, and Lord Rosse saw spiral structure. Today its appearance is generally compared to a comet's. Indeed, the nebula does resemble a comet with a split nucleus, parabolic hood, and a tail that fans from south to east. (Anyone who has seen the sunward fan of Periodic Comet Encke will appreciate how closely M78 resembles that feature.)

But there are "comets" within this "comet." Most striking is the long tail of material flowing southeast of the southern nucleus. Look at this tail under magnification and you will see it has two parts: a short, stubby fan associated with one of the 10th-magnitude stars and a long, graceful tail associated with a magnitude 13.5 star. The nucleus's northern component has a roughly 14th-magnitude companion to the east, around which diffuse material streams out to form a "flame" of nebulosity.

As I alluded to earlier, M78 is not without certain mysteries. When I first looked at it through the eyepiece, I expected, based on several written accounts, to see nebulosity surrounding two bright stars. But there were three obvious stars (as noted in the NGC description): the two dominant 10th-magnitude stars and a fainter one. Then I read in Luginbuhl and Skiff's book *Observing Handbook and Catalogue of Deep-Sky Objects* that the third star shines at magnitude 13.5. That seemed to match my visual impression, but when I returned to M78 a month later, the third star appeared fainter than 13.5, perhaps by as much as a magnitude. In a 1975 paper in *Astrophysical Journal* (vol. 196, p. 489), K. Strom et al. note that the star's measured magnitude was 13.1. Could it be a variable star?

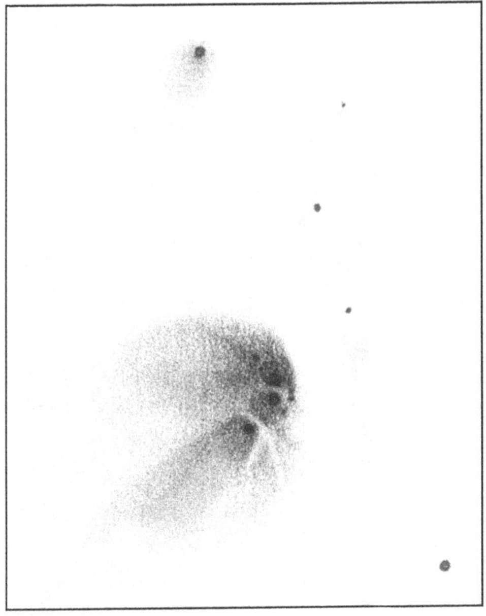

Here's something else to investigate. The nebula has a very faint southern extension, superposed on which appears to be a vein of dark nebulosity forming what I call the "skeleton's wrist." Actually, the wrist itself is not difficult to make out; it is the lane of darkness that separates the two southeast "comet tails." The wrist, however, then branches into two extremely tenuous dark wisps: one cuts sharply to the south and the other, more difficult, extension goes southwest – the skeleton's thumb and forefinger. I saw these other features only once in several sessions. Can you see them?

While you are in the neighborhood of M78, check out these interesting NGC objects. Refer to the finder chart for their locations. NGC 2071 to the northeast is an easy nebulosity, though it has a lower surface brightness than M78. At first glance, the nebula appears round, but it is actually extended to the south and slightly east. See if you can detect it in binoculars. More intriguing is NGC 2067, which I noticed only by circumstance. On one very gusty night, the Genesis was jittering slightly, setting the field in motion enough for my peripheral vision to catch sight of sharp-edged NGC 2067 to the northwest. There is yet another nebulosity – the minuscule (1.5′ × 1′) NGC 2064 to the southwest, which I have only suspected. Thus it is not included in the drawing.

Now for one of the most recent fantastic and unique amateur discoveries in recent times – a deep-sky discovery made just a few arcminutes southwest of M78. It's called McNeil's nebula and is a variable phenomenon, which only proves the words of prospectors, as told by legendary filmmaker Robert Flaherty (1854–1951): "Go out looking for one thing, and that's all you'll ever find."

The celestial prospector who struck it rich was Jay McNeil of West Paducah, Kentucky. On the night of January 23, 2004, he imaged a 2°-wide region of sky centered on M78 from his backyard – using a CCD mounted on nothing more than a 3-inch refractor. On one of the images, he spotted a diminutive 1′ glow shining at about 15th magnitude in M78 (see the accompanying European Southern Observatory image).

What made McNeil's observation so special is that it wasn't so much the image that led to the discovery but his 20 years of experience behind the eyepiece. When McNeil first scanned the discovery image, he immediately realized that something was amiss. "The new nebula was much brighter than any of the other objects I had observed visually around M78," he said, "and I have scrutinized that region on many a clear night."

Unlike the brighter regions of M78, however, McNeil's nebula flares into view only when its illuminating star (V1647 Orionis, possibly an FU Orionis-type variable)

sudden infall of matter onto the surface of the star from an orbiting disk of gas.

More recent x-ray observations have revealed that not only did the object flare to prominence, and thereby illuminate the surrounding nebula, but it is a *protostar* rotating 30 times faster than the Sun (about once a day) – nearly fast enough to tear itself apart. But what we actually appear to be witnessing is just the opposite: a protostar about five times the size of the Sun in the process of pulling itself together. The surrounding disk of matter that feeds the star – and occasionally causes the star and its surrounding nebula to burst forth from obscurity to prominence – is plentiful enough to nourish the newborn for at least another million years, when it might "turn on" as a star. The initial outburst died down in early 2006, but then V1647 Orionis erupted again in 2008, and remained bright well into 2012. But as Joe Philip Ninan and his colleagues reported in the July 2012 *Astronomer's Telegram* (#4237), V1647 Orionis has begun a slow dimming in brightness from the long-term optical and near-infrared photometric observations.

experiences an outburst. The period of the outbursts is unknown, but it is suspected to be very long. A search of past images revealed the nebula erupting in 1966, but none but McNeil had noticed it before.

Professional astronomers have turned much attention to this new nebula. Chandra x-ray observations, for instance, have shown that the young, illuminating star (buried deep in the nebula) has a coincident x-ray source that brightened fiftyfold in observations obtained before and just after the optical outburst. The x-ray data are strong evidence that the probable cause of the outburst was the

M79

NGC 1904
Type: Globular Cluster
Con: Lepus

RA: 05^{h}24.2^m
Dec: -24°31.5″
Mag: 7.7
Diam: 10′
SB: 12.6
Dist: ~42,100 light-years
Disc: Pierre Méchain, 1780

MESSIER: [Observed December 17, 1780] Nebula without a star lying below Lepus, and on the same parallel as a sixth-magnitude star. Seen by M. Méchain on 26 October 1780. M. Messier looked for it the following 17 December. This is a fine nebula; the center is bright, the nebulosity slightly diffuse. Its position was determined from fourth-magnitude ε Leporis.

NGC: Globular cluster, pretty large, extremely rich, extremely compressed, well resolved into stars.

THE NEXT TIME YOU ARE UNDER A VERY dark sky in a southerly locale, try the following. Using only your binoculars, visualize a line from Alpha (α) to Beta (β) Leporis, and extend it toward the southwest about 4 1/2° (about the width of two fingers held at arm's length). There you will see a magnitude 5.5 star (Herschel 3752). If you are properly dark-adapted, you might see a slightly dimmer "star" to the east - globular cluster M79. Try also to see it with the unaided eye; you will need to spend about 15 minutes or more trying. Do try if you're under a dark sky, because I certainly surprised myself by seeing it!

Like many globulars, M79 has an integrated spectral type of F5 and an average metallicity (each of its stars, on average, having about 1/40 the amount of metal per unit hydrogen as the Sun). Aside from its fair distance from the Sun, M79 also lies a significant distance from the Galactic center in the halo (~61,000 light-years distant). And one of its significant features is a tidal tail that points in the direction of the Galactic center. In 2010, globular cluster researchers J. A. Carballo-Bello and D. Martinez-Delgado of the Instituto de Astrofísica de Canarias (IAC) obtained photometry with the 2.2-meter European Southern Observatory telescope that reveals a distinct main sequence of a metal-poor stellar population, consistent with the presence of a stellar system of very low surface brightness that might belong to an unknown tidal stream in the Milky Way.

But some researchers have proposed that M79 belongs to the Canis Major Dwarf

M79

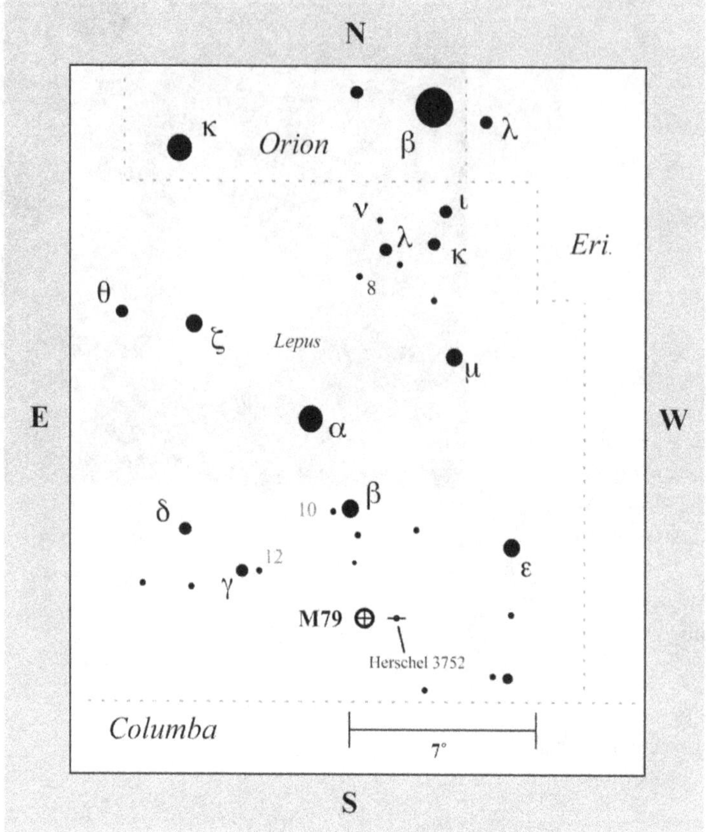

undergone a core collapse.

To find this fascinating globular, use the chart at the back of this book and the one here to find its location. Try first using binoculars. It shouldn't be that difficult under a dark sky, because its light compresses into a tiny 6′ disk. You'll be looking for a roughly 8th-magnitude fuzzy star bracketed nicely by two 9th-magnitude stars.

Even in the 4-inch at low power, M79 is surprisingly small and tight, so tight that I once swept over it, mistaking the globular for a star. It is hard to believe that this seemingly insignificant patch of light spans

Galaxy – an irregular system discovered in 2003 and thought to be the closest galaxy to our Milky Way. At 25,000 light-years distant, it appears to be having a deathly close encounter with the Milky Way, which it may not survive. If true, it will lend support to the theory that galaxies evolve over time by consuming their smaller neighbors.

M79 has a very large population of possible blue stragglers at very large distances from the cluster center. This goes against the Hubble Space Telescope's finding in 2007 that there are 39 bright blue stragglers in the cluster's core, which appear highly segregated. The vast majority of the cluster's heavy stars (binaries) appear to have already sunk to the core. Thus, the cluster system has already

some 120 light-years of space. If you consider how rapidly the globular is approaching us – at about 130 miles per second – it is easy to imagine it trying to rush into view to vie for our attention.

At low power, the cluster contains a smattering of fairly bright members against a blazing core, which was first described by both Smyth and Webb in the nineteenth century. The cluster's boxy sides are obvious in both the inner and outer regions. The outermost reaches can be resolved at moderate power, even though the individual stars shine between 13th and 14th magnitude at their brightest.

The tightly wound center is a demon for small apertures; full resolution occurs at magnitude 16.2, but you can see star clumps. Any

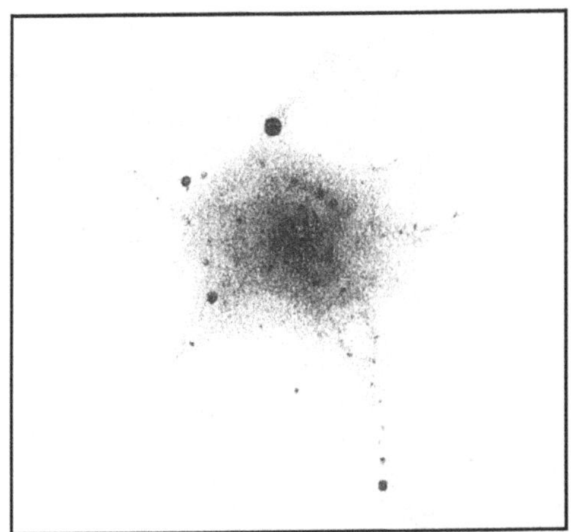

proper attempt at probing its depths requires high magnification, keen averted vision, and lots of patience.

If you spend enough time studying the nuclear region with high power, you might notice an asymmetry in that the bright western side fades gradually and subtly toward the east. Part of the asymmetry results from the presence of a nice string of glittering stars on the northwest side of the nucleus. Furthermore, there's a gap between them, resulting in fewer stars on that side of the cluster. Two smaller stumps of starlight to the west and southwest add further lopsidedness to the interior.

The cluster also shows five "arms" extending in various directions, which take on a distinctive starfish pattern with a 12th-magnitude star punctuating its northern fringe. I recall, as a teenager, seeing this shape through my 4 1/2-inch reflector and lightheartedly thinking how much it resembled a fly in a melting ice cube. One frigid evening many years later, as Lepus hopped above some distant fir trees at Oak Ridge Station in Harvard, Massachusetts, I turned a 16-inch reflector onto the globular and resolved it right to the core.

At first, the starfish pattern may look fuzzy, but with increased magnification you may be able to resolve each arm into faint streams of stars. Most intriguing is the southwestern arm, which seems to consist of a series of thinning stellar arcs that drip toward a bright star (remember the melting ice cube?).

As Mario's image shows, all these arms develop into crazy patterns of extremely faint outlying stars. I have yet to see these wild swirling patterns in the 4-inch. Perhaps one way to see them would be to use very high magnification and focus on each arm individually. I expect each search would consume hours. But if you feel up to the task, you could piece together a drawing that would rival the best CCD images.

M80

NGC 6093
Type: Globular Cluster
Con: Scorpius

RA: $16^h17.0^m$
Dec: –22°58.5′
Mag: 7.3
Diam: 10′
SB: 12.3
Dist: 32,600 light-years
Disc: Messier, 1781

MESSIER: [Observed January 4, 1781] Nebula without a star in Scorpius, between the stars *g* [now ρ Ophiuchi] and δ. It was compared with *g* to determine its position. This nebula is circular; the center is bright and resembles the nucleus of a small comet, surrounded by nebulosity. M. Méchain saw it on 27 January 1781.

NGC: Remarkable globular cluster, very bright, large, very much brighter in the middle (variable star), readily resolved, contains stars of the 14th magnitude and fainter.

IN AN ARTICLE TITLED "AN OPENING in the Heavens" in the 1785 edition of *Philosophical Transactions*, William Herschel shared his impressions of a 4°-wide swath of dark nebulosity above the Scorpion and its fanciful relationship with M80: "It is remarkable that [M80], which is one of the richest and most compressed clusters of small stars I remember to have seen, is on the western border of it and would almost authorise a suspicion that the stars of which it is composed, were collected from that place and had left the vacancy."

M80 in fact lies in a very busy region on the western bank of the Milky Way band. It is conveniently placed nearly 5° northwest of the ruddy 1st-magnitude star Antares, or Alpha (α) Scorpii – or about 2° almost due west of 4th-magnitude Rho (ρ) Ophiuchi, which might be the best way to find it. Just center Rho in your telescope at low power and make a slow and gentle sweep to the west.

Like M79, M80 is a tiny globular cluster with an extremely dense core and a spherical halo. The owner of a rich-field telescope will be truly rewarded by the sight of M80's thick pack of 7th-magnitude starlight glistening from amid the galactic pandemonium.

I believe that if it weren't a member of the Messier list, this fascinating object would not get the attention it deserves, given the powerful pull of brilliant M4 nearby. On the

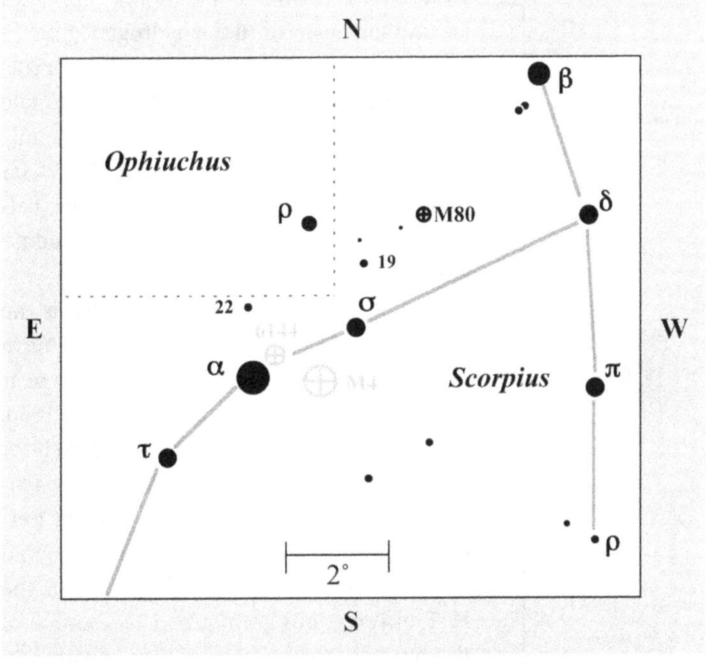

surface, M80 seems average enough; it has an average metallicity for a globular cluster and a typical integrated spectral type of F6. We see it hobbling toward us at ~5 miles per second. But of the 150 or so globular star clusters known, M80 is one of the densest. As the accompanying Hubble Space Telescope image shows, the cluster is jam-packed with hundreds of thousands of stars, all held together by their mutual gravitational attraction across a generous 120 light-years of space. Every star visible in this image is either more highly evolved than, or in a few rare cases more massive than, the Sun. Globular star clusters especially harbor red giant stars near the end of their lives. But ultraviolet images with the HST have also shown a large population of "blue stragglers" in the core of M80.

These stars appear to be unusually young and more massive than the other stars in a globular cluster. However, stellar collisions can occur in

dense stellar regions like the core of M80 and, in some cases, the collisions can result in the merger of two stars. This produces an unusually massive single star, which mimics a normal, young star (a blue straggler). M80 was previously not known to contain blue stragglers but is now known to contain more than twice as many as any other globular cluster surveyed with the HST. Based on the number of blue stragglers, the stellar collision rate in the core of M80 appears to be exceptionally high.

But if such close encounters were that common, astronomers would also expect to find a significant nova population as well. Novae erupt when a close companion star transfers fresh hydrogen fuel to a burned-out white dwarf. The infall of matter accumulates around the dwarf until the hydrogen ignites a thermonuclear explosion on the dwarf's surface, which we see as a nova. Yet the Hubble ultraviolet observations have revealed only three novalike close binary stars in M80 – far fewer than what the theoretical stellar collision rate is expected to be in such a dense environment.

One of the novae the HST detected is the remnant outburst of a 7th-magnitude nova (known as T Scorpii). This star flared to near naked-eye visibility from M80's core in 1860. The new star outshone the globular itself for a few days before it rapidly faded into obscurity. T Scorpii was one of only two (or perhaps three) novae that have been observed in a globular cluster. As reported in 1995, the HST observations pinpointed the source of the burst to a very blue candidate within 1″ of the astrometric position reported in 1860. Astronomers now believe, beyond any doubt, that this star is a cluster member. It was the first recovery of a quiescent old nova in a globular cluster core.

Through the 3-inch at moderate and high powers, M80 appears as a tiny wad of yellow light, like a swab of wet cotton, irregular at the edges. The core is tight but has some traces of glittering starlight around it. At 23× in the 4-inch, the core seems to glow with a softer yellow light that gradually fades to a silver sheen in the outer envelope.

Because the cluster appears so tiny – just 9′ in apparent diameter – I almost immediately switch to high power, which starts to resolve some cluster members, even though the brightest stars shine between 13th and 14th magnitude. The outer hood seems to scintillate, like flecks of mica. The high density of the cluster makes it difficult to resolve individual stars; high power is a must. At 130×, the core appears fractured, broken into at least three bright shards separated by delicate dark rifts.

When I'm just relaxing and enjoying the view, I like to imagine that M80's fractured core is the tragic result of that energetic blast – a crystalline sphere shattered by powerful shock waves. Now look closely at the nucleus. Does the brightness seem to favor the northeast, where one broad river of stars merges with another? The challenge is to see the dark bay between the rivers.

M81

NGC 3031
Type: Spiral Galaxy (SA(s)ab)
Con: Ursa Major

RA: $09^h55.6^m$
Dec: +69°04′
Mag: 6.9
Dim: 24.0′ × 13.0′
SB: 13.0
Dist: 12 million light-years
Disc: Johann Elert Bode, 1774

MESSIER: [Observed February 9, 1781] Nebula close to the ear of Ursa Major, on the parallel of star *d*, which is of fourth to fifth magnitude. Its position was determined from this star. This nebula is slightly oval, the center is light, and it is easily visible with a simple three-and-a-half-foot refractor. It was discovered from Berlin by M. Bode on 31 December 1774, and by M. Méchain during August 1779.

NGC: Remarkable! Extremely bright, extremely large, extended toward position angle 156°, gradually then suddenly very much brighter in the middle to a bright nucleus.

M82

NGC 3034
Type: Irregular Galaxy (IO)
Con: Ursa Major

RA: $09^h55.8^m$
Dec: +69°41′
Mag: 8.4
Dim: 12.0′ × 5.6′
SB: 12.8
Dist: 12 million light-years
Disc: Johann Elert Bode, 1774

MESSIER: [Observed February 9, 1781] Nebula without a star, near the previous one [M81]. Both appear in the same telescopic field, the latter is less conspicuous than the former. Its light is faint and elongated. At its eastern end there is a telescopic star. Seen from Berlin by M. Bode on 31 December 1774, and by M. Méchain during August 1779.

NGC: Very bright, very large, very much extended (ray).

UNQUESTIONABLY, M81 AND M82 ARE the most popular close pair of galaxies in the heavens, and they hold a special place in my heart. Whenever I see them, my mind conjures up pleasing memories of two warm summer nights spent under the stars with close friends. One evening, about a decade ago, nova discoverer Peter Collins and I were strolling down a dark country road in Concord, Massachusetts – an old haunt of literary legends Thoreau, Emerson, and Alcott. As we approached a dark forest of oak and elm, we noticed the Great Bear just lurching over the treetops. A minute of quiet contemplation passed before Peter raised his binoculars, which he usually had handy, and trained them on M81 and M82. I had always thought them to be only telescopic objects. Yet Peter seemed to be giving them an appreciative stare. He then handed me the glasses. Sure enough, there they were, glowing weakly like two spirits fading into the distance. He then asked if I could see them without the binoculars. Impossible, I thought, so I didn't give it much effort.

That night I did not see the galaxies without aid. Here in Hawaii, after years of trying from other locations, I have finally succeeded with M81. I estimated the galaxy's brightness with the naked eye to be magnitude 6.8. Now I have learned that Brent Archinal independently sighted this galaxy without optical aid, while on vacation under dark western skies at high altitude (12,000 feet), as did Becky

Ramotowski from New Mexico, also at high altitude. I hope that these observations inspire you to try to broaden your naked-eye horizon by seeking out this Milky Way neighbor.

Some catalogues underestimate M81's brightness by a magnitude or more. And as for its distance, M81 and the equally distant galaxy M83 in Hydra are the farthest objects in nature detectable with the naked eye. At a distance of 12 million light-years, M81 is about four times as far away as both the Triangulum Galaxy (M33) and the Andromeda Galaxy (M31).

M81 and M82 lie about 2° east of 24 Ursae Majoris – a magnitude 4.5 star about a fist width (~10°) northwest of Alpha (α) Ursae Majoris – which marks the Bear's ears. If you live under a dark sky, use binoculars first to locate the galaxy pair. The telescopic view is splendid: M81's egg-shaped, 7th-magnitude oval, almost the apparent size of the full moon, sailing past the smaller, fainter (magnitude 8.4), cigar-shaped ellipse of M82. These two galaxies, separated by a mere 38′, voyaged past one another about 300 million years ago with ominous consequences; now the galaxies are moving apart.

You can see visible evidence of this through large telescopes. The highly disheveled appearance of M82 stands in marked contrast with its properly groomed spiral neighbor. Here is an extragalactic radical, with spiked "hairs" bristling off a cigar-shaped body tattooed with dark matter. A very unusual galaxy, M82 appears in red-sensitive photographs to have a midsection that is bursting at the seams. Long filaments stream out at right angles from the galaxy's central region. The filaments can be traced out about 34,000 light-years; the galaxy itself is only 42,000 light-years long!

Although M82 was once believed to be experiencing a violent explosion, astronomers now believe its central region is the site of intense starburst activity – containing perhaps 40 or so supernovae in the early stages of expansion. We see this starburst activity nearly edge-on. The first starburst episodes probably began 40 million years ago, after M81, which is 10 times more massive than M82, ploughed past M82 like a speeding truck whizzing past a bicyclist.

During that encounter, a strong wake of gravity would have smashed into M82, causing interstellar clouds to collapse and trigger new star formation. M82's disrupted appearance probably results from interstellar material being gravitationally dragged away from its disk during that encounter. Newly formed supernovae, which usually flare up during episodes of star formation, could have also blasted material out of the galaxy's plane. As M81 sailed farther away from M82, the jostled interstellar medium of M82 would have gradually fallen back onto its parent galaxy, triggering yet more starburst episodes, which we may still be witnessing today.

Throughout the galaxy's center, young stars are being born 10 times faster than they are inside our entire Milky Way Galaxy. The Hubble Space Telescope image here reveals the resulting huge concentration of young stars carved into the gas and dust at the galaxy's center. The fierce galactic superwind generated from these stars compresses enough gas to make millions more stars.

Only the HST has resolved M82's young stars, congregated by the dozens to make bright "starburst clumps," in the galaxy's central region. Most of the other pale, fuzzy white "stars" sprinkled around the body of M82 are actually individual star clusters about 20

light-years across and contain up to a million stars.

The HST also recently discovered over 100 freshly formed (young) globular clusters, which probably formed during an encounter with M81. Owing to the encounter, M82 also displays two extended bright structures along the major axis and a fan-shaped plume along the minor axis to the southeast. Diffuse x-ray emission has been detected out to a distance of ~23,000 light-years below the plane of the galaxy toward the south, while a curious detached and bright arc of x-ray emission appears to the east of this southern flow. This latter feature is rather mysterious, as it does not appear associated with any of the tidal streamers around M82, though it could be associated with a superwind driven by supernova explosions in the starburst.

The starburst activity seems concentrated in the central 500 parsecs of M82, with a star-formation rate of ~10 solar masses per year. Astronomers expect the starburst to subside, probably in a few tens of millions of years.

M81, on the other hand, is a spiral galaxy similar to the Andromeda Galaxy (M31) or Milky Way but with thinner, more regular arms. Dust lanes thread through the galaxy's amorphous bulge and close to the nucleus. The bulge gives way to something that resembles a grand-design spiral with two arms, but these consist of spiral fragments that branch into many secondary filaments. The dust is generally on the inside of the fragments. These arms begin to resolve out at a blue magnitude of 18, making it capable of detection with amateur CCDs. Normal novae, Cepheid variables, and planetary nebulae have also been identified in the resolved stellar population.

M81's Seyfert/LINER nucleus is one of the best candidates for a low-luminosity active galactic nucleus, which harbors a black hole of 70 million solar masses. The black hole is about 15 times the mass of the Milky Way's black hole.

The Hubble Space Telescope image here, taken from data between 2004 and 2006, is the sharpest image ever taken of M81. It shows the galaxy tilted at an oblique angle to our line of sight, giving a "bird's-eye view" of the spiral structure. The spiral arms are comprised primarily of young, hot stars that formed in the past few million years. They also host a population of stars formed in an episode of star formation that started about 600 million years ago. As with many galaxies, the stellar population of M81's bulge contains much older, redder stars. M81 may be undergoing a

surge of star formation along the spiral arms as a result of its close encounter with M82 about 300 million years ago.

To find this extragalactic pair, use the chart at the back of the book to first locate 24 Ursae Majoris. If you live under light-polluted skies, you may need binoculars to identify 24 Ursae Majoris. It also forms the northeastern apex of a roughly 3°-wide acute triangle with the similarly bright stars Rho (ρ) and Omicron (o) Ursae Majoris; actually, this marks the position of Omicron[1] and Omicron[2] Ursae Majoris. M81 and M82 are only about 2° east-southeast of 24 Ursae Majoris. You can also starhop to the galaxies using the more detailed chart here.

From 24 Ursae Majoris, move 25′ southeast to magnitude 7.5 Star *a*. Next, move about 35′ farther to the southeast to 6th-magnitude Star *b*. Bright M81 lies about 1 1/4° east-southeast of Star *b*. M82 is only about 35′ due north of M81.

At 23× in the 4-inch, M81 looks enormously larger and more robust than M82. Immediately noticeable is M81's needle-sharp nucleus, which shines with a pale yellow light. Two 11th-magnitude stars burn just south of the core and can easily be mistaken for supernovae. Ironically, in 1993, a supernova in M81 blazed to prominence just west of these stars. Low power also reveals a mysterious "bar" of light crossing the nucleus, but this feature is probably nothing more than an enhancement of the spiral arms seen obliquely along the galaxy's major axis. Indeed, with a prolonged look, the bar seems to point toward M82, which can force you to

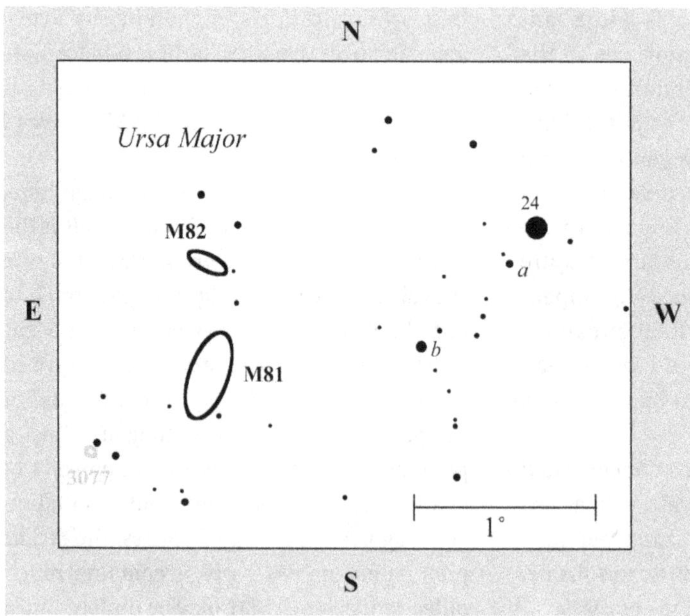

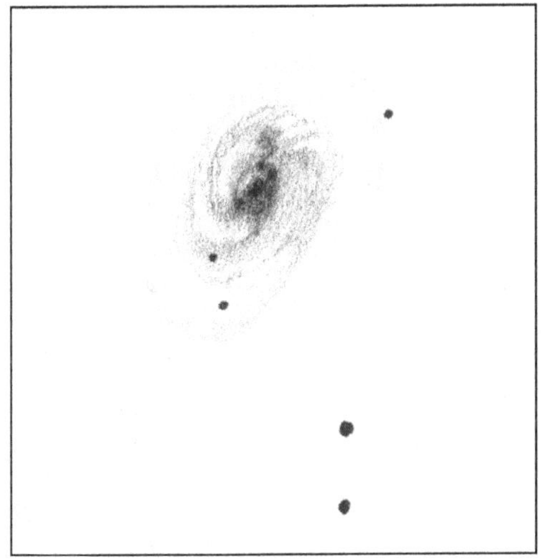

look at that neighbor galaxy. When I do that, my averted vision suddenly picks up swooping spiral arms on either side of M81! This is the magic of peripheral vision.

Something odd happens when I change to 72×. M82 increases in grandeur, while M81 loses some of its luster (because its faint outer arms have been overmagnified). The biggest surprise comes at 130×, when M81's core is transformed into a misty spring of light caressed by dark vapors, especially to the southeast, where a prominent lash of darkness abuts this region. Is this a paler version of M64's "black eye"? The bright bar slicing through the nucleus is now contained in a tiny inner ring of nebulosity, the southwestern half of which appears brighter than the northeastern half.

Delicate wisps of spiral arms surround the core, and together they look like a still photograph of a rotating lawn sprinkler. The most difficult detail to coax out (but the most rewarding once you see it) is the feathered texture of the spiral arm that lies midway between the outer, northeast arm and the burning core. This texture shows best with moderate power. Thus, to pick out the finest details of M81, you really have to study the outer arms at low power, the core at high power, and the middle arms at moderate magnification.

At moderate power in the 4-inch, M82 is a stunning sight. It reminds me of a ghostly starship, cracked and floating through an interstellar graveyard. As Luginbuhl and Skiff note, the western half of the galaxy is distinctly brighter than the eastern half, though the brightest segment of the fragmented hull lies just east of a central darkening. The far eastern half fades ever so gradually, and it looks as if the galaxy is slowly vanishing before your eyes, being consumed by the universe. Better yet, you can imagine this ghost ship snagged on a reef of stars – the brighter western half of the ship's hull sticking above water and the fainter part submerged in the shallows. When I really concentrate on the faint eastern extension, the brighter western half flares to prominence in my peripheral vision. It is really a very spooky galaxy.

As the drawing shows, M82 is highly complex and irregular. It takes time and patience to ferret out the details. Most impressive are the energetic-looking bursts of starlight running lengthwise through most of the galaxy. D'Arrest thought M82 scintillated "as if with innumerable brilliant points." At moderate power, an obvious elliptical halo surrounds the galaxy's brightest parts. The core is very angular – a sight created by wedge-like lanes of dust. Dark matter prevails on opposing ends of the galaxy. But look closely at the galaxy's extreme western end. Do you see a notch and a kink of stars punctuated with a stellar period? Now, let your eye relax and use keen averted vision. Do you see the very faint "explosions" emanating from the center of the galaxy? Try with high power.

At low power, M81 and M82 share the field of view with two companions, one large

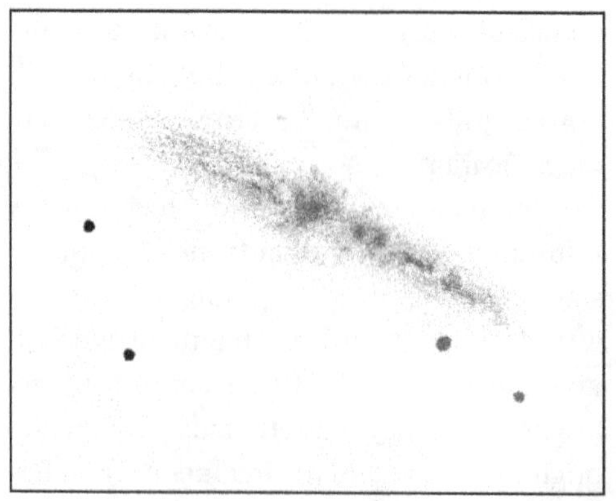

and one small. The large one, NGC 3077, is a 10th-magnitude elliptical galaxy 45′ to the southeast of M81 and sports a hurricane-like center. About 1 1/2° southwest of M81 is smaller NGC 2976, a 10th-magnitude spiral seen nearly edge-on. Note the little curl on the side away from M81.

In the summer of 1989, I was invited to look at Comet Okazaki–Levy–Rudenko through the 18-inch Clark refractor at Amherst College. The comet floated past M81 and M82 like a specter in a long white gown, and I felt like I was transported into Wilkie Collins's novel *The Woman in White,* just when Walter Hartright first encountered the specter: "I was far too seriously startled by the suddenness with which this extraordinary apparition stood before me, in the dead of night and in that lonely place, to ask what she wanted."

On March 26, 1992 UT, F. Garcia of Lugo, Spain, discovered a supernova (SN 1993J) in M81; the 10th-magnitude transient object appeared 5′ southwest of the nucleus. Two confirmed supernovae have occurred in M82: SN 2004am and SN 2008iz; a possible supernova may have been detected in 1986 (SN 1986D), though it may be an H II region.

M83

Southern Pinwheel
NGC 5236
Type: Mixed Spiral Galaxy (SAB(s)c)
Con: Hydra

RA: $13^h37.0^m$
Dec: -29°52′
Mag: 7.6
SB: 13.2
Dim: 15.5′ × 13.0′
Dist: 15 million light-years
Disc: Nicolas-Louis de Lacaille, 1752

MESSIER: [Observed February 17, 1781] Nebula without a star, close to the head of Centaurus. It appears as a faint, even light, but is so difficult to see with the telescope that the slightest illumination of the micrometer's crosshairs causes it to disappear. It requires considerable concentration to be seen at all. It forms a triangle with two stars estimated to be of sixth and seventh magnitude. Its position has been determined from the stars *i*, *k*, and *h*, in the head of Centaurus. M. de la Caille had already detected this nebula. See the end of this catalogue.

NGC: A very remarkable object. William and John Herschel found it very bright, very large, extended toward position angle 55°, especially bright in the middle to a nucleus. Seen as a three-branched spiral by Lassell.

WITHOUT QUESTION, M83 IS ONE OF MY favorite galaxies in the Messier list. It's bright, close, and a mesmerizing sight through the telescope. Its details seem unending, owing to their subtle complexity. In other words, though bright and impressive, its details pose a challenge, which should inspire you on many nights of observing this extragalactic treasure.

When Nicolas-Louis de Lacaille discovered this object from the Southern Hemisphere in 1752 – through a telescope that served no better than a pair of binoculars – he saw it merely as a "small, shapeless nebula." Because of its low declination, Messier said that from Paris he could see it only with the greatest concentration through his larger telescopes. Not until William and John Herschel observed

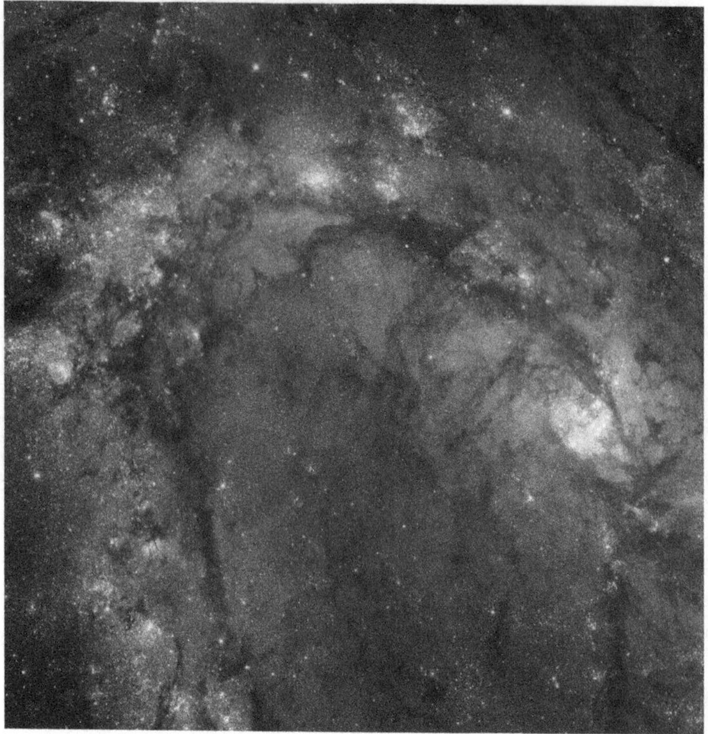

across the entire face of the central lens. M83 is a complex grand-design spiral with multiple arms: two bright ones and a multitude of low-surface-brightness fragments. Both the galaxy's nucleus and its arms are presently undergoing a burst of star formation, possibly triggered by tidal interactions with one of its traveling companions.

In 2009, the Space Telescope Science Institute released a stunning Hubble Space Telescope image of the ongoing starbursts along one of the major spiral arms extending northeastward from the nucleus (the small, bright region at right center), before looping southward. The new stars (only 1 million to 10 million years old) have formed largely in clusters along the dark dust lanes. As these newborn stars shed their natal dust wraps, bubbles of glowing hydrogen gas blossom forth, driven by fierce stellar winds from the young, hot cluster stars within.

As for the starbursts at the core, they are most likely caused by the thick bar, which funnels material to the galaxy's center along an arc near the core. There, the HST found the remains of about 60 supernovae, five times more than were known previously in this region.

In a 2012 paper in *Astrophysical Journal* (vol. 752, p. 47), J. Piqueras López (INTA-CSIC, Spain) and colleagues note that M83 is "one of the most studied galaxies given its

it with superior instruments (John from the Southern Hemisphere) do we see anyone referring to M83 as "a remarkable object." William Lassell (1799–1880), an amateur astronomer and brewer, first noticed it as a "3-branched spiral."

M83 belongs to a loose gaggle of galaxies known as the Centaurus A/M83 Group, which itself belongs to the Virgo Supercluster. In true physical extent, M83 is as large as our own Milky Way (~100,000 light-years across). Unseen, however, is an extensive gas disk component that contains 80 percent of its mass outside the visible disk.

This gorgeous, nearly face-on mixed spiral is one of the closest of its kind with a thick and prominent bar. Significant dust lanes emerge from the nucleus, line the bar, and then follow the inside of the two bright arms; other weaker dust patches spread in spiral patterns

proximity, orientation, and particular complexity. Nonetheless, many aspects of the central regions remain controversial." They found a complex situation where the gas and stellar kinematics are totally unrelated. "Supernova explosions play an important role in shaping the gas kinematics," they say, "dominated by shocks and inflows at scales of tens of parsecs that make them unsuitable to derive general dynamical properties. We propose that the location of the nucleus of M83 is unlikely to be related to the off-center 'optical nucleus.'" The researchers conclude that the optical nucleus is a gravitationally bound massive star cluster with a mass of about 11 million solar masses, formed by a past starburst.

That same year, Roberto Soria of Curtin University and his colleagues discovered that a previously unknown x-ray source in the nuclear region entered an ultraluminous state between August 2009 and December 2010. It was first seen with Chandra on December 23, 2010, and remained ultraluminous through December 2011. The luminosity and spectral properties are consistent with accretion powered by a black hole with a mass of about 40–100 solar masses, the researchers say, adding: "These optical observations suggest that the donor star is a low-mass star undergoing Roche lobe overflow, and that the blue optical emission seen during the outburst is coming from an irradiated accretion disk."

M83 is powerfully condensed at low power in the 4-inch, easy in binoculars, and just visible to the naked eye. It is a true showpiece for small telescopes, displaying so much structure – bright knots, dark lanes, and oddly shaped arms – that it begs for high power.

Despite its brightness, M83 might pose a challenge for observers at mid-northern latitudes. It resides in a fairly nondescript part of Hydra, the longest constellation in the sky, at a low southern declination (–30°). Start by using the chart at the back of this book to locate 1, 2, and 3 Centauri – a tight equilateral triangle of bright stars about 20° (two fist widths) south and slightly east of 1st-magnitude Alpha (α) Virginis (Spica). Now refer to the chart here. Notice that the galaxy is 3 1/2° northwest of 3 Centauri and just about 30′ southwest of magnitude 5.5 Star *b*. Adding to the wealth of the telescopic vista is a beautiful string of 10th-magnitude stars bordering the galaxy to the southeast and two fainter stars bracketing its farthest spiral arms.

Through the 4-inch at 23×, the galaxy immediately looks elongated southeast to northwest, with obvious patches on either side of the nucleus. One patch in the outer, southeast arm is glorious, and, with time, several star-forming regions (nebulae) pop in and out of view all over the galaxy. You can spend many enjoyable hours just hunting them down.

To see the most detail, alternate between medium and high power. I use 72× to see the faintest details in the arms and 130× to pick out structure in the nuclear region, and I'm amazed at how incredibly detailed the nucleus and its surroundings appear in the 4-inch – how well defined even the tiniest bits of dark matter appear. The nucleus itself is tiny, looking like a star trapped in a maelstrom of shimmering light.

While concentrating on the nucleus, I notice that a bright knot in the southeastern arm becomes clearly apparent, as does another knot equidistant from the nucleus in the northwestern arm. The knots are oriented in a row that is skewed about 20° from the galaxy's main "bar," which appears to extend across the major axis. The bar is composed of numerous nebulous patches along the galaxy's major axis and ultimately branches off the axis to form the stunning spiral arms. The sight reminds me of the great controversy over

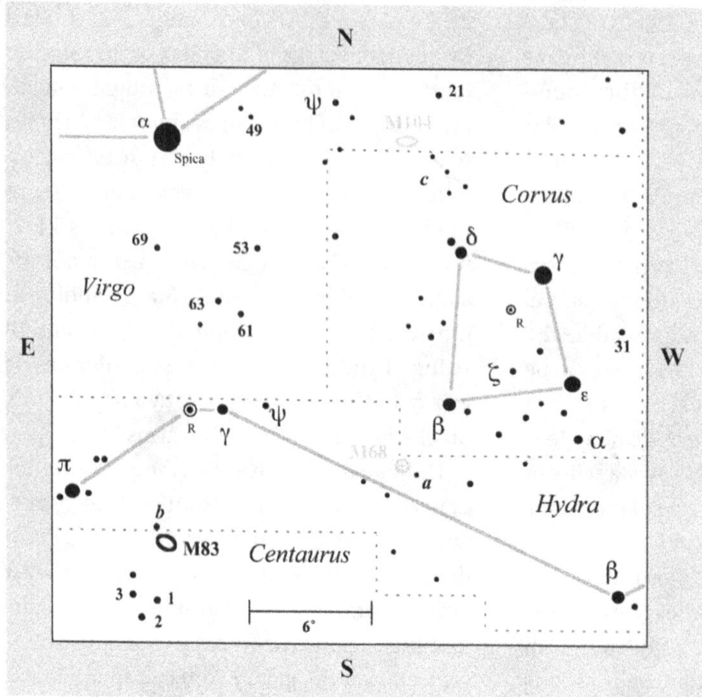

Lowell's canals on Mars: in imperfect seeing, the bar looks like a straight line of light, but under excellent conditions, the line breaks up into individual spots.

Once you can follow some of the spiral patterns (and this takes time), try focusing not on the bright arms themselves but on the dark matter between them. The challenge here is to make sense of the innermost region – that maelstrom. I spent hours doing this, to confirm and reconfirm what I was seeing. Take it piecemeal – first look southwest of the nucleus, where the arms appear brighter than those to the northeast.

There is one beautiful dark lane, which I think is the most striking feature in the galaxy, just to the northeast of the nucleus. It looks like a black canal whose sandy banks are being illuminated by starlight. On the south side, use averted vision to see a broad, arcing band of patchy nebulosity between the nuclear pool and the closest spiral arm. Now spend time concentrating on that arm. When I do this, I discover that the nebulosity separates into two arms; one is more tightly wound than the other and lies closer to the nucleus. You need acute vision and resolution to separate these arms.

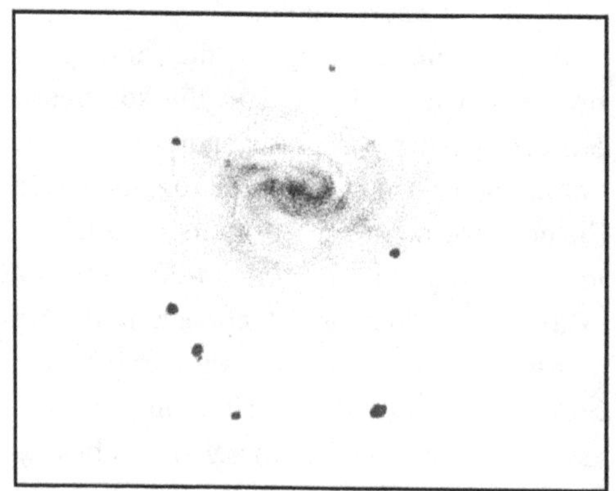

The vast and complex network of spiral structure exhibited by M83 could possibly be explained, in part, by tidal warping from a collision with tiny NGC 5253, an elliptical galaxy 10 times less massive than M83 and that now lies about 2° to the southwest. Six supernovae are known to have occurred in M83, and they are as follows, with their magnitudes in parentheses (it seems like one is overdue to erupt): SN 1923A (14.0), SN 1945B (14.2), SN 1950B (14.5), SN 1957D (15.0), SN 1968L (11.9), and SN 1983N (12.5).

M84

NGC 4374
Type: Elliptical Galaxy (E1)
Con: Virgo

RA: $12^h25.1^m$
Dec: +12°53′
Mag: 9.1
Dim: 5.1′ × 4.1′
SB: 12.3
Dist: 55 million light-years
Disc: Charles Messier, 1781

MESSIER: [Observed March 18, 1781] Nebula without a star in Virgo. The center is pretty bright, surrounded by slight nebulosity. Its brightness and appearance are similar to those of numbers 59 and 60 in this catalogue.

NGC: Very bright, pretty large, round, pretty suddenly brighter in the middle, mottled.

M86

NGC 4406
Type: Elliptical/Spiral Galaxy (E3)
Con: Virgo

RA: $12^h26.2^m$
Dec: +12°57′
Mag: 8.9
Dim: 12.0′ × 9.3′
SB: 13.9
Dist: 53 million light-years
Disc: Charles Messier, 1781

MESSIER: [Observed March 18, 1781] Nebula without a star in Virgo, on the same parallel as number 84 above, and very close to it. They have the same appearance, and both appear in the same telescopic field.

NGC: Very bright, large, round, gradually brighter in the middle to a nucleus, mottled.

M86

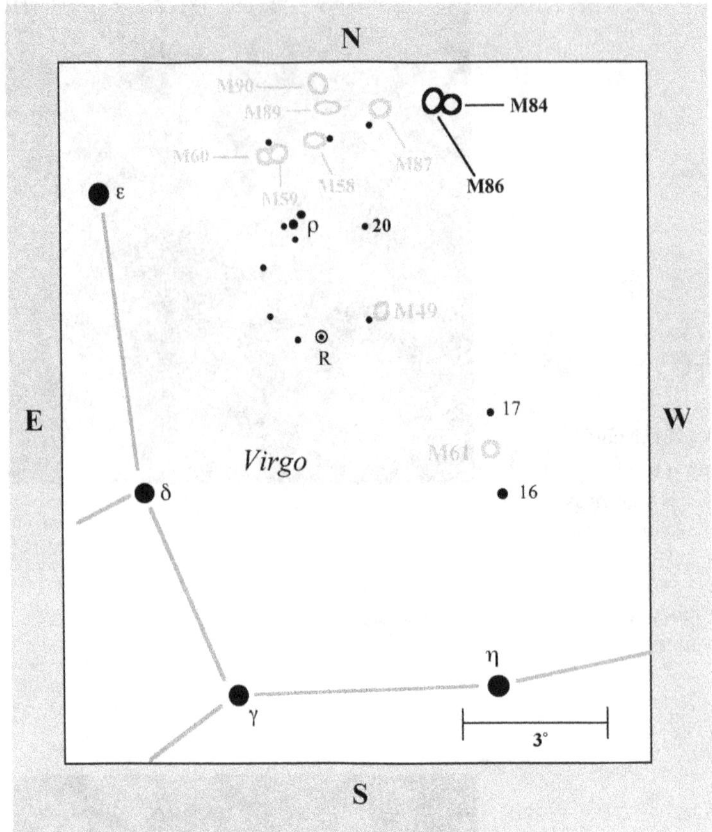

Cluster. It was once believed to be only a foreground object, residing about halfway between us and the Virgo Cluster, but this is no longer accepted.

In fact, both M84 and M86 are giant elliptical radio galaxies that not only form a true physical pair but their outskirts overlap! One of the more prominent ellipticals near the center of the Virgo Cluster, M84 has a compact nuclear source and two warped dust lanes perpendicular to a radio jet that streams away from the nucleus in a north–south direction. Hubble Space Telescope data suggest this gas disk surrounds a black hole with a mass of ~1.5 billion solar masses. M84's outer halo appears smooth and pockmarked with prominent globular clusters, and its stars appear to be very old, about 11.8 billion years.

By comparison, M86 is a normal elliptical E3 galaxy. Its smooth-looking surface gets gradually fainter the farther you look away from the core. There is no evidence of an outer envelope, and M86 does not appear to be interacting with its tiny companion NGC 4402, located about 10′ north of its core. M86 has a high velocity relative to the average cluster velocity and is probably on a radial orbit almost parallel to our line of sight, passing close to the cluster core about every 5 billion years.

Careful studies of M86's core have revealed a slight dimming relative to its surroundings,

M84 AND M86 IN VIRGO SEEM TO HAVE A lot in common: they are separated by a mere 20′, they both shine at around 9th magnitude, and they look about the same size. (So, it makes sense to discuss them together, leaving M85 to follow.) But note that M86 is also larger in apparent diameter and more diffuse, so it has a lower surface brightness.

M84 is an E1 system – a round giant elliptical that resides in the Virgo cluster of galaxies, the nearest of the large extragalactic populations. It stretches across ~82,000 light-years of space. By contrast, M86 is an E3 system, meaning it's more elliptical, and a tad closer. Nevertheless, it dwarfs M84, being ~186,000 light-years in true physical extent. M86 appears to lie at the very heart of the Virgo

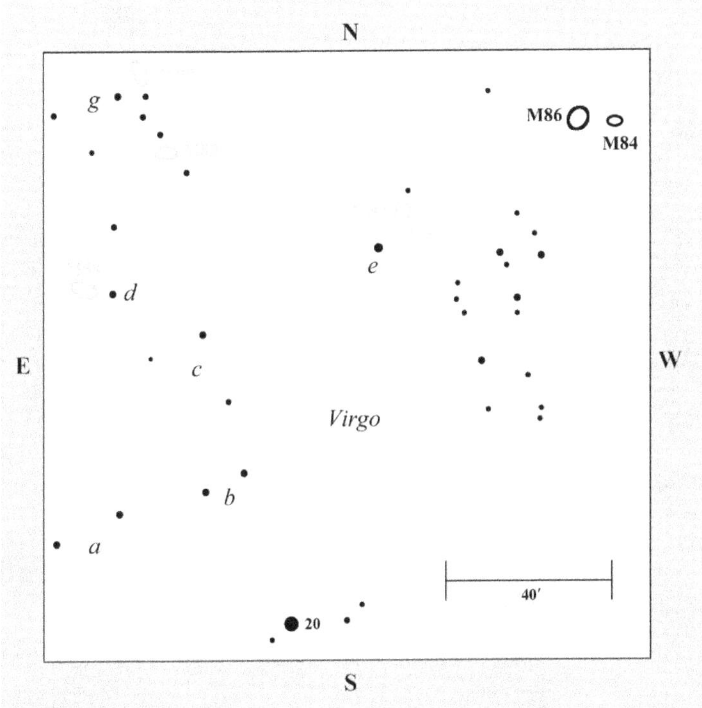

Indeed, earlier research by Jeffrey D. P. Kenney of Yale University and his colleagues, reported in 2008 in *Astrophysical Journal* (vol. 687, p. 69), interpreted the nearly 400,000-light-year-long hydrogen-alpha filaments that clearly connect M86 with NGC 4438 as strong evidence for a previously unrecognized collision between them. Spectroscopy of selected regions shows a fairly smooth velocity gradient between M86 and NGC 4438, consistent with the collision scenario. "Such a collision would impart significant energy into the interstellar medium of M86, probably heating the gas and acting to prevent the gas from cooling to form stars," they said, adding, "We propose that cool gas stripped from NGC 4438 during the collision and deposited in its wake is heated by shocks, ram pressure drag, or thermal conduction, producing most of the H-alpha filaments."

In another recent study, released in 2012 by the Subaru Observatory and published the same year in *Astrophysical Journal* (vol. 757, p. 184), Hong Soo Park (Seoul National University) and colleagues produced the first spectroscopic study of M86's globular clusters. Radial velocities determined for 25 of them reveal "a hint of rotation" in the system that strongly suggests the existence of an extended dark matter halo. The clusters show a wide age distribution from 4 billion to 15 billion years, another indication that M86 has

which could suggest a nuclear dust ring or torus seen face-on, though other studies have contested this. In 1999, x-ray observations revealed a weak nuclear source with a plume of diffuse emission extending toward the northwest – presumably resulting from the stripping of the hot gas as it falls into the core cluster's surroundings.

In a 2010 paper in *Astronomy and Astrophysics* (vol. 518, p. L45), H. L. Gomez (Cardiff University) and colleagues announced that they found three dust components to M86: emission from the central region, a dust lane extending north to south, and a bright emission feature 32,600 light-years to the southeast. They estimate that tidal interactions between M86 and another companion, NGC 4438 (about 25′ to the east-northeast), has stripped NGC 4438 of about ~1 million solar masses of dust.

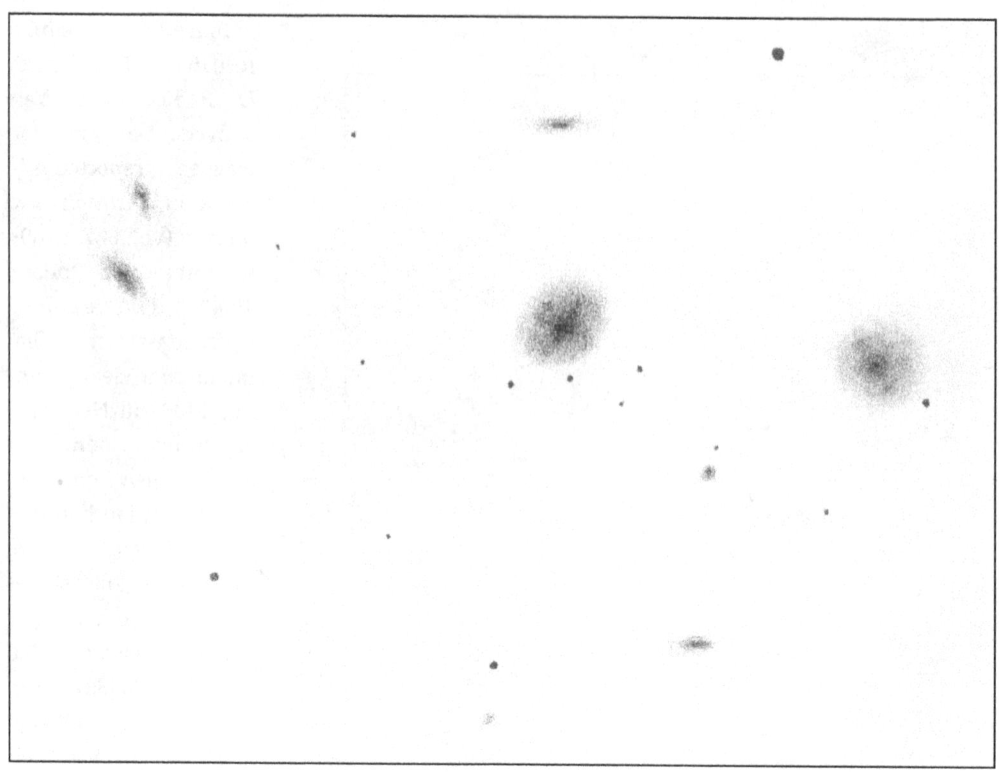

perhaps had mergers or strong tidal interactions in the past.

To find this curious pair, use the star charts here to find Rho (ρ) Virginis, then 20 Virginis a little more than 2° to the west. From 20 Virginis, move about 40′ northeast to Pair *b*, then 30′ to Triangle *c*; both are comprised of mostly 8th-magnitude stars. Now move 1° to 8th-magnitude Star *e*, just north-northwest of which should be M87. M84 and M86 lie about 1 1/2° to the west-northwest (no other galaxies should interfere with your view). You're looking for a double "nebula."

In the 4-inch, both exhibit softly glowing disks with condensed centers that gradually give way to diffuse outer halos. But there are subtle differences. With low power, M86 appears slightly larger and brighter than M84. Curiously, my eye is first drawn to M84, whose stellar nucleus and tightly packed core appear more conspicuous than those of M86. Moderate power does little to improve the image. But 130× reveals a 14th-magnitude star southwest of M84's nucleus, which enhances the prominence of the galaxy's core. Do you see another star or condensation equidistant from and northeast of the nucleus?

Now use high power to concentrate on M86. It, too, has a sharp nucleus, but its core is more diffuse than that of M84. I see a very curious feature in this nuclear haze, appearing as a four-armed windmill with broken blades – glazing an otherwise porcelain-pure oval.

A minefield of galaxies, the Virgo Cluster must have posed quite a problem for Messier and his contemporaries as they swept this region for comets. So many nebulous glows went unrecorded by these skilled observers that I wonder if, in trying to plot them, utter confusion was the culprit more than the optical limitations of their telescopes!

For the patient observer under a dark sky, the region centered on M84 and M86 just might be one of the richest galaxy fields visible in small telescopes. Although few of the objects display impressive details (most look like fuzzy patches of varying brightnesses and sizes), they nevertheless have an enchanting aura about them. What the Virgo Cloud cannot offer in quality, it makes up for in quantity. Fully 10 galaxies fill a 1° field centered on M84 and M86.

And the entire Virgo Cloud – which is more properly referred to as the Coma–Virgo Cloud because the greatest concentration of galaxies in the region huddles close to the border of these two constellations – contains at least 10 times that many galaxies within reach of a 4-inch under dark skies. But not all are easy to see. It takes a dedicated effort, for example, to sight each individual object shown in the accompanying drawings.

M85

M85

NGC 4382
Type: Lenticular Galaxy (SA0^+(s) pec)
Con: Coma Berenices

RA: $12^h25.4^m$
Dec: +18°11′
Mag: 9.1
Dim: 7.5′ × 5.7′
SB: 13.0
Dist: 56 million light-years
Disc: Pierre Méchain, 1781

MESSIER: [Observed March 18, 1781] Nebula without a star, above and close to the ear of corn in Virgo, between two stars in Coma Berenices, Flamsteed 11 and 14. This nebula is very faint. M. Méchain determined its position on 4 March 1781.

NGC: Very bright, pretty large, round, with a brighter in the middle, star to the northwest.

ALTHOUGH M85 LIES IN COMA BERENICES, it belongs to the Virgo cloud of galaxies. This peculiar lenticular system has some similarities to M84 in Virgo. M84 and M85 both are ~55 million light-years distant, and both shine at 9th magnitude. They have masses of 500 billion and 400 billion solar masses, respectively. But unlike M84, M85 is a gas-poor galaxy that displays low-contrast spiral structure.

M85 is a radio-quiet, lenticular galaxy, oriented nearly face-on. In many images, it has an extremely bright, though diffuse, nucleus nested in smooth whorls that form a lens. However, an HST unsharp-masked image suggests the presence of a dust lane running across the nucleus along the minor axis.

M85 has a companion, NGC 4394, ~8′ to the northeast, and it forms a strongly interacting pair with M85. Despite its tempting treat nearby, M85 has most likely already consumed another companion, which induced an episode of recent star formation. This interaction might explain M85's bluer hue when compared with other more typical lenticular galaxies. If true, the color would come from enhanced star formation in the disk as opposed to star formation in the nucleus. Further evidence is a warp in the outer reaches of the galaxy's quite extensive outer envelope. What's more, M85 has a system of star clusters unlike any other galaxy studied to date. Hundreds of them are fainter and more extended than typical globular clusters, and have no local analog.

One mystery prevails. Many galaxies, including the Milky Way, harbor a massive black hole at their heart. M85 may be an exception. In a 2011 paper in *Astrophysical Journal* (vol. 741, p. 38), Eric Peng of the

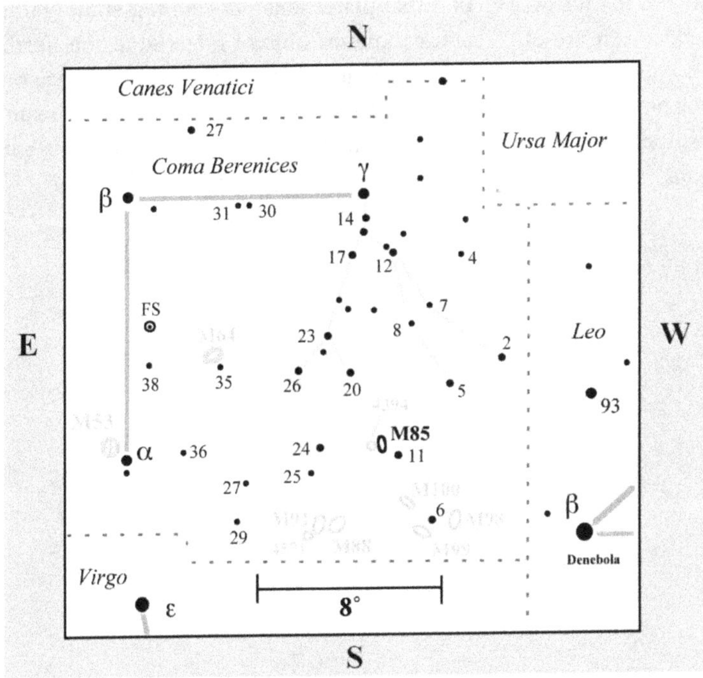

University of Michigan and his colleagues explain how they used the spectrograph on board the Hubble Space Telescope to examine the orbital velocity of stars near M85's diffuse nucleus. They expected to find a black hole in the mass range of 300 million to 2 billion solar masses.

The sign would appear as a specific shift in color that relates to the orbital speed of the stars – blue as stars approached us in their orbit around the core and red as they moved away from our line of sight. The more massive the central object, the greater the measured shift. But the expected shift was not observed, given a galaxy of this size, which spans about 120,000 light-years across (a little larger than the Milky Way). The researchers explained that M85's black hole is much smaller than predicted, perhaps with a mass of about 50 to 80 million solar masses. This might explain why the galaxy is so "quiet."

To find this peculiar galaxy, use the chart at the back of this book or the one here to locate 11 Comae Berenices; M85 is just a little more than 1° east of that 5th-magnitude star. It's also 5° north of M84. I really like part of Heinrich d'Arrest's description of the galaxy's surroundings, where he says it is "abundant with true and false nebulae." What a grand way to describe the visual chaos of the Virgo Cluster.

Through the 4-inch at 23×, the galaxy appears as a powerfully glowing oval mass with a distinct core. It shares a high-power field of view with two 10th-magnitude stars and the 11th-magnitude barred spiral NGC 4394 to the northeast. M85 forms a strongly interacting pair with NGC 4394. See if you notice a slight blue coloration to M85.

The telescopic view at high power is reminiscent of photographs I have seen of this pair taken with large instruments! A glorious

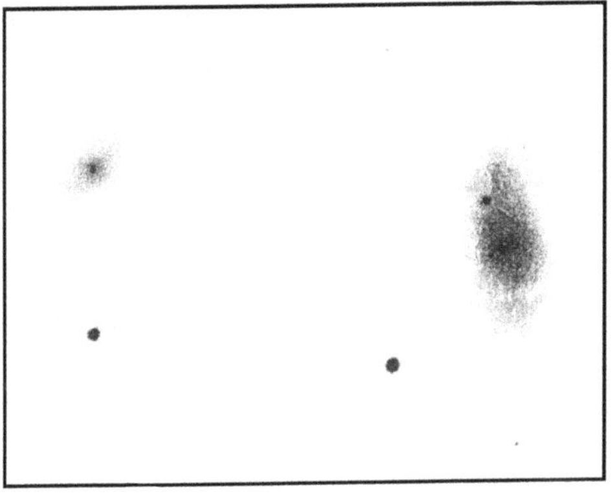

magnitude 12.5 star blazes just to the north-northeast of the galaxy's nucleus. An arc of bright material to the south caught me by surprise. The only reference I have found that refers to this peculiar yet distinct feature is *Burnham's Celestial Handbook*, which says of it that plates taken at Palomar show "faint elongations or tufts of material at the north and south tips of the system, either the vague beginning of a spiral pattern or the last surviving remnant of one." Do you also see a bar running through the galaxy's major axis?

M87

NGC 4486
Type: Elliptical Galaxy (cD pec)
Con: Virgo

RA: $12^h30.8^m$
Dec: +12°24′
Mag: 8.6
Dim: 7.1′ × 7.1′
SB: 12.7
Dist: 55 million light-years
Disc: Messier, 1781

MESSIER: [Observed March 18, 1781] Nebula without a star in Virgo, below and quite close to an eighth-magnitude star. The star has the same right ascension as the nebula, and its declination is 13°42′21″ north. This nebula appears to have the same brightness as the two nebulae numbers 84 and 86.

NGC: Very bright, very large, round, much brighter in the middle, third of three [the easternmost].

A MONSTROUS BALL OF ENERGY, THE giant elliptical galaxy M87 measures ~115,000 light-years across and has a total mass of nearly 800 billion solar masses, making it one of the more massive galaxies known. It is also one of the brighter ellipticals in the Virgo Cluster and one of the more luminous of all known ellipticals. M87 is easily located by drawing a line from M58 to M84; M87 lies almost equidistant between them, being slightly closer to M84 and just 6′ south of an 8th-magnitude star.

The central member of the Virgo Cluster, M87 is a cosmic powerhouse with a very strong central radio source and a furious jet of matter blasting out from its nuclear region. In fact, it is the central radio source in the Virgo Cluster. After studying Hubble Space Telescope images of M87, astronomers now believe that a black hole with a mass of 3 billion solar masses lurks in the galaxy's nucleus, within a zone about 120 light-years across. It is the largest black hole known. High-resolution terrestrial images reveal a beaded jet of hot ionized gas, called plasma, though the brightest beads are barely within the visual range of large-aperture amateur instruments. Barbara Wilson and I spied evidence of the jet in her 20-inch reflector during a Texas Star Party.

The jet extends nearly 5,000 light-years from M87 and points toward M84. It was once believed to be an intergalactic bridge between them, but the mysterious feature probably has a more interesting origin, the voracious supermassive black hole lurking at the center of M87, whose nuclear region

M87

is swarming with billions of rapidly orbiting stars. The disk around that rapidly spinning black hole has magnetic field lines that entrap ionized gas falling toward the black hole. These particles, along with radiation, flow rapidly away from the black hole along the magnetic field lines. The rotational energy of the spinning accretion disk adds momentum to the outflowing jet.

The black hole is already thought to have ingested clouds of gas (about 5 billion solar masses worth) and some whole stars as well. If you consider that M87's jets are only about as massive as the Sun, they may be the ejected remains of a star that was sucked into the black hole. In 2009, the Hubble Space Telescope observed a flare-up in M87's jet – a blob of matter 214 light-years from the galaxy's core and embedded in that powerful, narrow beam of hot gas – that outshone M87's brilliant core.

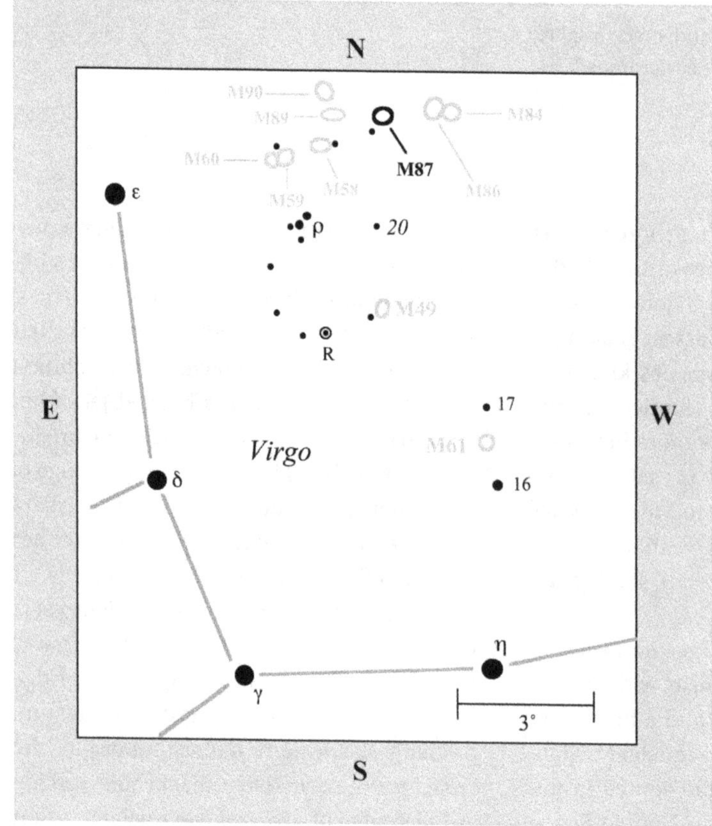

As the Hubbell press release stated: "Despite the many observations by Hubble and other telescopes, astronomers are not sure what is causing the brightening. One of the simplest explanations is that the jet is hitting a dust lane or gas cloud and then glows as a result of the collision. Another possibility is that the jet's magnetic field lines are squeezed together, unleashing a large amount of energy. This phenomenon is similar to how solar flares develop on the Sun and is even a mechanism for creating Earth's auroras."

To find the galaxy, use the star charts here to

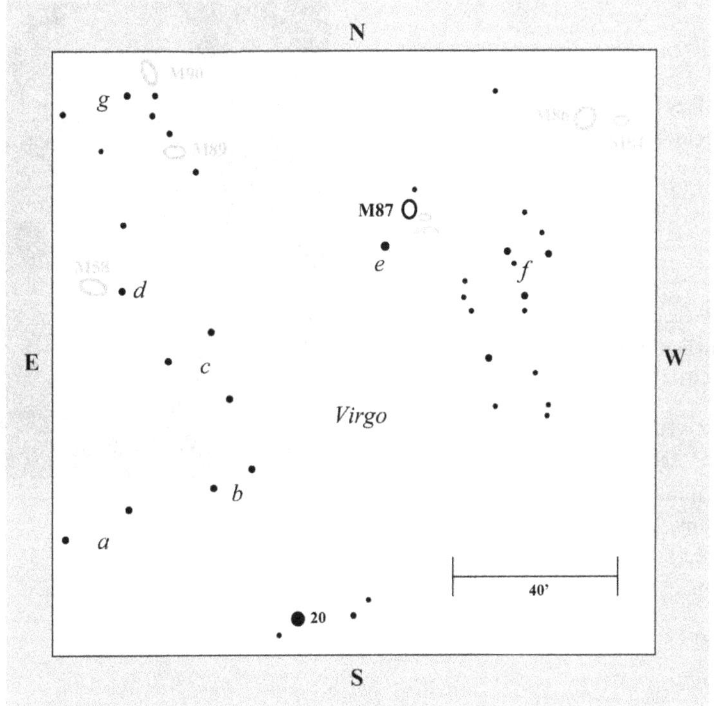

find Rho (ρ) Virginis, then 20 Virginis a little more than 2° to the west. From 20 Virginis, move about 40′ northeast to Pair *b*, then 30′ to Triangle *c*; both are comprised mostly of 8th-magnitude stars. Now move 1° northwest to 8th-magnitude Star *e*, just north-northwest of which should be M87.

At 23×, the magnitude 8.6 elliptical can be seen glowing to the east of a fine telescopic asterism that looks a bit like Leo, the Lion. The galaxy itself resembles an unresolved globular cluster, one with a bright spherical shell that gradually condenses inward. I cannot see a sharp nucleus or much other detail, even at high power.

Don't miss tiny NGC 4478, an 11th-magnitude elliptical 10′ southwest of M87. At 23×, the galaxy looks merely like a star, but higher magnification brings it clearly into view. I also spotted a close double star to the west, which resembles the light of a galaxy.

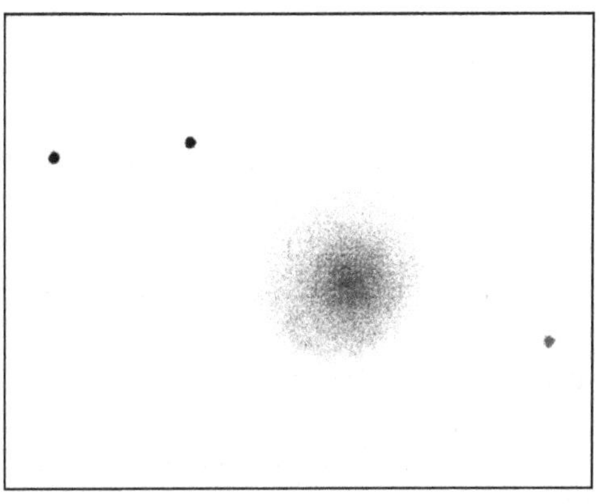

M88

NGC 4501
Type: Spiral Galaxy (SA(rs)b)
Con: Coma Berenices

RA: $12^h32.0^m$
Dec: +14°25′
Mag: 9.6
SB: 12.6
Dim: 6.1′ × 2.8′
Dist: ~62 million light-years
Disc: Messier, 1781

MESSIER: [Observed March 18, 1781] Nebula without a star in Virgo, between two faint stars and a sixth-magnitude star, which all appear in the same telescopic field as the nebula. Its luminosity is one of the faintest and it resembles that reported for number 58, also in Virgo.

NGC: Bright, very large, very much extended.

JOHN MALLAS APTLY DESCRIBED M88 as resembling the Andromeda Galaxy (M31) depicted in a very small-scale photograph. Lying about 4 1/2° southeast of 11 Comae Berenices, M88 is one of the more conspicuous spiral systems in the Virgo cluster of galaxies. Although it shines at only magnitude 9.5, its compact size and 30° tilt from edge-on give it good surface brightness.

Stretching across 110,000 light-years of space, M88 is one of the largest spirals in the Virgo Cluster region. It has well-organized dust lanes that form a prominent spiral pattern with many filamentary arms studded with knots. Beyond the dust, young and hot star clusters and hydrogen-rich star-forming regions line the arms. To the southwest are a pair of detached outer arms. The galaxy's dust lanes project keenly against its bright spherical bulge and disk pattern. On red Hubble Space Telescope images, spiral-like dust lanes wind down to a very small and intense Seyfert nucleus.

M88 is not only far from the core of the Virgo Cluster but also receding from us at quite a clip (~1,400 miles per second) – fast enough to leave some astronomers doubting M88's relationship with the Virgo Cluster. Indeed, the system's high redshift lies near the upper limit for inclusion in the cluster. That said, consider now that M91, which has nearly the same angular size as M88 and lies only about 40′ from it in the sky, is moving away from us at only 227 miles per hour, assuring its membership in the Virgo Cluster.

Some say this circumstantial evidence is enough to warrant M88's membership in the Virgo Cluster, but while M88 for now is allowed to be a member, some astronomers wonder if it's just a projection effect. No

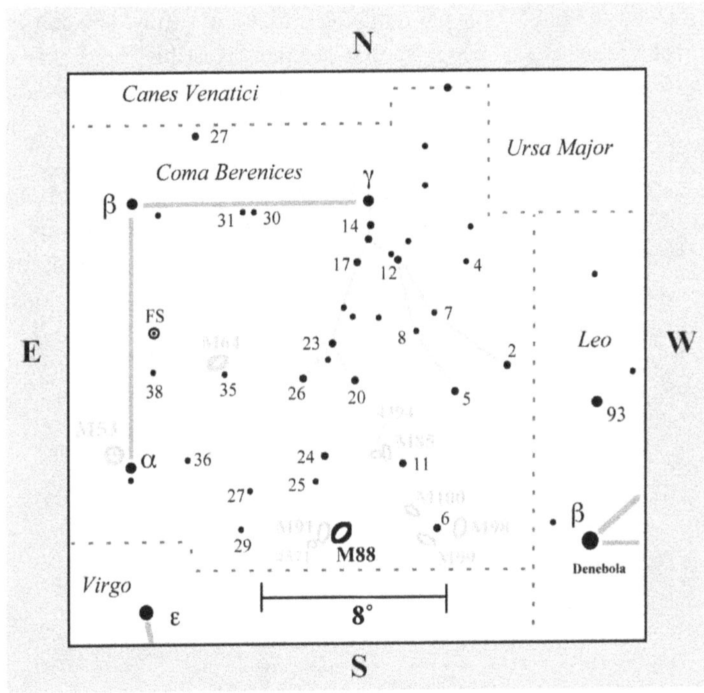

matter, membership seems inevitable given that M88 is rushing headlong toward M87 at the core of the Virgo Cluster and is expected to make a close approach to it in about 200 million to 300 million years.

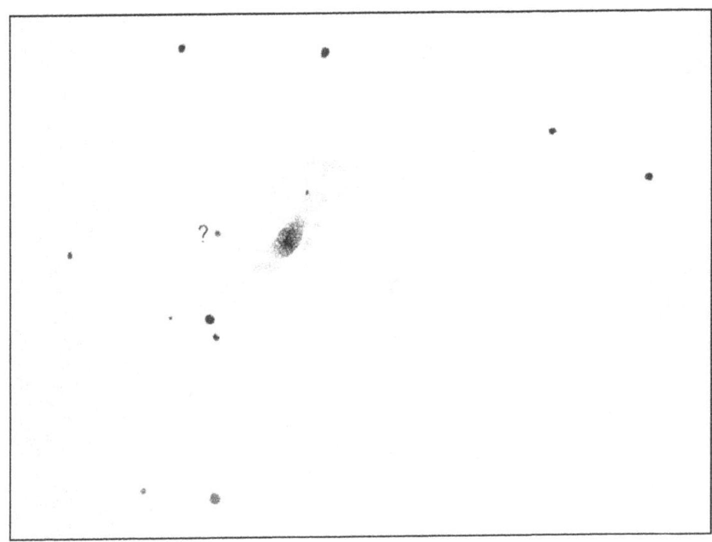

To find this wonder, draw an imaginary line between Epsilon (ε) Virginis and Beta (β) Leonis. M88 lies midway and a tad northeast along that line. It's also midway between the 6th-magnitude stars 29 and 6 Comae Berenices.

At 23×, the galaxy displays a definite spindle shape and is easily picked out from the galactic morass. No other galaxies near it shine as brightly, and a close pair of 12th-magnitude stars (separated by 30″) touch the southeast extremity of the nebulous glow. An 11th-magnitude star is visible to the north.

Moderate power reveals a strong suggestion of spiral structure along the western and eastern rims. It took me a few nights and many hours to plot the fainter arms closer to the nucleus. Although I could detect them – and, in moments of excellent seeing, they appeared obvious – I had difficulty translating those impressions to paper. The problem is that most of the brighter inner detail shines with a uniform glow that is broken in places by dark matter.

The one interesting feature of this galaxy, however, seems to be a mystery. When I made my first drawing of M88, it showed the galaxy's pinpoint nucleus and outer spiral arms. I noted the galaxy's position between the three foreground stars mentioned previously.

In a drawing I made 10 nights later, however, I included a magnitude 13.5 star southeast of the nucleus (indicated with a question mark in the drawing), which I had apparently not seen the first time. Moreover, the photographs I have seen of M88, including the accompanying one, do not show any star in or near that position. Was this "new" star a supernova? An asteroid passing by? Or is it a foreground variable star? Give the galaxy a look and see what you think.

On May 29, 1999 UT, the Lick Observatory Supernova Search team discovered a supernova in M88. SN 1999cl shined at magnitude 16.4 when discovered 46″ west and 23″ north of the galaxy's center.

M89

NGC 4552
Type: Elliptical Galaxy (E0–1)
Con: Virgo

RA: $12^h35.7^m$
Dec: +12°33′
Mag: 9.8
Dim: 3.4′ × 3.4′
SB: 12.3
Dist: 50 million light-years
Disc: Charles Messier, 1781

MESSIER: [Observed March 18, 1781] Nebula without a star in Virgo, not far from and on the same parallel as the nebula number 87, described above. Its light is extremely faint and diffuse, and it is visible only with difficulty.

NGC: Pretty bright, pretty small, round, gradually much brighter in the middle.

M90

NGC 4569
Type: Mixed Spiral Galaxy (SAB(rs)ab)
Con: Virgo

RA: $12^h36.8^m$
Dec: +13°10′
Mag: 9.5
Dim: 10.5′ × 4.4′
SB: 13.6
Dist: 40 million light-years
Disc: Charles Messier, 1781

MESSIER: [Observed March 18, 1781] Nebula without a star in Virgo. Its light is as faint as the previous one, number 89.

NGC: Pretty large, brighter in the middle to a nucleus.

M90

The Virgo Cluster harbors some 250 large galaxies and perhaps a thousand or more smaller systems. One distinguishing characteristic of this particular galaxy cluster is its smattering of spirals loosely dispersed among the more amorphous ellipticals. This peculiarity is nicely represented by the pairing of M89 and M90, an elliptical and a spiral, respectively. Separated by only 40′, these galaxies should fit in the low-power field of almost any amateur instrument. Although their magnitudes are similar, M90 is bigger, bolder, and more interesting than its counterpart.

M90 is one of the larger spirals in the Virgo Cluster, though it may be of low density. It may also be nearly perfectly spherical. Though unusual, its shape could reflect an instance of near perfect geometry, one where we see an elliptical system looking directly down its major axis. At a glance, we're looking at a modest system only about 50,000 light-years in true physical extent. But images of M89 made in 1979 with the UK Schmidt telescope by Australian astronomer David Malin of the Anglo Australian Observatory reveal that the galaxy is much larger than was previously believed.

The entirety of M89 is surrounded by a faint but significant outer envelope that reaches out to about 150,000 light-years in distance. It was the first galaxy to be discovered with such a feature. Malin's discovery implies that a significant portion of M89's mass exists in this low-luminosity halo. This discovery would make M89's size comparable to that of M90, which has a diameter of about 125,000 light-years and a mass of 80 billion solar masses.

An extensive system of globular star clusters surrounds M89. A 2006 survey estimates that there are about 2,000 globulars compared with the estimated 150–200 thought to surround the Milky Way.

At optical wavelengths, an arc-shaped dust lane obscures M89's nucleus. At x-ray, ultraviolet, and radio wavelengths, however, the central point source reveals itself to be variable in nature. The galaxy also displays streams of filaments to the northeast, while a gas flow extends farther to the southwest. This outflow of hot particles penetrates deep into the galaxy's outer envelope.

In a 2006 paper in *Astrophysical Journal* (vol. 648, p. 947), N. Machacek and colleagues report that Chandra X-ray Telescope observations of the nuclear outflow show two ringlike features ~5,500 light-years in diameter, which appear to be shocks. The central source is most likely a low-luminosity active galactic nucleus. The researchers also found that the central region's gas temperature (out to about ~325 light-years or so) is hotter than anywhere else in the galaxy, suggesting that we may be directly observing the reheating of the galaxy's interstellar medium by the nuclear activity. Hubble Space Telescope images reveal the nuclear dust ring.

Hubble Space Telescope observations also indicate a supermassive black hole with a mass of about 700 million solar masses. Chandra results, combined with new theoretical calculations, suggest a supermassive black hole spinning extremely rapidly at its core. Visible in the Chandra images are pairs of huge bubbles or cavities in the galaxy's hot gaseous atmosphere, created in each case by jets shooting out of the black hole. But other observations have indicated that the core activity in M89 is caused by the remains of a dwarf galaxy that M89 absorbed or disrupted long ago.

M90, on the other hand, is among the earliest of the Sab-type galaxies, displaying two smooth and massive regular arms extending from a large central bulge. Massive dust lanes in the outer region wind closely inward,

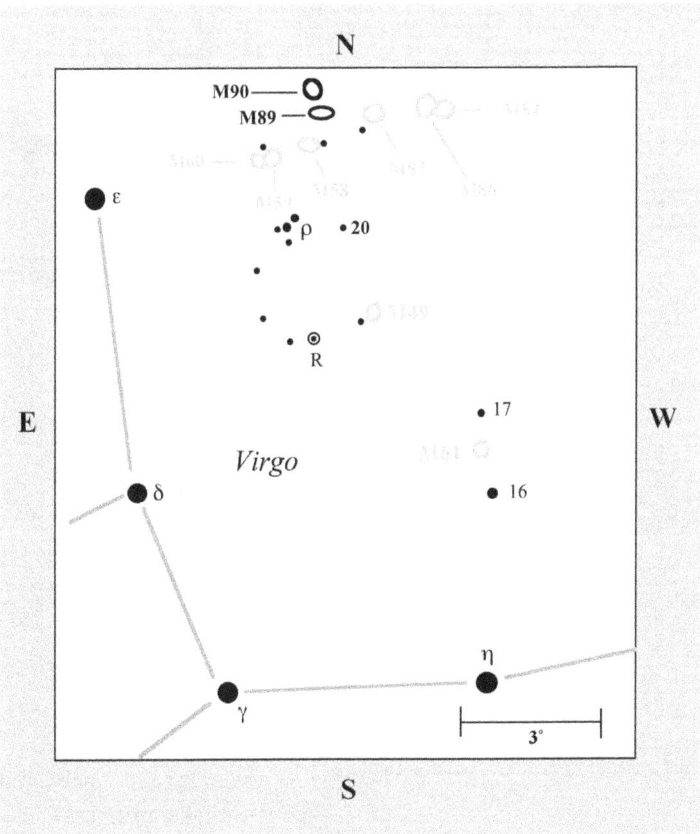

but with no definite spiral pattern. The dusty chaos in the region is most likely a result of the galaxy's high inclination. At the core is a very small but extremely bright nucleus that suggests a central starburst dominated by hot A-type supergiants in a young star cluster.

M90 has also been classed as "anemic," because of its small gas content. Recent radio observations show strong signatures of stripping by the intragalactic gas and large radio lobes with vertical magnetic fields extending more than 65,000 light-years from the disk – a strange combination of features for a spiral galaxy. The nature of these outflows is still being debated. Suggestions include a nucleus powered by a black hole, gas stripping by the intracluster medium, and a central starburst-driven outflow.

A giant 26,000-light-year outflow of hydrogen-alpha gas has been found perpendicular to M90's disk. At the 2006 International Astronomical Union Symposium held in Prague, Czech Republic, S. Ry and colleagues shared how they inspected emission from M90's lobes and found that most likely they are powered not by the active galactic nucleus but by a nuclear starburst and superwind-type outflows that occurred 30 million years ago.

Recent International Ultraviolet Explorer satellite observations and optical studies of M90 demonstrate that its nucleus is a young star cluster 6 magnitudes more luminous than the core of 30 Doradus in the Large Magellanic Cloud. Hubble Space Telescope data show that the ultraviolet spectrum is nearly an

M90

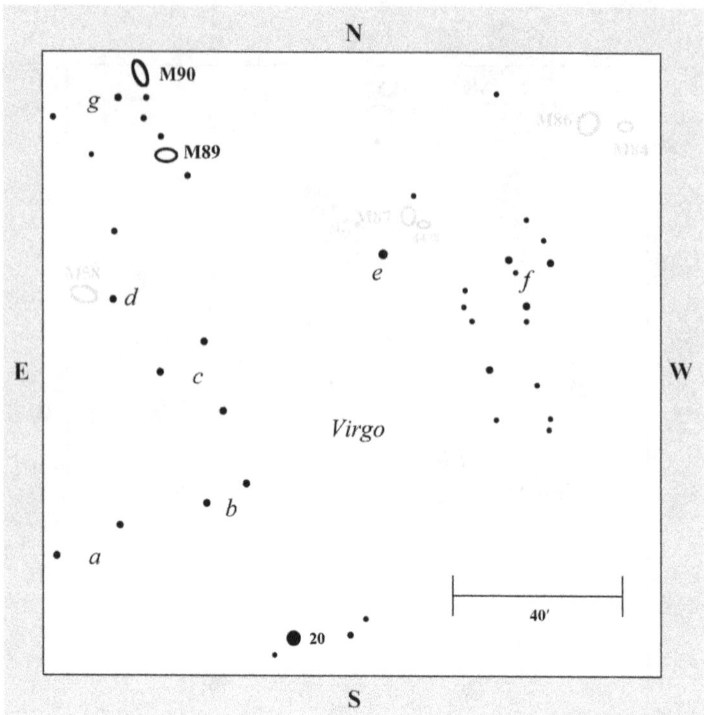

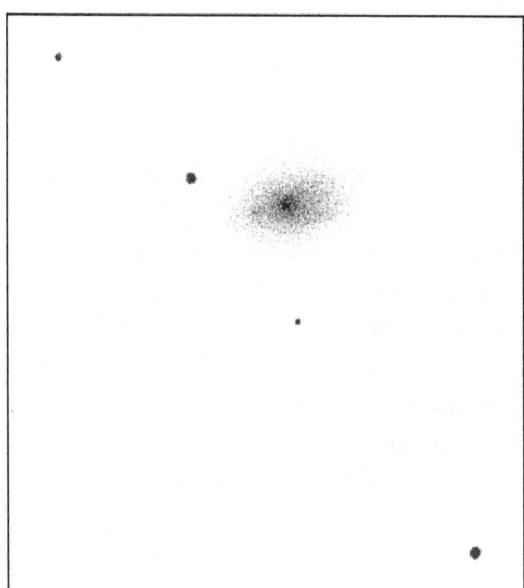

exact match to that of the young starburst cluster NGC 1741-B, which has a likely age of 4 million to 5 million years. M90 is approaching us at 193 miles per second and may be destined to leave the gravitational bounds of that cluster; some sources have speculated it may already have left the cluster and is now a considerable distance nearer to us.

To find M89 and M90, use the star charts here to find Rho (ρ) Virginis, then 20 Virginis a little more than 2° to the west. From 20 Virginis, move about 40′ northeast to Pair *b* and then 30′ to Triangle *c*; both are comprised mostly of 8th-magnitude stars. Next hop ~30′ northeast to Star *d*, with M58 just to its east. M89 is only about 50′ to the north-northwest. M90 is about 40′ north-northeast of M89; a zigzagging chain *g* of faint stars connects the two galaxies (*g*).

At 23×, M89's slightly swollen starlike disk is easy to overlook, especially with mighty M90 to its northeast. Yet I find the elliptical strikingly beautiful at 72×, its silvery light pure and pleasing to the eye. M89 does have an extremely sharp nucleus surrounded by

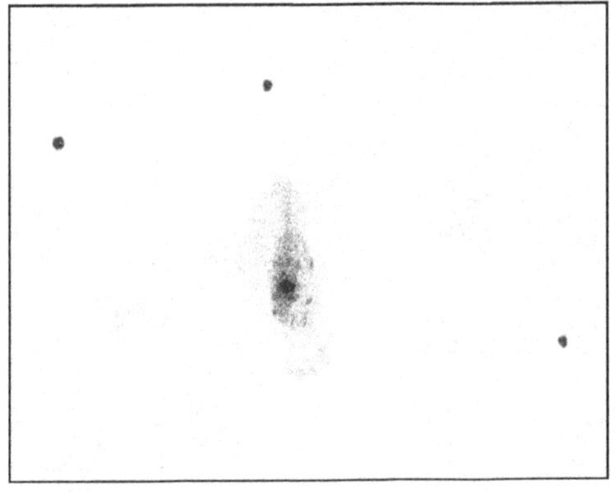

a misty glow; high power brings out suggestions of irregularities in the halo. For instance, there seems to be, though I am not certain of this, a knot to the east, which makes that side of the galaxy look brighter than the other. That knot might be a very faint star.

M90, on the other hand, is a strong, oval glow - a dimly shaped spiral whose major axis is elongated slightly southwest to northeast. But close inspection shows a bar of light across the nucleus with a north–south orientation. Clawlike arcs of light curve off to the west from each end of the nuclear region, giving it the shape of a crab. The outer spiral arms make a slightly warped S-shaped pattern. Skiff noticed that the western flank of the halo bulges in the center and looks flattened to the east. The nuclear region also looks mottled, either with faint starlight or dark nebulosity.

M91

Missing Messier Object
NGC 4548
Type: Barred Spiral Galaxy (SB(rs)b)
Con: Coma Berenices

RA: $12^h35.4^m$
Dec: +14°30′
Mag: 10.2
SB: 13.3
Dim: 5.0′ × 4.1′
Dist: ~53 million light-years
Disc: Charles Messier, 1781

MESSIER: [Observed March 18, 1781] Nebula without a star in Virgo, above the preceding one, number 90. Its light is even fainter than that one. *Note:* The constellation of Virgo, and above all the northern wing, is one of the constellations that contains the most nebulae. This catalogue includes 13 that have been detected, namely numbers 49, 58, 59, 60, 61, 84, 85, 86, 87, 88, 89, 90, and 91. All these nebulae appear to be without stars. They can be seen only under extremely good skies, and close to meridian passage. Most of these nebulae have been pointed out to me by M. Méchain.

NGC: Bright, large, little extended, a little brighter in the middle.

"Of Messier's 14 nebulae in [the Coma-Virgo] region, 13 are easily identified today, but M91 cannot be found. ... There is simply no bright nebula omitted by Messier that could conceivably be identified with M91." Thus wrote a perplexed Owen Gingerich in a chapter he contributed to Mallas and Kreimer's *Messier Album*. Gingerich opened a Pandora's box of possibilities. Was M91 a comet, as John Herschel, Harlow Shapley, and many others had proposed? Is it NGC 4571, the closest galaxy to Messier's recorded position, as William Herschel deduced? Could it have been a duplicate observation of M58 (a simple mistake made by Messier), as Gingerich suggests? Perhaps it is just as Heinrich d'Arrest supposed: "The nebulae M91 in this position, no longer exists in the heavens."

Whatever one chooses to believe, without question the most widely accepted explanation today is that Messier's ninety-first catalogue entry was actually NGC 4548, a 10th-magnitude barred spiral galaxy in Coma

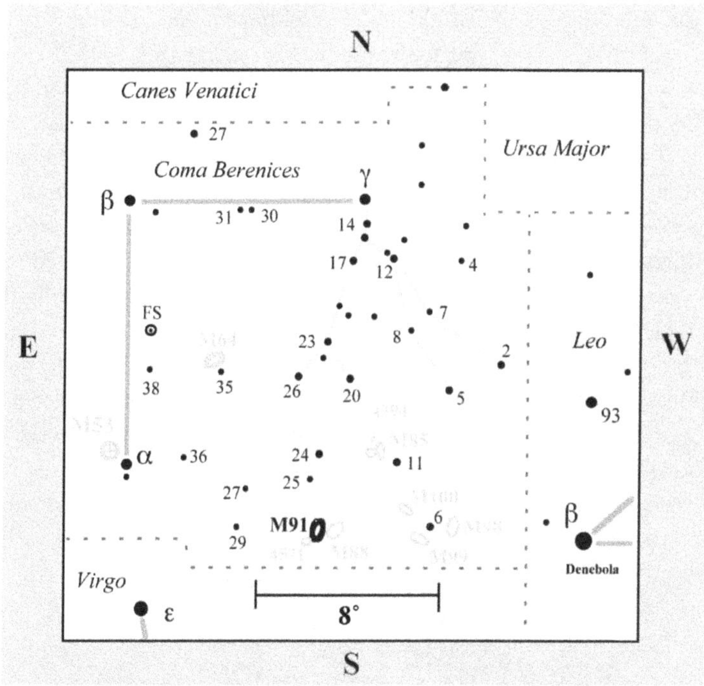

Berenices – so argued W. C. Williams of Fort Worth, Texas, in the December 1969 issue of *Sky & Telescope*. He described how Messier probably made a mistake in both reduction and plotting. By applying a mathematical correction to Messier's right ascension and declination for the object, Williams encountered NGC 4548, about 4 1/2° northwest of magnitude 4.9 Rho (ρ) Virginis, and deduced that this was the missing Messier object, M91. And that is the object we seek out today as we check off the Messier objects.

M91 is a pretty typical barred spiral some 80,000 light-years across in true physical extent. Its modest glow is threaded by a prominent bar that connects to two arc-like structures – one on either end of the bar – that form an incomplete inner pseudo-ring. These arcs are the bright inner segments of the galaxy's two major arms, which contain at least nine prominent H II regions.

After the arms wrap past the bar and its bright, diffuse nucleus, the southeast arm then becomes very open and quickly becomes diffuse and frayed, while the northwest arm wraps more tightly around the inner region; it also remains intact for much longer. The east arm has many more knots in its inner part than the west arm. The outer disk is asymmetric, with the southeast side having a higher surface brightness than the northwest side. The arm structure

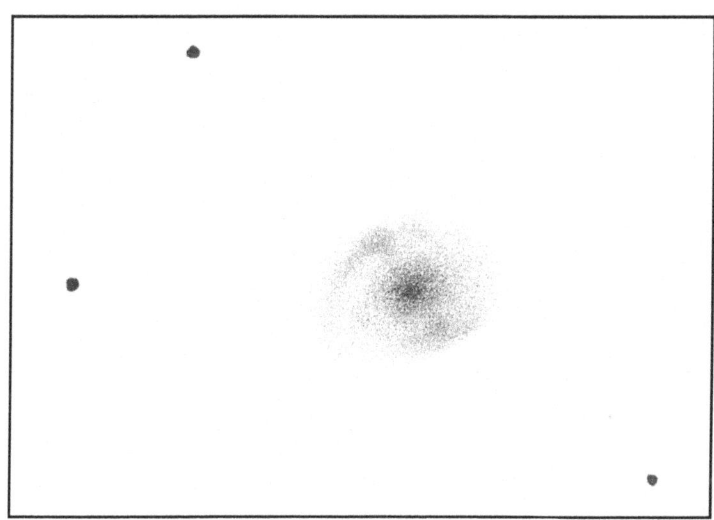

splits into several armlets in the outer disk to create an overall global spiral pattern.

To find this wonder, draw an imaginary line between Epsilon (ε) Virginis and Beta (β) Leonis. M88 lies midway and a tad northeast along that line. It's also midway between the 6th-magnitude stars 29 and 6 Comae Berenices. M91 is just 50′ to the north-northeast of M88.

When I use low power to look at the field containing M91, I am immediately impressed by how obvious this galaxy appears compared with NGC 4571 – a smaller, 11th-magnitude spiral 1/2° to the southeast, which would certainly have been beyond Messier's visual grasp. NGC 4571 is faint, even under superb Hawaiian skies through superb optics! Also, would not Messier have noted the bright star 30′ to its southeast?

At 23×, M91 is a very modest glow, definitely out of round – a broad spiral shape with a bright center. Medium power begins to display its complexities, such as a central bar, knots, and hazy arcs outlining the apparent halo. High power, and persistence, are required to trace out the arms (I had to really breathe hard and use averted vision). Interestingly, Glyn Jones thought a 12-inch was required to see the central bar! But look at my pencil rendering. Perhaps I really did not see the physical bar in my 4-inch; maybe my eye was merely connecting the dots – the starlike nucleus and the knots at the ends of the bar. Study it with your scope and see what you think.

M92

NGC 6341
Type: Globular Cluster
Con: Hercules

RA: $17^h17.1^m$
Dec: +43°08′
Mag: 6.5
Diam: 14′
SB: 12.2
Dist: 27,100 light-years
Disc: Johann Elert Bode, 1777

MESSIER: [Observed March 18, 1781] A fine, conspicuous nebula, very bright, between the knee and left leg of Hercules. It is clearly visible in a one-foot refractor. It does not contain any stars. The center is clear and bright, surrounded by nebulosity, and resembles the nucleus of a large comet. Its light and size are very similar to those of the nebula in the belt of Hercules. See number 13 in this catalogue. Its position was determined by comparing it directly with the fourth-magnitude star σ Herculis. The nebula and the star lie on the same parallel.

NGC: Globular cluster of stars, very bright, very large, extremely compressed in the middle, well resolved, faint stars.

JUST AS GLOBULAR CLUSTER M28 IN Sagittarius is commonly overlooked in favor of the more dazzling globular M22 nearby, so M92 in Hercules plays second fiddle to the much vaunted M13. As Robert Burnham Jr. noted, if M92 were in any other constellation, it would be considered a showpiece. Located 5° southwest of Iota (ι) Herculis, this magnitude 6.4 bolus is a surprisingly easy naked-eye catch, even though it is about a magnitude fainter than M13. One night, I discovered that the two Hercules globulars could be seen simultaneously with the unaided eye! Try this from your observing site, and see if you surprise yourself. Both globulars lie about the same distance from us, so you are really seeing a difference in physical size, reflected in the objects' brightnesses. M92 is not only 20 light-years smaller in true physical extent than M13 but 6′ smaller in apparent diameter.

M92 is a stunning inner halo globular cluster system near the farthest point in its orbit from the Galactic center, some 32,300 light-years distant. Astronomers believe it has already made 16 perigalactic transits in its

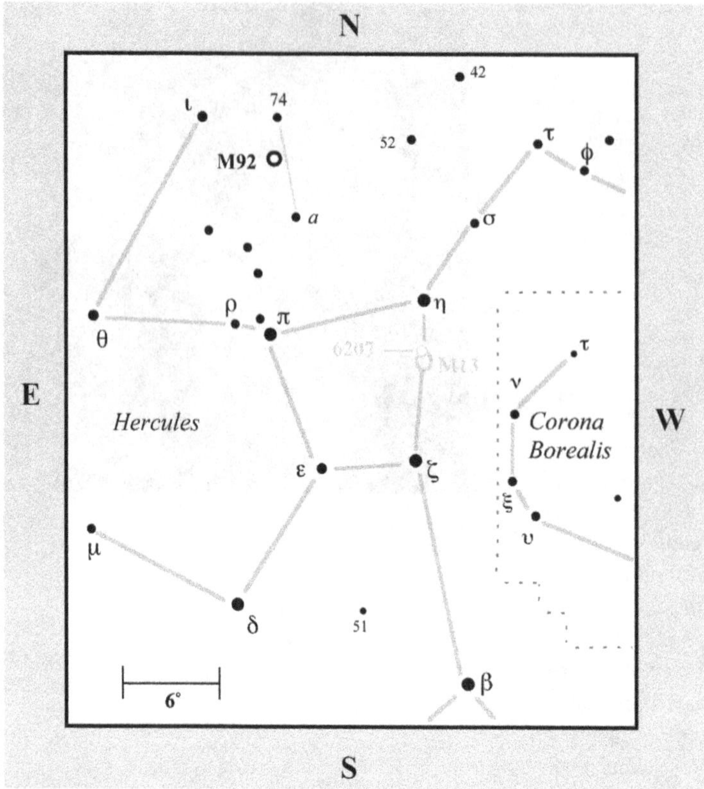

triggered everywhere at the same time in our Galaxy," as opposed to merging with other systems over a longer time frame.

To find M92, use the chart here to locate 4th-magnitude Iota (ι) Herculis, then 5th-magnitude 74 Herculis ~ 3° to the west. M92 is roughly midway between 74 Herculis and similarly bright Star *a*, about 8° to the south-southwest.

At 23×, M92 displays a very tight core and some definition in the outer halo; that is, it looks faintly granular and elongated north to south. But overall the globular appears much more compact than M13. Resolution begins at 72× (its brightest stars shine around 12th magnitude!), while 130× and averted vision transform the cluster into myriad points of light that appear to mushroom from a very tight core. The region immediately surrounding the core is not as uniform as it might first appear. In fact, I find it strongly asymmetrical to the north and shaped somewhat like a lobster claw. Full resolution will be achieved if you can see to 15th magnitude.

Close examination of the void between the claw's pincers reveals a very faint arc of stars enclosing the darkness to the west, so the void now looks like a keyhole. Streams of stars flow out of the core and nearly wrap themselves around the cluster's northeastern part, which looks rather heart shaped. These streamers

lifetime. Because of these passages, the cluster has periodically been stripped of its stars. M92 exhibits long tidal tails, especially in the northeast–southwest direction along its orbit. The cluster has an integrated spectral type of F2 and is metal-poor, with each of its stars (on average) having about 1/200 as much metal per unit hydrogen as the Sun. In a 2010 paper in *Publications of the Astronomical Society of the Pacific* (vol. 122, p. 911), A. di Cecco and colleagues report that they used Hubble Space Telescope optical photometry to determine that M92 is about 11 billion years old. Earlier HST near-infrared and spectroscopic observations by Jae-Woo Lee (University of North Carolina, Chapel Hill) and colleagues indicate results "consistent with the idea that the globular cluster formation must have been

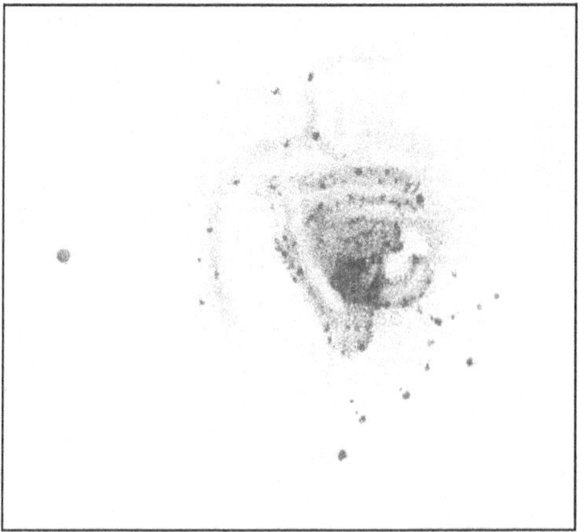

remind me of the wings of stars flowing from the core of M13, but in M92 they are flowing in the opposite direction.

Not everyone will discern all of the features I have described here. The stars in M92 vary so much in brightness and are in such complex arrangements that I can only liken the view to a Rorschach test, the inkblot test used in psychology: at one moment, I see a face-on spiral galaxy (Lord Rosse, by the way, saw a spiral contour to the nucleus); at another, I see two rings of starlight perpendicular to each other. I blink and the lobster claw materializes.

Finally, to add to the confusion, M92 is riddled with dark lanes. In fact, when Isaac Roberts looked at his photographs of M92 taken through a 20-inch telescope, he believed the cluster was involved with dense dark matter that almost prevented stars from being seen through it. Aside from the keyhole, I believe I have found one other obvious dark pattern, a wishbone, in the northeast part of the cluster.

M93

NGC 2447
Type: Open Cluster
Con: Puppis

RA: $07^h44.5^m$
Dec: -23°51.2′
Mag: 6.2
Diam: 10′
Dist: 3,400 light-years
Disc: Messier, 1781

MESSIER: [Observed March 20, 1781] Cluster of faint stars, without nebulosity, between Canis Major and the prow of Argo Navis.

NGC: Cluster, large, pretty rich, little compressed, with 8th- to 13th-magnitude stars.

THERE ARE THREE REASONS WHY I consider this open cluster in Puppis a stunning visual treat. First, its stars are like tiny gems scattered in a rich field of even tinier gems; these are subtle beauties, as delicate as pearls on white satin. Second, the core of the cluster has a distinct arrowhead shape, much like that of M48 in Hydra. Third, in binoculars, M93 appears asymmetrical, and I can see a "cat's eye" pattern, similar to that in M4.

The cluster lies almost exactly in the Galactic plane, though it is reddened by less than 0.05 magnitude. It's of moderate size, extending across 10 light-years of space. It has a Trumpler class of I3r, meaning it's a rich cluster with a strong central concentration comprised of bright and faint stars, which are mainly solar type. M93 also contains a healthy population of red giants. A 2MASS survey determined a distance of about 3,400 light-years and an age of about 400 million years.

Perched fairly low in the sky from temperate latitudes, M93 sits about 1 1/2° northwest of Xi (ξ) Puppis, a 3rd-magnitude yellow supergiant star 7° northeast of Eta (η) Canis Majoris, in the tail of the Great Dog. Despite its low position in the sky as seen from mid-northern latitudes, the late Walter Scott Houston said he could detect it from his home in New England without optical aid.

Telescopically, the cluster fires the imagination. Smyth fancied a starfish pattern to the stars. (Smyth also shared a somewhat amusing story of how Chevallier d'Angos of the Grand Master's Observatory in Malta once mistook M93 for a comet, which caused Baron de Zach to call such astronomical blunders "Angosiades.") Kenneth Glyn Jones visualized a butterfly with open wings, and Brian Skiff saw three rows of stars emanating from the center like "trident spikes." To me, M93's V shape is part of a sunlit spider resting

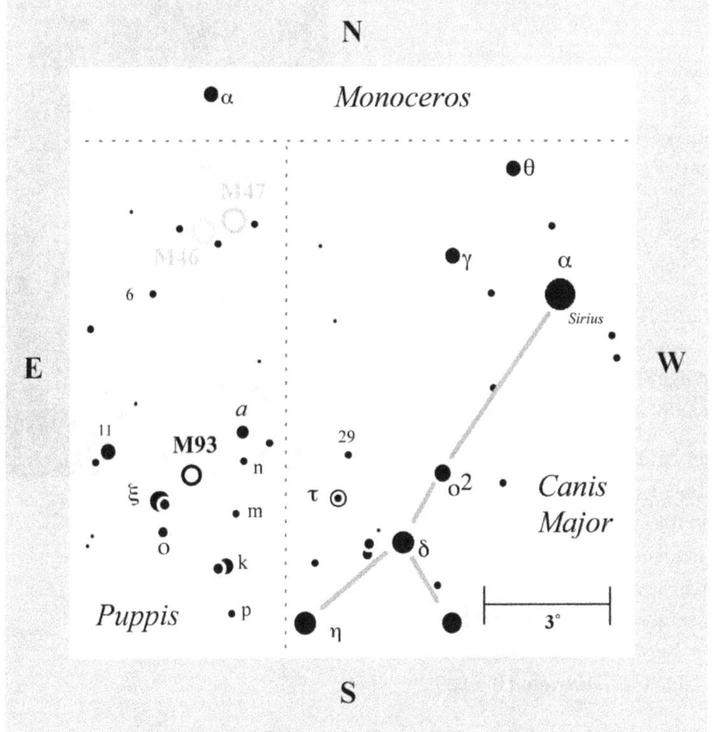

in the center of its dew-laden web, as you can see from my sketch.

M93 contains about 80 stars in a 10′ field. Do you get the impression that the two brightest stars in the southwest portion of the arrowhead appear yellowish orange? I see this only at 23×. Be sure to look all around the arrowhead because a swarm of fainter stars appears to form a glittering spherical halo around it. The sight is almost mystical.

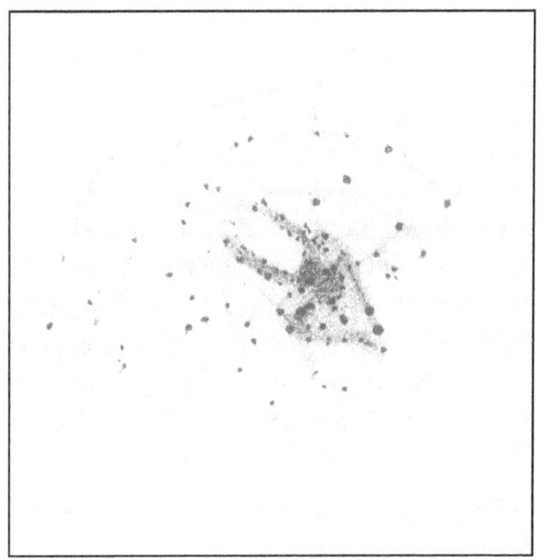

M94

Croc's Eye Galaxy
NGC 4736
Type: Spiral Galaxy ((R)SA(r)ab)
Con: Canes Venatici

RA: $12^h50.9^m$
Dec: +41°07′
Mag: 8.2
Dim: 13.2′ × 1.4′
SB: 13.5
Dist: ~17 million light-years
Disc: Pierre Méchain, 1781

MESSIER: [Observed March 24, 1781] Nebula without a star, above Cor Caroli [α Canum Venaticorum], on the same parallel as the sixth-magnitude star Flamsteed 8 Canum Venaticorum. The center is bright and the nebulosity not very diffuse. It resembles the nebula that lies below Lepus, number 79, but this one is finer and brighter. M. Méchain discovered it on 22 March 1781.

NGC: Very bright, large, irregularly round; very suddenly very much brighter in the middle to an extremely bright nucleus; mottled.

M94 IS A VERY SOFT AND BEAUTIFUL glow that adorns the high northern sky between the two brightest stars in Canes Venatici, the Hunting Dogs, lying between Boötes and Ursa Major. Smyth found it to be a "comet-like Nebula; a fine, pale-white object with evident symptoms of being a compressed cluster of small stars." Like John Herschel before him, he noticed that the object brightened toward the core, much like a globular cluster, so one can understand the confusion. In 1855, however, Lord Rosse discovered "a dark ring round the nucleus; then bright ring exterior to this. The annulus is not perfect, but broken up and patchy, and the object will probably turn out to be a spiral. Much faint outlying nebulosity."

How prophetic. Despite its placid appearance, M94 is an early-type spiral with ring structures. Unknown to our visual pioneers, we now know that the galaxy may have experienced a violent explosion only 10 million years ago – a cleansing event that disgorged millions of solar masses of material out of its disk or nucleus. Episodes of violent behavior may be typical for spiral systems; if so, we are seeing this mundane-looking spiral in a quiescent state. Its disk is oriented nearly face-on

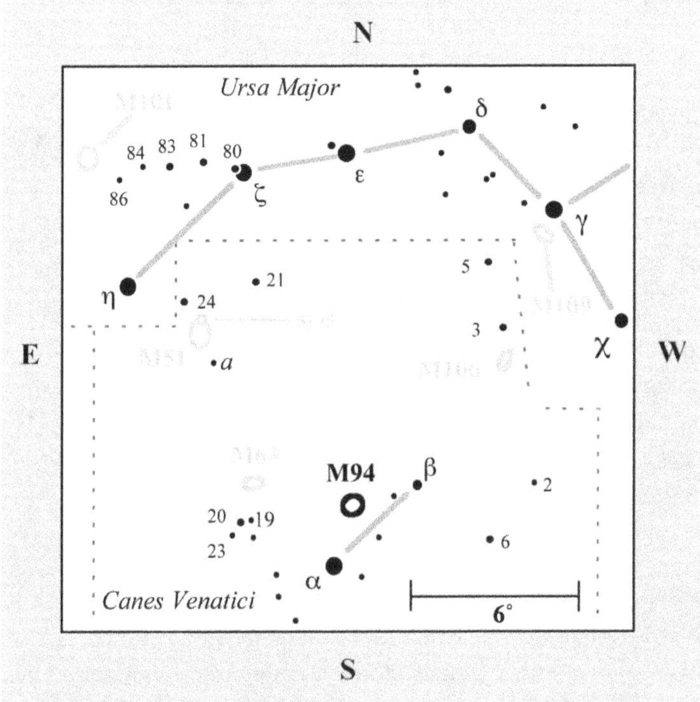

to us, and photographs through large instruments reveal its tightly wrapped arms, reminiscent of those in satellite views of a tropical hurricane.

Indeed, M94, a moderate-sized system some 65,000 light-years in true physical extent, harbors a poststarburst nucleus that hosts the closest example of a LINER 2 nucleus. Ultraviolet-light images by the Hubble Space Telescope reveal two bright point sources in the nuclear region and possible bow shocks, leading researchers to believe that two objects are in the final stages of merging in the nucleus. One or both of these sources may be a very compact star cluster or supermassive black hole, created during the merger event.

While this activity blossoms in the nuclear region, M94's tightly wound spiral arms appear smooth, with no evidence of recent star formation. Furthermore, no H II (star-forming) regions have been detected in the galaxy's innermost arms, which arc outside a nuclear bar. A pair of molecular arcs extend from the ends of the nuclear bar mostly toward the east but partly toward the west. Recent imaging has resolved the mysterious nuclear region into an expanding ring of intense star-formation activity. The ring is 50″ from the nucleus, is relatively blue, and has an extremely high surface brightness.

Ignacio Trujillo of the Instituto de Astrofísica de Canarias and his colleagues reported in a 2009 paper in *Astrophysical Journal* (vol. 704, p. 618) the results of a study of M94's outer region, which was long believed to form a closed stellar ring system. They found instead that the outer ring is actually a spiral arm structure that contains ~23 percent of the total stellar mass budget of the galaxy and contributes ~10 percent of the new stars created, showing that this region of the galaxy is also active. "In fact," the researchers say, "the specific star formation rate (SFR) of the outer disk ... is a factor of ~2 larger ... than in the inner disk. We have explored different scenarios to explain the enhanced star formation in the outer disk. We find that the inner disk (if considered as an oval distortion) can dynamically create a spiral arm structure in the outer disk which triggers the observed relatively high SFR as well as an inner ring similar to what is found in this galaxy."

This latest discovery was inspired by a CCD image of M94 taken in 2006 by amateur astronomer R. Jay Ganaby of New Mexico. When astronomers saw Ganaby's enhanced photo, which showed the galaxy's phantom outer ring as two giant spiral structures, they literally took to the sky and imaged the galaxy using NASA's Galaxy Evolution Explorer (ultraviolet light), 2MASS (infrared), and the Spitzer Space Telescope (infrared), as well as the Sloan Digital Sky Survey (optical) on the ground. A comparison of all these images led them to the discovery that M94's outer ring was indeed formed by two giant spiral arms extending from the galaxy's inner regions.

This compact galaxy looms brightly about 3° northwest of the double star Alpha (α) Canum Venaticorum, or 1 1/2° northeast of the midpoint between Alpha and Beta (β) Canum Venaticorum. At 23×, the galaxy can be seen in the neck of a Cygnus- or Grus-like asterism, shining with an almost starlike radiance, though a careful look will reveal that it is surrounded by a much larger, albeit fainter, glow.

M94 is a most intriguing sight at moderate to high powers because, with time, it seems to peer right back at you, like a hypnotic eye. A near opposite of the famous Black Eye Galaxy (M64), whose central region is lined with dark material, M94 sports a faintly luminous ring that surrounds a sharp nucleus. This ring reminds me of the yellow circle around a crocodile's eye – a discriminating, all-the-better-to-see-you-with marking. (Can you make out the southern curl of this ring?) The crocodile metaphor is also fitting because, as described earlier, this is a "violent" galaxy.

The 48-inch Schmidt camera on Palomar Mountain recorded another faint elliptical ring outside M94, perhaps debris blown out from the bellowing episode. It measures about 15′ in diameter (M94 is only 11.2′ in diameter on its major axis), has a sharp inner edge, and reveals a major axis skewed about 30° from that of M94. One would think that this ring would be out of reach of amateur instruments, so why bother?

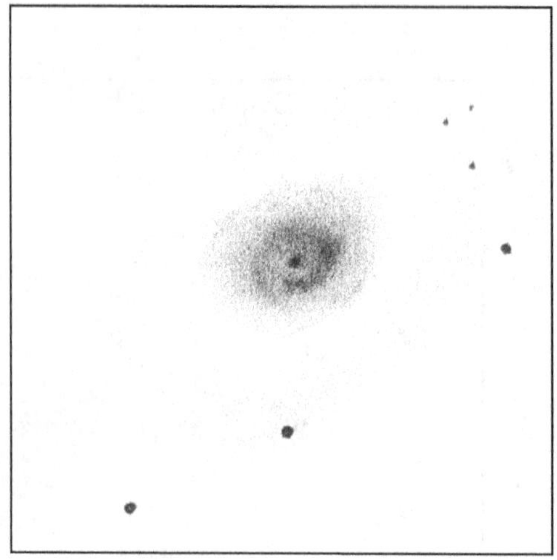

Well, when I swept the 4-inch back and forth across that field several times at 23×, I convinced myself that this outer ring might be at the visual threshold. My drawing of its location, extent, and orientation appears to be consistent with the one in the 48-inch image. Am I simply being misled by the power of suggestion? Or is this ring real and within reach of amateur instruments? Many times an object discovered with large telescopes turns out to be visible in small telescopes once knowledge of it exists. I'll let you be the judge.

Speaking of possible illusions created by the power of suggestion, Heinrich d'Arrest thought M94's circular disk and central star were similar to those of a planetary nebula. He even saw a "sky-blue" color!

M95

NGC 3351
Type: Barred Spiral Galaxy (SB(r)b)
Con: Leo

RA: $10^h44.0^m$
Dec: +11°42′
Mag: 9.7
Dim: 7.8′ × 4.6′
SB: 13.5
Dist: ~33 million light-years
Disc: Pierre Méchain, 1781

MESSIER: [Observed March 24, 1781] Nebula without a star in Leo, above the star *l* [53 Leonis]. Its light is very faint.

NGC: Bright, large, round, pretty gradually much brighter in the middle to a nucleus.

M96

NGC 3368
Type: Mixed Spiral Galaxy (SAB(rs)ab)
Con: Leo

RA: $10^h46.8^m$
Dec: +11°49′
Mag: 9.2
Dim: 6.9′ × 4.6′
SB: 12.9
Dist: ~35 million light-years
Disc: Pierre Méchain, 1781

MESSIER: [Observed March 24, 1781] Nebula without a star in Leo, near the preceding one [M95]. This one is less conspicuous. Both are on the same parallel as Regulus. They resemble the nebulae in Virgo, numbers 84 and 86. M. Méchain saw both on 20 March 1781.

NGC: Very bright, very large, little extended, very suddenly very much brighter in the middle, mottled.

M96

M95 AND M96, TWO SPIRAL SYSTEMS IN Leo, fit comfortably together in the field of view of 7 × 50 binoculars, separated by only 42′. Huddling with them in the same field is the elliptical galaxy M105. They lie 8° west of another Leo galaxy trio, consisting of M65, M66, and NGC 3628. All belong to a small cluster of galaxies known as the Leo Galaxy Group, whose members are speeding away from us at 800 miles per second.

M95 is a broken-ring spiral some 70,000 light-years across, seen almost face-on. Nearly circular arms spring from the ends of a very strong bar and wind slightly outward, almost touching the opposite bar and arm after a revolution of about 180°. They give the impression of a ring, but they are clearly broken, being composed of two separate spiral segments, from which dust lanes emerge. Multiple low-surface-brightness outer arms branch into separate fragments that can be traced for only about one-quarter of a revolution.

The ring is a region of starburst activity, which appears as blue knots where intense star formation is taking place. The blue ring measures about 20″ across and surrounds a redder nucleus of about 2″. As is normal for barred systems, the starburst appears to be the result of an inflow of gas toward the nucleus from the arm, which drives gas toward the star-forming regions.

M96 is another broken-ring barred spiral galaxy, the same physical size as M95 but inclined by about 45°. It has a small, bright nucleus in a broad, diffuse bar with many dark lanes. And, like M95, it has a bar that draws in a high concentration of gas toward the nucleus bar. But the activity there may also result from tidal interactions with other objects in the group.

M96 displays a curious morphology; namely, it's a one-armed barred spiral rather than a grand-design two-armed system. This arm contains many small H II regions. It emerges from the disk from the southwest and then coils outward to the right and up. Furthermore, M96's gas disk appears inclined to the symmetry plane of the stellar disk and may be of external origin.

The more we study galaxies and their components, the more we learn. For instance, astronomers expect any galaxy with a massive bulge to host a central supermassive black hole. The mass of the black hole also appears directly related to the mass of the bulge, providing evidence that the two have evolved together over time. As explained in a 2010 paper in the *Monthly Notices of the Royal Astronomical Society* (vol. 40, p. 646), N. Novak (Max-Planck-Institut für extraterrestrische Physik) and colleagues note that the massive bulge of M96 is expected to harbor a supermassive black hole with a mass of about 76 million solar masses.

M96 has been the focus of cosmologists' attention of late. Astronomers used the Hubble Space Telescope to probe the galaxy's spiral arms in search of Cepheid variable stars. The period of a Cepheid's variability is strictly related to the intrinsic luminosity of the star. By comparing how bright a Cepheid looks with how luminous it really is, astronomers can determine the star's distance. Using this stellar yardstick on Cepheids in M96 and comparing them with those in the nearby Large Magellanic Cloud, astronomers determined that M96 lies 38 million light-years away – nearly 60 percent farther away than previously thought. This implies that all the galaxies in the Leo Group might be that much more distant than earlier calculations had put them.

You can find M95 and M96 lurking about 9° east of Regulus (Alpha [α] Leonis) and 1 1/2° northwest of 5th-magnitude 53 Leonis. Although I did not estimate their magnitudes,

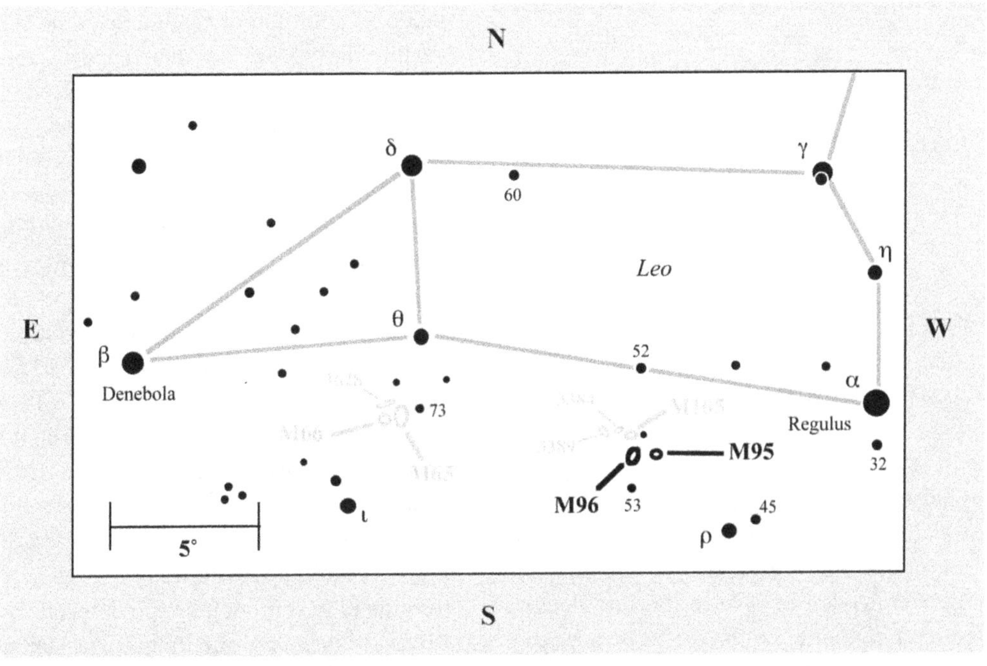

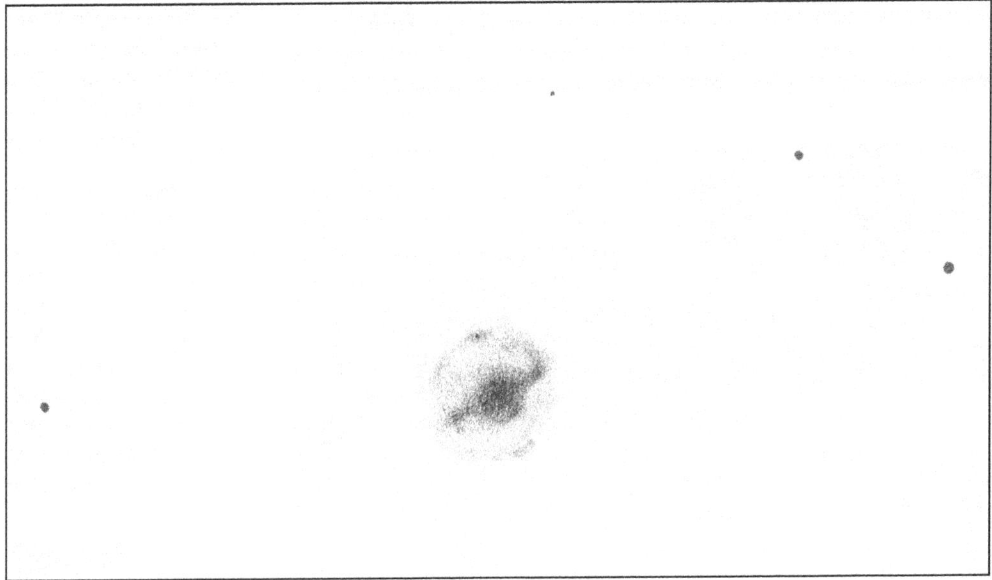

my gut feeling is that both are brighter than the values listed for them (9.7 for M95 and 9.2 for M96), perhaps by a half magnitude or more. At low power, M95 appears larger and more diffuse than its neighbor M96, whose core is much tighter.

M95 is a remarkable object that reminds me of Darth Vader's Death Ship in *Star Wars*

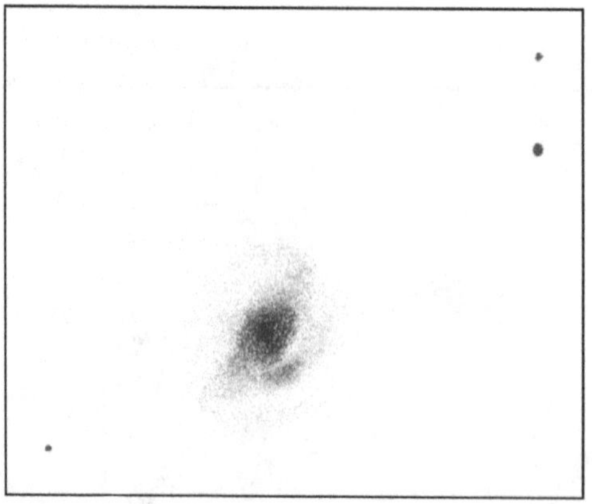

because it has a round center with bars extending out to arcing "wings." It shows hints of irregularity even at a glance, but high power is needed to resolve the fine details hiding in the inner region. At 130×, the central bar shows up with little difficulty, and it seems to have knots at each end. John Mallas, using his 4-inch f/15 refractor, could not see the bar. With its central disk, narrow arms, and beads of light, the galaxy looks a little like an image of Saturn with nearly edge-on rings. But this illusion vanishes with a longer look. The bar stretches away from the center and connects to a faintly luminous, slightly beaded ring that completely encircles the bar. Now the galaxy looks like an aerial view of an island covered by clouds and surrounded by a coral atoll. An impressive knot adorns the northern part of the ring.

The core of M96 is much more condensed than that of M95. A tiny nucleus lies within the strong central glow, but high power is needed to bring it out well. After studying the galaxy's faint textures at 130×, I noticed a detached arc of light just southwest of the elliptical nucleus. It appears to be the brightest part of a separate arm that branches off the southeastern side of a warped central bar; the southeastern tip of this bar curls westward, while the northwestern tip curls eastward. So the galaxy looks like a barred spiral that is on the verge of becoming fully spiral. See if you start to feel watched the longer you look at M96; I find that its shape resembles that of an eye – a piercing, almost sinister one at that!

Both M95 and M96 are known to have had one supernova to date: SN 2012aw appeared 60″ west and 115″ south of M95's center and shined at 15th magnitude at discovery. And Mirko Villi of Forli, Italy, discovered SN 1998bu at 13th magnitude, 4″ east and 55″ north of M96's center.

M97

Owl Nebula
NGC 3587
Type: Planetary Nebula
Con: Ursa Major

RA: $11^h14.8^m$
Dec: +55°01′
Mag: 9.8
Diam: 202″
Dist: ~1,300 light-years
Disc: Pierre Méchain, 1781

MESSIER: [Observed March 24, 1781] Nebula in Ursa Major, near β. M. Méchain reports that it is difficult to see, especially when the micrometer crosshairs are illuminated. Its light is faint, and it is without any stars. M. Méchain first saw it on 16 February 1781, and the position given is the one that he reported. Close to this nebula he saw another, whose position has not yet been determined, as well as a third, which is close to γ Ursae Majoris.

NGC: Very remarkable, planetary nebula, very bright, very large, round, very gradually then very suddenly brighter in the middle, 150 seconds in diameter.

THIS DISTINCTIVE PLANETARY NEBULA, so strongly reminiscent of the face of an owl (at least in images), is a real challenge for small telescopes because its faint light is spread over an area roughly five times the size of Jupiter's apparent disk. Thus, to see the Owl Nebula well, one needs to be under a dark sky.

In good conditions, M97 is one of the finer examples of a planetary nebula for a 4-inch telescope, displaying a pale but intricate shell of nebulosity. You have to keep a fixed gaze, though, to see its delicate details. It earned the "Owl" moniker from a sketch made in 1848 by Lord Rosse, who through his 72-inch telescope spied the two dark holes that look like eyes peering spookily from a round face. The Owl is certainly challenging, especially from a city, but I can recall seeing it many years ago from Cambridge, Massachusetts, with a 4-inch reflector. And John Mallas said he definitely saw the Owl's eyes with a 4-inch refractor in the dark skies over Arizona.

M97 sports a classical circular shape with a complex structure consisting of three concentric shells: a faint outer halo, a circular middle

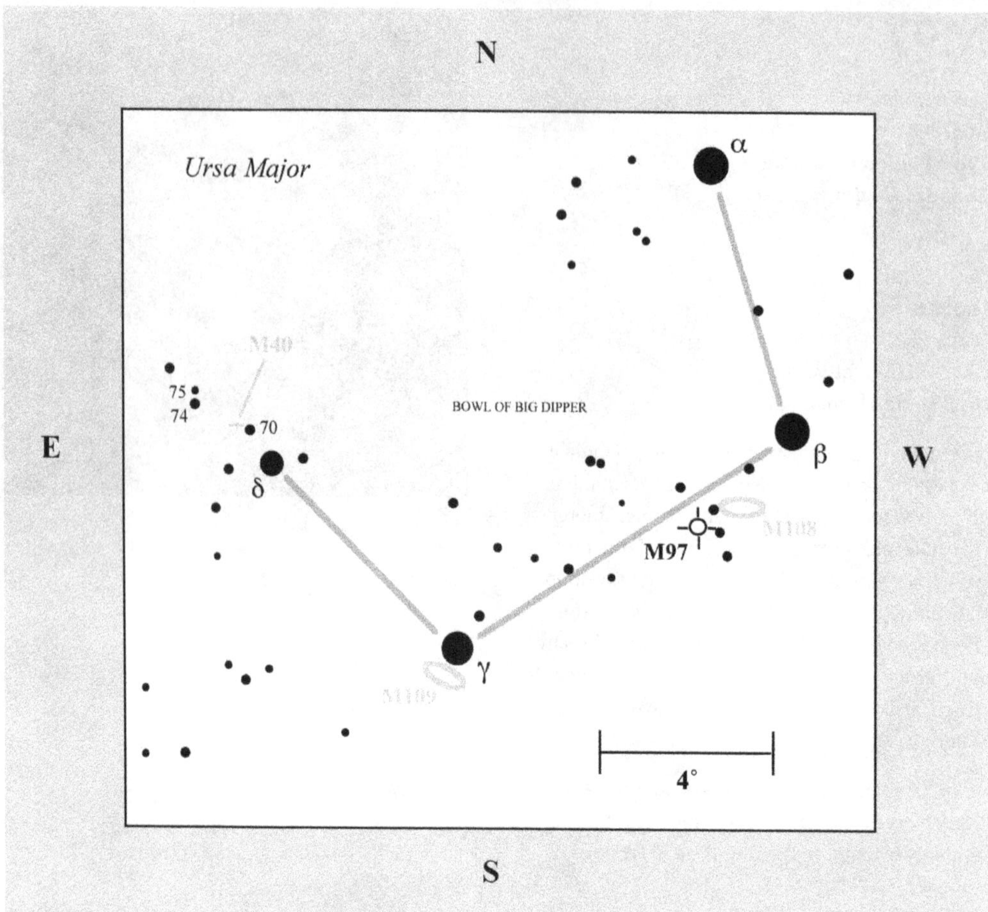

shell, and a roughly elliptical inner shell. As reported in the June 2003 *Astronomical Journal* (vol. 125, p. 3213), researchers from the University of Illinois at Urbana-Champaign, the Instituto de Astrofisica de Canarias in Spain, and Williams College in Williamstown, Massachusetts, used observations made with several telescopes to model how the Owl evolved. They concluded that the 6,000-year-old nebula formed in three wind-driven episodes. First, normal stellar winds created the outer shell after fusion in the parent star's core ceased. Second, a superwind later drove even more gas and dust outward to form the middle shell, though some say this shell is comprised of two components. And third, an even faster stellar wind finally compressed the superwind to form the inner shell and a bipolar cavity – a barrel-like component oriented at a 45° angle to the line of sight and whose matter-poor ends create the Owl's eyes.

The nebula's 16th-magnitude central star has a solar mass between 0.55 and 0.6 and appears to be in a fairly advanced stage of evolution. The solar mass of the nebula itself is much lower, on the order of 0.13. X-ray emission from M97 was first reported in 1989, and recent ROSAT observations confirm only one related source, the hot central star, with a

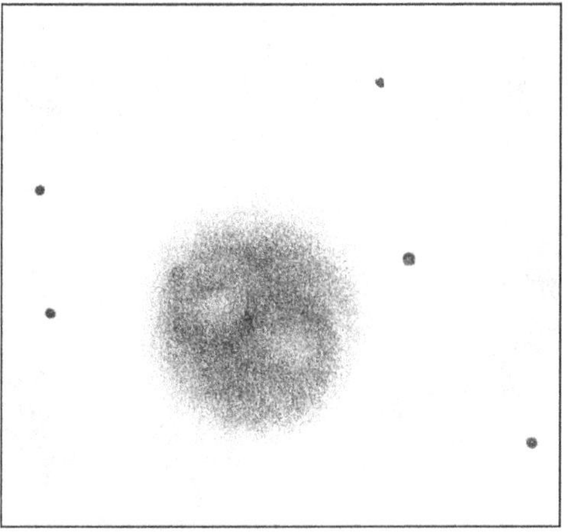

temperature between 100,000 and 200,000 K, which emits soft x-rays.

From Hawaii, M97 is just visible in 7 × 35 binoculars. The field is easily located 2° southeast of Merak (Beta [β] Ursae Majoris), a blue A1 main-sequence star in the bowl of the Big Dipper. In binoculars, the Owl forms the eastern arm of an asterism resembling the Southern Cross (Crux). Through the Genesis at 23×, the nebula is a strong ball of light surrounded by a thinner, weaker shell, which is slightly off-center - reminiscent of a composite print out of registration.

To see the innermost details of the Owl, you have to experiment with different eyepieces and determine which one gives you the highest magnification without diminishing the object's brightness. Jumping back and forth from 72× to 130× works best for me. At medium power, the glow from the Owl promptly appears as a nonuniform disk. Every now and then, something at the core pops into view, but I have seen it only occasionally with averted vision and it must be an illusion.

The Owl's eyes can be essentially inferred at 72× by sweeping your gaze back and forth across the disk. When I do this, I see a small smoke ring, like the Ring Nebula. At high power, two stars appear on the southern side, and I catch hints of the central brightening. The northern side of the planetary is brighter, slightly off-axis, and strongest toward the northeast. The eyes themselves are rather difficult to resolve, and it takes a lot of time behind the eyepiece to figure out their orientation. Rather than trying to see dark holes, try instead for the dim bar of light that separates them.

Don't be fooled, however, by the brighter northern outline of the eye sockets, which can trick you into believing that it is the bar. This pseudo-bar might be responsible for an optical illusion, which I call the "black-lash phenomenon." With low power and strong peripheral vision, a mysterious curved lash of darkness arcs across the glowing disk and wavers in and out of view. When I change the tilt of my head, the axis of this phantom lash also changes. I have also seen it flip-flop between two different positions. By the way, I have not read an account of anyone seeing color (outside of gray or silver) in this planetary. What do reflector owners have to say about this?

Coincidentally, when I finished observing M97 on the third and final night, I looked skyward and heard the hoot of a Pueo, the Hawaiian short-eared owl.

M98

NGC 4192
Type: Mixed Spiral Galaxy (SAB(s)ab)
Con: Coma Berenices

RA: $12^h13.8^m$
Dec: +14°54′
Mag: 10.1
SB: 13.2
Dim: 9.1′ × 2.1′
Dist: ~52 million light-years
Disc: Pierre Méchain, 1781

MESSIER: [Observed April 13, 1781] Nebula without a star, whose luminosity is extremely faint, above the northern wing of Virgo, and on the same parallel as and close to the fifth-magnitude star Flamsteed 6 in Coma Berenices. M. Méchain saw it on 15 March 1781.

NGC: Bright, very large, very much extended along position angle 152°, very suddenly very much brighter in the middle.

WHAT A REFRESHING SIGHT! THE nearly edge-on galaxy M98 in Coma Berenices offers the small-telescope user more detail to hunt down than most of the face-on spiral and elliptical systems in this region. Its pale, slender disk, hooklike extensions, and dark dust make it a soothing yet enticing sight. Like M86, this metropolis of 130 billion solar masses, extending 140,000 light-years across, lies in the great Coma–Virgo galaxy cloud. But unlike most of the other members of the cloud, which are receding from us at several hundred miles per second, M98 is one of only five galaxies in the cluster having a negative velocity relative to the Local Group, appearing to be approaching us at 142 miles per second.

This mixed spiral galaxy is highly inclined to our line of sight (~83°). Nevertheless, we can see a very small and extremely bright nucleus (partially hidden by a dark lane) embedded in a smooth bar with resolved spiral arcs near its edge. The galaxy's smooth outer arms form a pseudo-ring, and H II regions dapple the arms.

M98's inner dust lanes cross the bulge in such a way that astronomers early on concluded this was a sign of a strong tidal disturbance. Later observations suggested that the galaxy has an active galactic nucleus with ongoing star formation.

In a 2009 paper in the *Monthly Notices of the Royal Astronomical Society* (vol. 400, p. 2098), Sebastian Perez (Universidad de Chile) and his colleagues discovered shocks (possibly driven by a radio jet) in the nuclear region as well as x-ray heating from the active galactic nucleus. But the jury is still out on

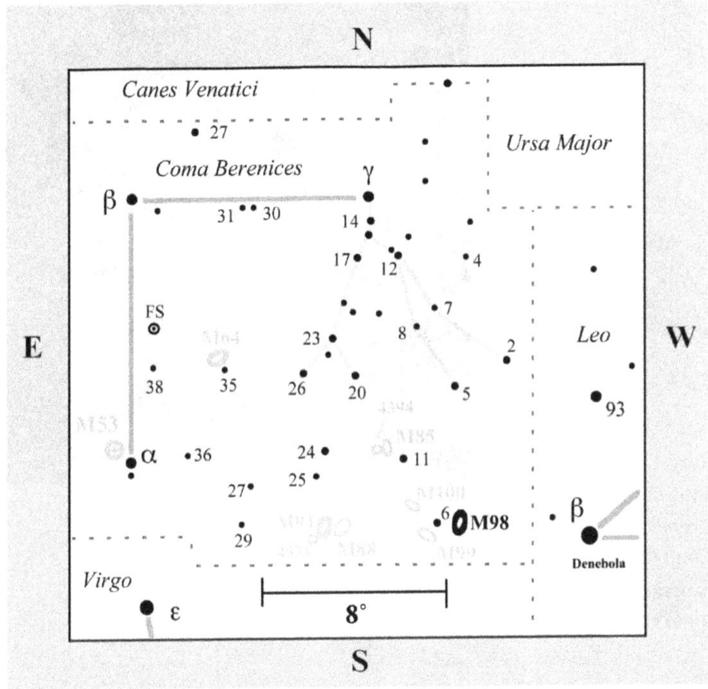

whether the outflows are created by superwinds driven from nuclear star clusters or if a highly reddened black hole lurks within the dusty inner heart of M98.

M98, a 10th-magnitude spiral galaxy, can be found just 1/2° east of 5th-magnitude 6 Comae Berenices, so you will have to use high magnification to exclude that star from the field of view. Through the 4-inch, the most obvious feature of the galaxy, aside from its highly elongated shape, is a fan of material southwest of the nucleus, which when studied more closely seems composed of bright and dark patches reminiscent of those in M108. The inner nuclear region has spiral structure that abruptly terminates along the eastern rim, yielding semicircular arms that look like pincers – much more prominent and dramatic than the crablike claws of M90.

These features of M98, created by a remarkable dustiness, are best seen with high power. Now switch to a low-power eyepiece. Do you see how the entire galaxy is surrounded by an elongated halo that stands out boldly against the dark background? For *Star Trek* fans, does this galaxy remind you of a Klingon vessel?

Admittedly, the first night I happened to view M98 turned out to be one of the clearest nights in this observing project. That evening, the Milky Way stood above the horizon like a god. It struck me as ironic that, after spending so many hours squinting at dim and distant M98, I could look away from the eyepiece to see our Galaxy spread so boldly across the sky that I thought it would crush me.

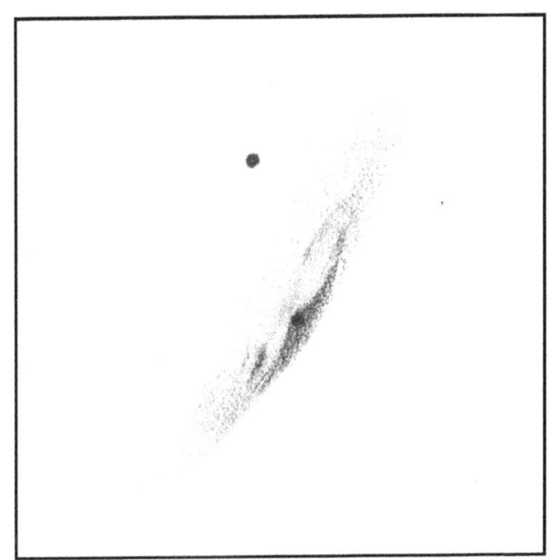

M99

NGC 4254
Type: Spiral Galaxy (SA(s)c)
Con: Coma Berenices

RA: $12^h18.8^m$
Dec: +14°25′
Mag: 9.9
Dim: 4.6′ × 4.3′
SB: 13.0
Dist: ~50 million light-years
Disc: Pierre Méchain, 1781

MESSIER: [Observed April 13, 1781] Nebula without a star, whose luminosity is very dim, but which is slightly clearer than the previous one [M98], lying on the northern wing of Virgo, and close to the same star, Flamsteed 6 in Coma Berenices. The nebula is between two stars of the seventh and eighth magnitude. M. Méchain saw it on 15 March 1781.

NGC: Very remarkable. William and John Herschel called it bright, large, round, gradually brighter in the middle, mottled. Rosse and William Lassell saw it as a three-branched spiral.

M100

The Mirror of M99
NGC 4321
Type: Mixed Spiral Galaxy (SAB(s)bc)
Con: Coma Berenices

RA: $12^h22.9^m$
Dec: 15°49′
Mag: 9.3
Dim: 6.2′ × 5.3′
SB: 13.0
Dist: 56 million light-years
Disc: Pierre Méchain, 1781

MESSIER: [Observed April 13, 1781] Nebula without a star, with the same luminosity as the preceding one [M99], lying within the ear of corn in Virgo. Seen by M. Méchain on 15 March 1781. These three nebulae, numbers 98, 99, and 100, are very difficult to recognize, because of the dimness of their light. They may be seen only under good conditions, and near meridian passage.

NGC: Very remarkable, pretty faint, very large, round, very gradually then pretty suddenly much brighter in the middle to a mottled nucleus. With the 26-inch Leander McCormick refractor, Lassell saw it as a two-branched spiral.

JUST 1 1/2° SOUTHEAST OF THE NEARLY edge-on spiral M98 in Coma Berenices looms the face-on pinwheel M99. The two galaxies present a nice morphological contrast together in the low-power field of view. Another way to locate M99 is to look 50′ southeast of the magnitude 5.1 star 6 Comae Berenices, 2° northeast of which beckons M100. Let's look at M99 first.

M99 is among the 10 largest spirals in the Virgo Cluster. In images made with large-aperture instruments, the 70,000-light-year-wide system shows two thin spiral arms that fly off to the west from the main central mass and one strong arm that arches far to the east. Several minor tufts of starlight also stray off the nuclear region to the west. Overall, the entire spiral structure looks chaotic and asymmetrical.

The two principal arms are of the grand design type. Many of the subsidiary arms are branched and studded with rich star-forming knots. In 1995, near-infrared images of M100 released by astronomers working on Mauna Kea in Hawaii and on La Palma in the Canary Islands revealed stubby spiral arms close to

M100

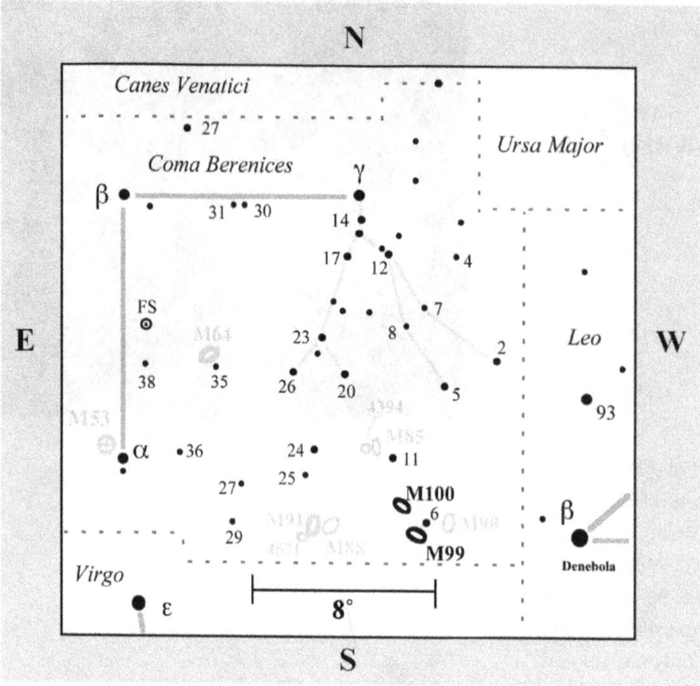

the nucleus that appear to be rotating in the opposite direction of arms farther out in the giant spiral. The astronomers suggested that these represent regions of new star formation, where hot young stars are being forced into independent orbits, especially along the southern arm, where much recent star formation has occurred.

Dust lanes line the major arms (as usual, generally on the inside edge) and also pervade the entire central lens. There is some circumnuclear star formation scattered between the dust spirals, but they do not appear associated with them.

Although M99 is not classified as a barred spiral galaxy, the molecular gas distribution suggests a weak bar with multiple molecular gas arms extending from it. In 2005, astronomers discovered a bridge of neutral-hydrogen gas 800,000 light-years long that appears to connect M99 with a curious region known as Virgo HI 21 – which may contain a galaxy's worth of dark matter! The gravity from this controversial "dark galaxy" may have drawn out this tail when M99 had a close encounter with it in the past. Then again, the tidal tail could have resulted from "galaxy harassment" as M99 entered the Virgo Cluster.

M99 is not alone in its struggles against tidal interactions. M100 appears to be a victim of tidal harassment as well. There is a lack of symmetry in the brightest features of M100's disk, which might suggest that the galaxy is being warped by its neighbors.

M100 is *the* brightest spiral in the Virgo Cluster. It possesses characteristics of both M51 and M101, displaying multiple arms, of which only its two grand-design arms are bright. Faint secondary arms lie between the principal arms on the east side of the nucleus, completely filling the space between the bright arm closest to the nucleus on the north side and the extension of the second main arm in the northeast quadrant after it has spiraled through 360°.

A "hot spot" galaxy, M101 harbors a nuclear ring that contains four distinct H II regions, where strong star formation is present. It's argued that this mixed spiral's very weak bar is driving vigorous star formation in a ring that surrounds the central disk.

The most surprising image of M100, however, came in late 1993, when the newly repaired Hubble Space Telescope turned its

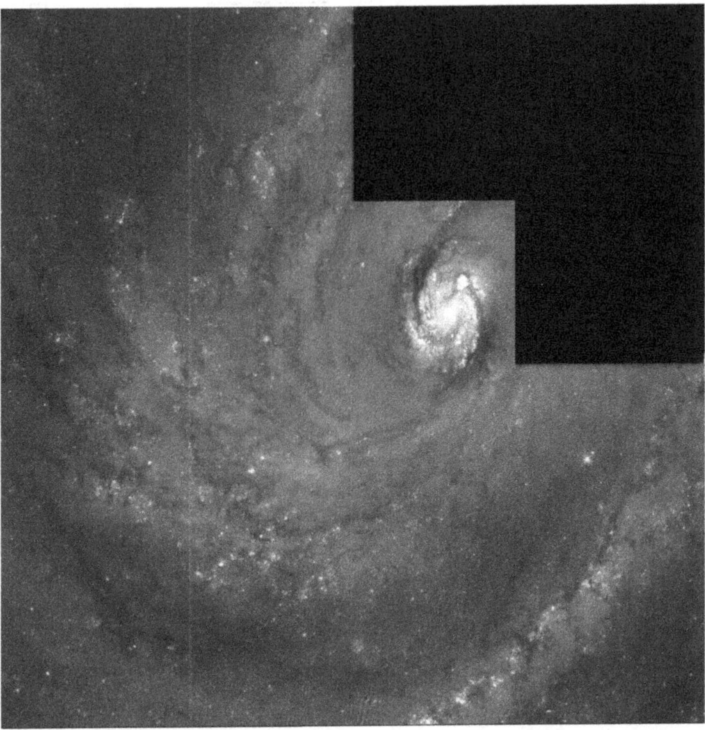

variables in the outskirts of the spiral, leading astronomers to refine the distance scale of the universe. For example, 41 million light-years had once been the generally accepted distance for M99. But the HST data have led to a recalculation of the distance to more like 56 million light-years.

The telescopic image of M99 in the 4-inch is but a pale ghost of that mighty spectacle. Yet details do emerge with effort. At 23×, a bead-like nucleus is set inside a round, unassuming outer glow. Increasing

Wide Field and Planetary Camera onto the galaxy's core, revealing a vortex of blue and yellow stars that looks like a galaxy within a galaxy. The HST also resolved about 20 Cepheid

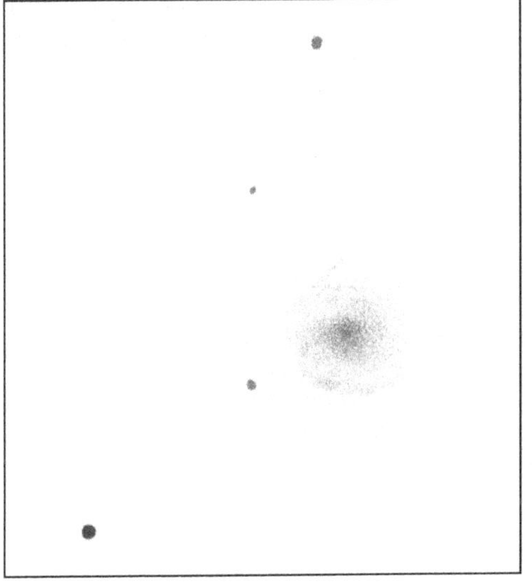

to moderate magnification transforms the galaxy into a subtle beauty. The two strongest arms materialize slowly, soon displaying graceful knots that look like diamonds balanced on a fingertip. This is a magnitude 9.8 galaxy with an angular size of 4.6′ × 4.3′ and a low overall surface brightness. The quality of your skies may well determine whether you see any spiral structure at all. Roger Clark, in his book *Visual Astronomy of the Deep Sky*, reported seeing "no hint" of spiral detail in his 8-inch Cassegrain under "moderate to good" sky conditions.

Astronomically, M100 is one of the larger spiral systems in the Coma–Virgo cloud of galaxies, comparable in size to our own galaxy. Through a telescope, however, M100 has little to offer. Although at magnitude 9.4 it is brighter than M99, it is nevertheless a more difficult object because its light is spread over an even larger surface area (6.2′ × 5.3′).

It remains a pale orb even with high magnification, though perseverance will show traces of ghostly arms.

Photographically, M100 is supremely textured, its spiral appendages interspersed with small patches of obscuring matter. The moral here is not to become too spoiled by photographs, especially of face-on galaxies, which are always much more transparent to the eye!

Just look at the drawings of the various galaxies throughout this book. I find it remarkable that I can look at an oblique galaxy such as M98, 55 million light-years away, and resolve more detail more easily than I can for a face-on galaxy like M74, which is 23 million light-years closer. This is similar to the effect one sees when looking at a planetary nebula's spherical shell of gas, where more light can be seen on the outer edge than in the center because more material is concentrated in our line of sight on the edge.

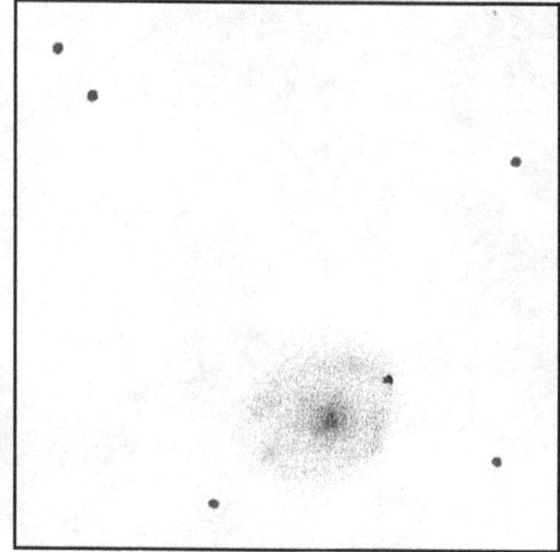

To date, M99 has had three known supernovae: SN 1967H, SN 1972Q, and SN 1986I. M100 has had five: SN 1901B, SN 1914A, SN 1959E, SN 1979C, and SN 2006X.

M101

The Pinwheel Galaxy
NGC 5457
Type: Mixed Spiral Galaxy (SAB(rs)cd)
Con: Ursa Major

RA: $14^h03.2^m$
Dec: +54°21′
Mag: 7.9
Dim: 26.0′ × 26.0′
SB: 14.8
Dist: 25 million light-years
Disc: Pierre Méchain, 1781

MESSIER: M101: [Observed March 27, 1781] Nebula without a star, very dark and extremely large, 6 to 7 minutes in diameter, between the left hand of Boötes and the tail of Ursa Major. Difficult to distinguish when the crosshairs are illuminated.

NGC: Pretty bright, very large, irregularly round, gradually then very suddenly much brighter in the middle to a small, bright nucleus.

M102

duplicate observation of M101 or =
NGC 5866?
Type: Lenticular Galaxy (SAO)
Con: Ursa Major? Draco?

RA: $15^h06.5^m$
Dec: +55°46′
Mag: 9.9
Dim: 6.6′ × 3.2′
SB: 13.1
Dist: 48 million light-years
Disc: William Herschel, 1788

MESSIER: M102: [No date listed] Nebula between the stars ο Boötis and ι Draconis. It is very faint. Close to it is a sixth-magnitude star.

W. HERSCHEL: [Observed May 5, 1788] Very bright, considerably large, extended, following 2nd star. (H I-215)

NGC: Same as M101, if duplicate observation. If NGC 5866: Very bright, considerably large, pretty much extended toward position angle 146°, gradually brighter in the middle.

M101 IS ONE OF MY FAVORITE FACE-ON spirals. Its numerous, far-flung arms and distinct asymmetry make it readily identifiable in images. Visually, its pale 9th-magnitude glow is diffuse and difficult to make out, yet it has much to offer if you are patient.

Although stunning, M101 is similar to M100 in that its spiral pattern is very chaotic; dust segments cross the entire inner lens of M101 and, though they define an overall spiral pattern, cannot be connected with one another to form continuous lanes. Still, the outer arms can be traced from a branching of two principal dust arms that begin in the nucleus. In fact, Many Sc galaxies are of the M101 type, with three main characteristics: (1) a small nucleus from which the spiral pattern emerges in dust lanes, at first, and farther out in luminous filaments; (2) multiple spiral arms, thin and highly branched in their outer regions; and (3) arms well resolved into individual stars and H II regions.

A giant spiral disk of stars, dust, and gas, M101 spans 170,000 light-years of space – nearly twice the diameter of the Milky Way. It contains at least one trillion stars, 100 billion of which are Sun-like. In 2006, the Space Telescope Science Institute released the largest and most detailed photo ever taken of M101. The galaxy's portrait is actually composed of 51 individual Hubble exposures, in addition to elements from images from

ground-based photos. The final composite image measures a whopping 16,000 by 12,000 pixels. As the image here shows, the galaxy's spiral arms are sprinkled with large regions of star-forming nebulae. These nebulae are areas of intense star formation within giant molecular-hydrogen clouds. Since the galaxy is 25 million light-years distant, we see its light as it shined at the beginning of Earth's Miocene Period, when mammals flourished and the mastodon and the first modern-day cats (like the saber-toothed tiger) first appeared on Earth.

M101 is easy to locate because you can starhop to it from the famous visual double star Alcor and Mizar in the handle of the Big Dipper. Southeast of this pair is a path of four 5th-magnitude stars (81, 83, 84, and 86 Ursae Majoris) that leads in the direction of M101, which lies 1 1/2° east-northeast of 86 Ursae Majoris at the end of the trail. The galaxy is visible in binoculars, so try them first before taking to the telescope.

N
Ursa Major
M101
84 83 81 80
86
δ
ε
ζ
γ
η
21
24
5
3
χ
E
W
a
β
2
20 19
23
6
α
Canes Venatici
6°
S

I wonder whether someone with young eyes could detect M101 without optical aid. At magnitude 7.7, it would be quite a challenge – extremely dark skies and a high-altitude site would probably be required – but the reward is obvious: this

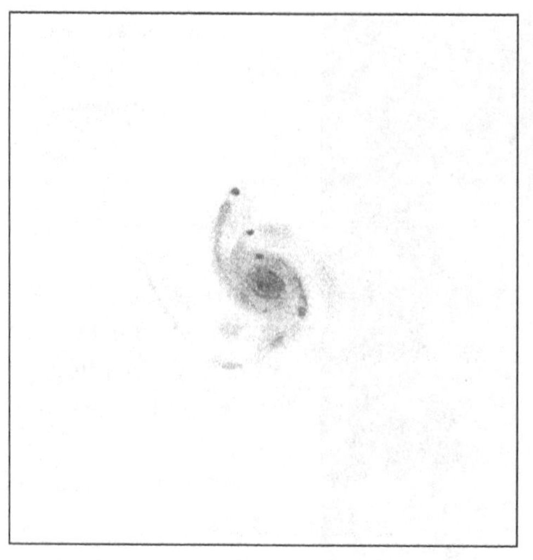

person would belong to a club of one, having the distinction of being the only human to see 17.5 million light-years into the universe with the naked eye.

Telescopically, at 23×, the galaxy's core is compact but slightly elliptical. Knots in phantom spirals emerge from the galactic mists, but moderate power is much more effective in coaxing them out. Low power is best saved for determining the full extent of the far-flung arms. High power, on the other hand, is your best bet for examining the inner region, which shows a sharp nucleus inside a pentagon-shaped core. Only two arms are very definite; beyond that, everything shifts in and out of view. You might have to alternate between moderate and low power to preserve your sanity as you try to keep track of their directions.

I needed several nights to study this one galaxy, knowing that persistence behind the eyepiece would enhance my view. And, indeed, the time was well spent, because I was able to trace the patterns of a few of its major arms. Still, the view through the eyepiece pales in comparison to the photographic image. And to think that, as Timothy Ferris points out in his book *Galaxies*, billions of stars and interstellar clouds between these arms are too faint to register in photographs. "It is as if we are flying on a landscape at night," he writes, "where the brightly lit cities catch our eye at the expense of the dark farmlands."

For a pleasant diversion, there is a pentagon of 11th-magnitude NGC galaxies 1° north of M101 whose pattern mimics the shape of M101's nuclear region. Another solitary 11th-magnitude galaxy, NGC 5474, lies about 45′ south-southeast of M101; its spectral glow vanishes, however, if the sky is at all hazy.

M101 has hosted three supernovae: SN 1909A (magnitude 12.1), SN 1951H (magnitude 17.5), and SN 1970G (magnitude 11.5).

Turning now to our discussion of M102, the unassuming lenticular galaxy NGC 5866 in Draco has long been at the center of a popular debate among deep-sky observers: whether it is M102 – one of the few ambiguous objects in Charles Messier's famous catalogue of nebulae and clusters. The history surrounding this debate is worth repeating. In Messier's catalogue, the one hundred second entry is described as:

> Nebula between the stars ο [Omicron] Bootis and ι [Iota] Draconis. It is very faint. Close to it is a sixth-magnitude star.

Omicron Boötis and Iota Draconis are separated by nearly 40°, making them unlikely markers for even the most skilled observers of Messier's day. Thus, something is clearly amiss with the description. Here's why.

On April 13, 1781, Messier made the last observation of the objects included in his final supplement of the catalogue. It was of M100, a "nebula without star" in Virgo, which Messier's contemporary Pierre Méchain had discovered the previous month. The supplement appeared in the French almanac *Connaissance des Temps* for 1784, which would be published in 1781. Méchain had supplied Messier with a number of new objects, and though Messier worked feverishly to observe all of them before his publishing deadline, time ran out, and three objects went unseen. Still, Messier appended these objects to the supplement as numbers 101, 102, and 103 under the heading: "By M. Méchain, which M. Messier has not yet seen." Furthermore, Méchain had provided Messier with a measured position for only entry 101.

After the supplement was printed, Méchain noticed an error – entry 102 was not a new object but rather a second observation of entry 101. After discussing the matter with Messier, Méchain drafted a letter of correction, which first appeared, in abbreviated form, in the *Histoire* of the Berlin Academy for 1782. A copy of the letter (dated May 6, 1783) was also sent to Johann Bernoulli, editor of the *Berliner Astronomisches Jahrbuch*, in Berlin. Bernoulli translated the original letter into German and printed it in 1786:

> On page 267 of the *Connaissance des Temps* for 1784 M. Messier lists under No. 102 a nebula which I have discovered between omicron Bootis and iota Draconis. This is nothing but an error. This nebula is the same as the preceding No. 101. In the list of my nebulous stars communicated to him M. Messier was confused due to an error in the sky-chart.

Mystery solved? It would seem so, but alternate theories exist.

At the heart of the M102 theory is the error on Messier's sky chart. Many have speculated on this error and have come to the same conclusion, which was first introduced by William Henry Smyth in 1844 in *A Cycle of Celestial Objects*. It can be found in his discussion of M102:

> A small but brightish nebula, on the belly of Draco, with four small stars spreading across the field, north of it. There may be a doubt as to whether this is the nebula discovered by Méchain in 1781, since Messier merely describes it as "very faint," and situated between *o* Bootis and *i* Draconis. But there must be some mistake here; the one being on the herdsman's leg and the other in the coil of the Dragon far above the head of Bootes, having 22° [sic] of declination and 44′ of time between them, a space full of all descriptions of celestial objects. But as the θ in the raised right hand of Bootes, if badly made, might be mistaken for an *omicron*, this is probably the object seen by Méchain, and [John Herschel's] 1910; it being the brightest nebula of five in that vicinity.

In addition to being the first to suggest that Messier confused the star θ Boötis for o Boötis, Smyth appears to have been the first to suggest that M102 is NGC 5866 (John Herschel's 1909, which Smyth misidentifies as Herschel's 1910), which is situated between θ Boötis and ι Draconis.

Smyth's solution is beautiful except for one fact: If Méchain had discovered NGC 5866, he was not aware of it. In the letter he sent to Bernoulli – the same one in which he corrects the error concerning M102 – Méchain included the descriptions of six nebulae he discovered in 1781, and NGC 5866 is not among them.

But mistaking θ Boötis for o Boötis, as Smyth suggests, can't be the only error, since M101 does not lie between θ Boötis and o Draconis.

Some have argued that Méchain's descriptions of M101 and M102 are suspiciously different, since he calls M101 a "nebula without star, very obscure and large" and M102 a "very faint" nebula near a "star of the 6th magnitude." The difference between a very obscure nebula and a very faint nebula is trivial. M101 is two magnitudes brighter than NGC 5866. M101 is a round glow nearly 30′ in diameter, while NGC 5866 is a mere 7′ × 3′ spindle of light. Through a small telescope, M101 is highly diffused, while NGC 5866 is highly condensed. The two objects are dramatically different in size and appearance.

So what is M102? If we accept Méchain's words, there was an error on Messier's star chart, Messier was confused, and M102 is a second observation of M101.

Here is how Messier presented Méchain's description of M101 in the 1784 *Connaissance des Temps*:

> Nebula without a star, very obscure and extremely large, 6 to 7 minutes in diameter, between the left hand of Bootes and the tail of Ursa Major. Difficult to distinguish when the crosshairs are illuminated.

Messier also included a position determined by Méchain on March 27, 1781, that is quite accurate in declination but about 1 3/4° west of its true position. As reported on the Students for the Exploration and Development of Space (SEDS) Web site (http://www.seds.org/messier/m/m102d.html), Messier's personal copy of the catalogue printed in the *Connaissance des Temps* for 1784 contains a penned position for M102 that falls almost midway between θ Boötis and ι Draconis. Just as Méchain had stated in his letter to Bernoulli, there was an error on Messier's star chart. Either θ Boötis was mislabeled ο Boötis or, in his haste, Messier simply mistook θ for ο.

Unfortunately, this discovery does not solve the mystery of M102, because nothing exists at the penned position. The closest object is the 12th-magnitude galaxy NGC 5687, which was too faint for Méchain to see. NGC 5866 is 3° east-northeast of the penned position, and M101, on the other hand, is more than 6° west of it.

If M102 is indeed M101, and if θ Boötis was mislabeled on Messier's star chart (which makes perfect sense because θ is in the "left hand of Bootes"), then one thing that does not make sense is the position of M102 that Messier penned in his personal copy of the *Connaissance des Temps*. Did Méchain mean that Messier was confused because he went in the wrong direction when he plotted Méchain's second observation?

Since θ Boötis is the brightest naked-eye star near M101, Messier likely would have used it as an offset star. Messier does use "the left hand of Bootes [θ Boötis]" to help position M101.

Given these facts, I plotted Messier's penned position of M102 on a star atlas and stepped back to study the scene. I was immediately struck by how well this position mirrored M101's position with respect to θ. If Messier's penned position of M102 is plotted west of θ Boötis instead of east of the star, then it falls almost right on top of the modern position of M101.

Do two positions of M101 actually exist? Yes. *Sky & Telescope* magazine senior editor Dennis di Cicco noticed that in *The Messier Album* by Mallas and Kreimer, in the facsimile of the Messier catalogue (as Messier himself published in the *Connaissance des Temps*), there are two right-ascension positions for each object – one in time, the other in degrees, minutes, and seconds. Unlike all the other objects, the two positions given for M101 differ from one another.

The 1781 position of M101 in time, when precessed to epoch 2000.0, is right ascension $13^h51^m33^s$, declination +54°19′09″. That position places M101 1 3/4° to the west of M101's modern position. It also places M101 a mere 40′ from the 6th-magnitude star 86 Ursae Majoris. This, of course, would explain why Messier, in his description of what he thought was M102, wrote: "Close to it is a sixth-magnitude star."

On the other hand, the 1781 position of M101 recorded in degrees, minutes, and seconds, when precessed to epoch 2000.0 and converted to time, is $13^h59^m55^s$, declination +54°20′25″. The NASA Extragalactic Database position for M101 (epoch 2000.0) is $14^h03^m13^s$, declination +54°20′57″ – a difference of only 1′ from the second, more refined position of M101 published in Messier's catalogue.

It seems reasonable that one good and viable solution to the mystery is that M102 is a second, more refined observation of M101.

But what about NGC 5866? On the SEDS Web site (http://messier.seds.org/m/m102d.html), Hartmut Frommert presents an equally intriguing argument for NGC 5866 to be M102. The full article and a supplement are well worth reading, and I encourage you to do so. In summary, Frommert has this to say:

> The object that really deserves the designation "Messier 102" should be identical to one of the two observed by Méchain and Messier, may they be identical or not. As nobody is still alive who has witnessed them during their observation and recording, we can currently not reconstruct what they actually observed. Méchain's description gives good evidence that the object M102 could be NGC 5866, which most probably everybody would believe if he had not retracted the discovery in the letter mentioned, or if this letter had stayed forgotten. It may depend on taste to speculate which was erroneous: The observation or the letter. Moreover, Messier has probably observed NGC 5866 and taken it for M102, but again made an error in data reduction. Once more, it is a question of taste if these facts entitle the lenticular galaxy NGC 5866 to bear the designation "M102."

NGC 5866 is a remarkable, nearly edge-on object (90,000 light-years in diameter) with a very bright ring and a dust lane – both of which are tilted against the equatorial plane of the elliptical halo – containing an extended halo of globular clusters. Indeed, the outer envelope simulates an elliptical galaxy (which are generally known to be rich in globular clusters), but the lens shows a narrow ring of dark material surrounding the large nucleus, so it has a classic lenticular form.

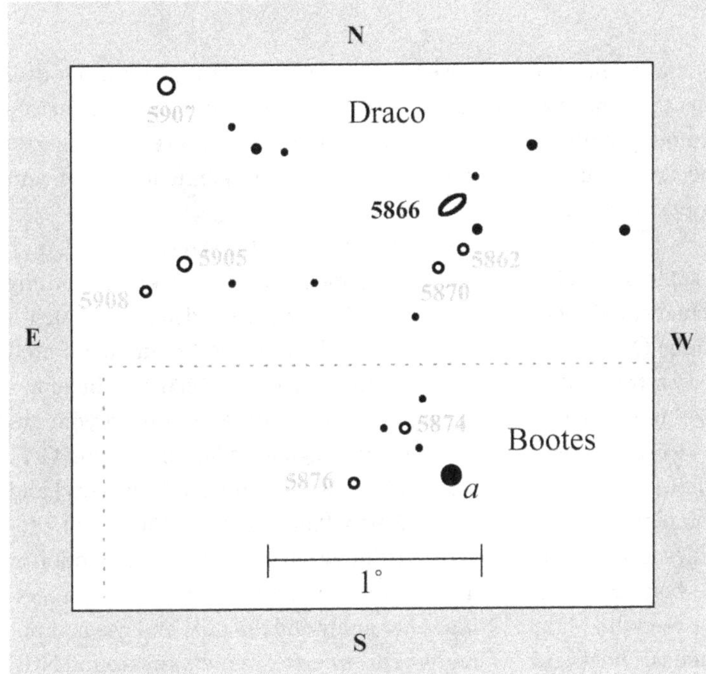

NGC 5866 has an unusually rapid gas rotation curve, with a measured maximum rate of 320 kilometers per second within about 3,000 light-years of the Galactic center. This is much faster than would be expected based on the light distribution of the bulge and the disk, and two possibilities

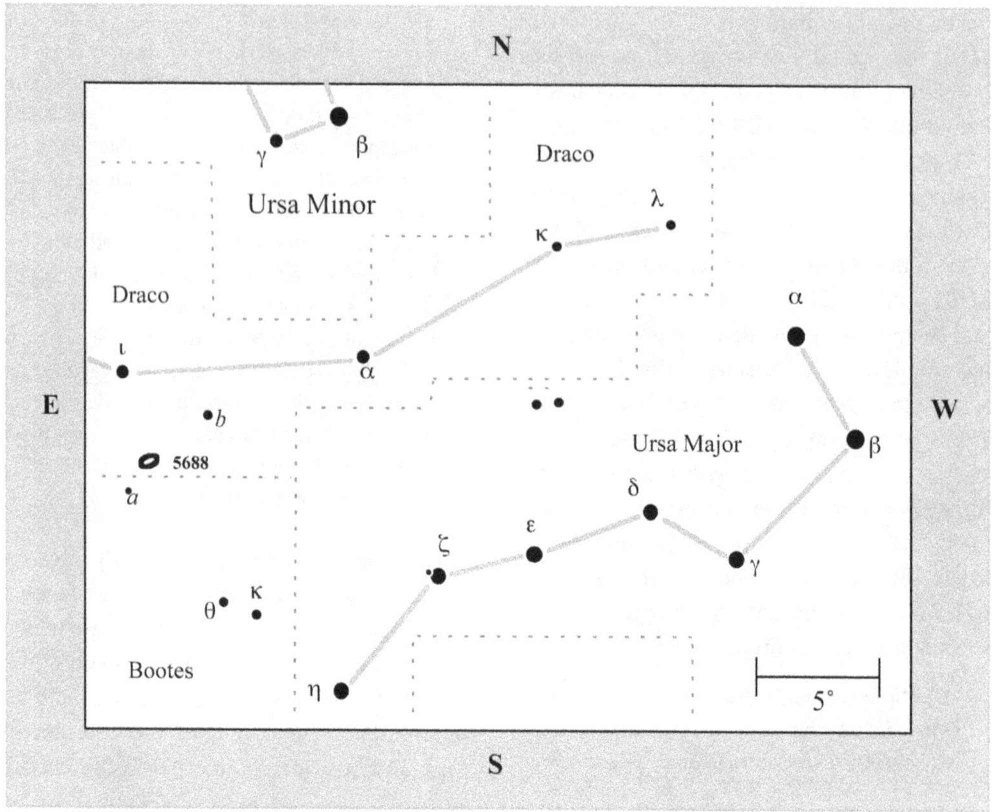

have been proposed. One is that within the disk there is a mass component that has not been accounted for, and the other possibility is a decoupled system located at the Galactic center that is the result of a past capture or merger process.

NGC 5866 is also remarkable for being immersed in a massive molecular cloud. NGC 5866 has two wide companions (NGC 5907 and NGC 5879). NGC 5907 is a thin, 10th-magnitude ellipse about 1 1/2° to the northeast, and NGC 5879 is an 11th-magnitude ellipse about 1 1/4° to the north-northeast. All three galaxies have very similar redshifts, which suggests that the galaxies form a wide physical group. The distance between them is approximately half that between the Milky Way and M31, so hypothetical backyard observers living on a planet within the disk of any one of these three galaxies certainly have two stunning naked-eye extragalactic wonders in their night skies to look upon and wonder about.

Otherwise, NGC 5866 is relatively isolated, and there appears to be no evidence for any tidal distortions. So the existence of such a large gas shell, which is bright in infrared light, indicates that NGC 5866 may be entering an era of star formation fueled with gas donated by its aging stellar population. G. K. Kacprzak (New Mexico State University) and G. A. Welch (Saint Mary's University) presented evidence to support this contention at a 2003 American Astronomical Society meeting. After analyzing the dust and gas components of the interstellar medium around NGC

5866, the researchers concluded that NGC 5866's x-ray-hot interstellar medium is likely being swept perhaps by a global wind powered by supernovae.

To find NGC 5866, first locate 3rd-magnitude Iota Draconis (Edasich). Now look for a 5th-magnitude star (*a*) just about 5° southwest. Use binoculars if you have to, but Star *a* should be easy to identify since it is the brightest star in the region. NGC 5866 is just a little more than 1° due north of Star *a* and 10′ northeast of a magnitude 7.5 star.

Once you get the field, try relaxing and seeing NGC 5866 with binoculars. I find that, under a dark sky, NGC 5866 is *just* visible with 7 × 50 binoculars with effort. You have to know exactly where to look. The galaxy is easier to see in my antique telescope, but it is not something that calls attention to itself in a sweep. Again, you have to know exactly where to look to see it.

In the 4-inch at 23×, it is immediately obvious as a condensed, elliptical glow with white, winglike extensions. At this low power, the galaxy's light is almost combined with that of a magnitude 11.5 star about 1.5′ to the northwest. The galaxy, however, stands out much more brightly. Its core is very small (~2′) and very compact.

The galaxy's spindle shape is nicely revealed at 72×. A number of fainter stars peek out from the background immediately surrounding the galaxy, adding a bit of dimension. It's a nice view at this power, and the lens looks extremely uniform and smooth.

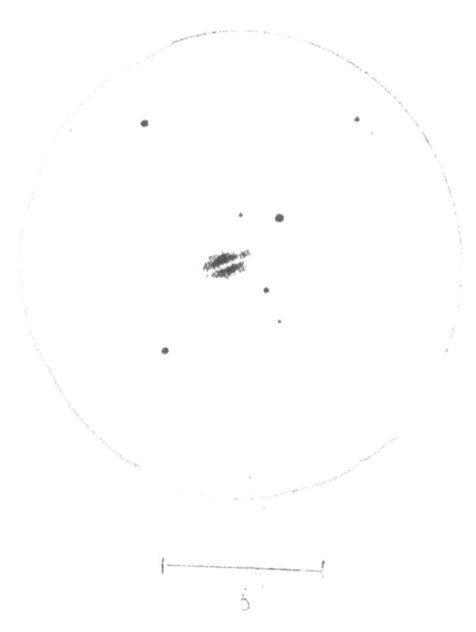

But an object as condensed as this one begs for higher powers. At 303×, the dust lane can be seen as a whisper of darkness – one that cleanly splits the central budge in two. I needed extreme averted vision to see that thin sliver of darkness extend into the disk. But, in time, see if the dark matter just *pops* into view. The central bulge at high power is itself interesting – a prominent ellipse of light that gradually becomes more circular to a point. In the modern scopes of today, with their superb optics and light-gathering power, NGC 5866 is, overall, a much more dramatic object than either M99 or M100 – or M101, for that matter.

M103

M103

NGC 581
Type: Open Cluster
Con: Cassiopeia

RA: $01^h33.4^m$
Dec: +60°39.5′
Mag: 7.4
Diam: 6′
Dist: 8,130 light-years
Disc: Pierre Méchain, 1781

MESSIER: [No date listed] Nebula between the stars ε and δ Draconis of the leg of Cassiopeia.

NGC: Cluster, pretty large, bright, round, rich, stars of 10th and 11th magnitude.

STARS IN AN OPEN CLUSTER ARE destined to be fickle. Without the "glue" of strong mutual gravitation to hold them together – as the hundreds of thousands of stars are held together in a globular cluster – they gradually separate from the main pack and form other alliances. Or, like the Sun, they travel solo through the emptiness of space.

When we look at a loose cluster of stars through a telescope, then, it is difficult to judge what state of association, or disassociation, it is in. The last object in Messier's original catalogue, M103 in Cassiopeia, is a case in point. Astronomer Harlow Shapley went so far as to call this stellar sprinkle a possible chance alignment of stars. But recent data suggest that M103 is a young, metal-rich cluster with as many as 172 stars in an area of only 6′ of arc.

The 25-million-year-old cluster has some 20 variable stars and several binary systems – one visual beauty, Struve 131, shines as the brightest star in the cluster but, alas, it is not physically related but rather a projection effect. By finding the cluster's brightest main-sequence stars and estimating their lifetimes, astronomers have found an evolved giant branch of stars, one of which has a reddish sheen as seen against the cluster's core. The cluster stretches across 14 light-years of space and is seen against the rich backdrop of the Cassiopeia Milky Way.

Located about 1° northeast of Delta (δ) Cassiopeia, the cluster can be seen under a dark sky in 7 × 35 binoculars, which also resolve the cluster's brighter stars, the brightest being magnitude 10.5. The view through the telescope is really best at low power, when the loose grouping appears more compact. When I place brilliant Delta on the western side of the low-power field of view, M103 lies between it and a congregation of four other open clusters to the east! What a dynamic field, even if these clusters are not related.

M103 is clearly the brightest of the five clusters, and it is easily distinguished by its

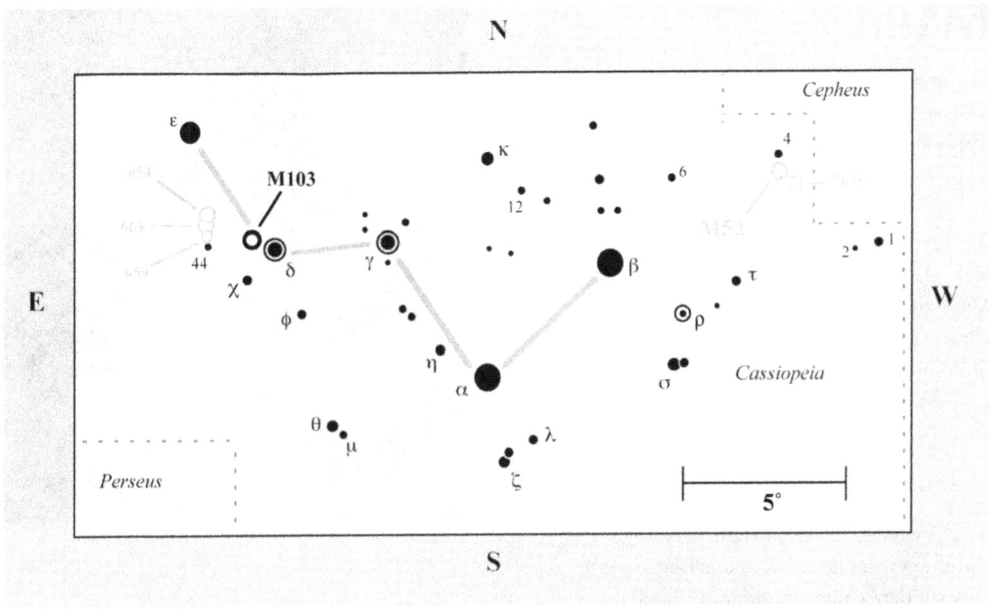

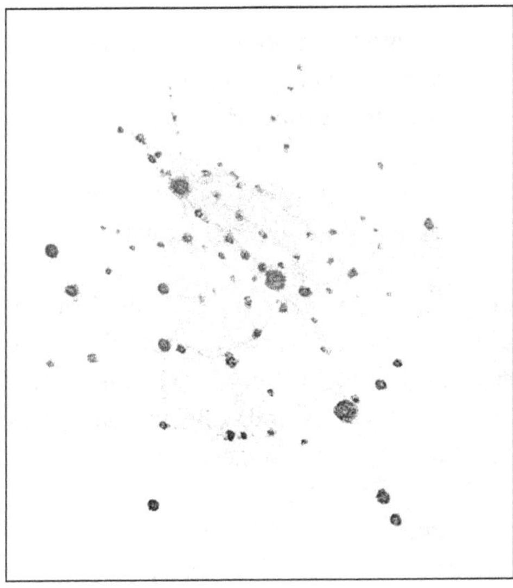

Christmas tree shape. The luminous star at the apex of the tree is Struve 131. Take the time to admire this unrelated system. It has a magnitude 7.3 primary and a magnitude 10.5 secondary 14″ away toward the southeast. Smyth saw their colors as straw and dusky blue.

Smyth also saw a red 8th-magnitude star southwest of the brightest star in the cluster. D'Arrest described this same star shining with a rose tint, but he recorded it as 10th magnitude. Luginbuhl and Skiff describe it as "very red" and shining at magnitude 8.5. Photometric measurements confirm that the star is fairly red. Curiously, I saw no reddish color at all, even though this is one of the cluster's most evolved stars.

M104

Sombrero Galaxy
NGC 4594
Type: Spiral Galaxy (SA(s)a)
Con: Virgo

RA: $12^h40.0^m$
Dec: –11°37′
Mag: 8.0
Dim: 7.1′ × 4.4′
SB: 11.6
Dist: 34 million light-years
Disc: Pierre Méchain, 1781

MESSIER: None

NGC: Remarkable, very bright, very large, extremely extended toward position angle 92°, very suddenly much brighter to a nucleus.

PICTURESQUE M104 IS AN ENIGMA. In the 1920s, Edwin Hubble began classifying spiral galaxies based on the size of their central bulge relative to their arms. A galaxy such as M74 in Pisces, for example, has a very tight nuclear region and wide-sweeping spiral arms. M96 is on the other side of the spectrum, with a large bulge but relatively insignificant arms. Like M96, M104 displays a dominant central bulge in long-exposure photographs, but this nearly edge-on system (~6° from the line of sight) also shows a preponderance of interstellar gas and dust, which is characteristic of a more evolved spiral galaxy, somewhere between M96 and M74 in the classification scheme.

In 1913, Vesto Slipher of Lowell Observatory became the first astronomer to detect rotation in a galaxy other than our own. By studying the spectrum of M104, he deduced not only that the galaxy was receding from us at 700 miles per second but that its disk was rotating: One side was moving toward us while the other side was moving away from us.

M104 is not as much a mystery today as it was in the early twentieth century. We now know it is one of the largest galaxies at the southern edge of the Virgo Galaxy Cluster. It is also one of the closest Sa-type galaxies. In detailed images, we see most or all of the galaxy's dust confined to a very thin plane consisting of knots and spiral segments, as well as clearly defined spiral arms. The galaxy may have formed in two stages. First, there was a brisk free-fall collapse accompanied by star formation so rapid that all the gas in the bulge changed into stars before completion of the collapse to the plane. Second, the disk formed on a longer time scale as the outer regions of the protogalaxy continued to collapse, probably even into modern times.

The Hubble Space Telescope image shown here is a reprocessed archival image that diminishes glare from the bulge and allows us

M104

to see dust lanes well into the bright central region. The image also easily resolves M104's rich system of globular clusters, estimated to be nearly 2,000 in number – 10 times as many as orbit the Milky Way. The ages of the clusters are similar to those of the clusters in the Milky Way, ranging from 10 billion to 15 billion years old.

Embedded in the bright core of M104 is a smaller disk, which is tilted relative to the large disk. X-ray emission suggests that there is material falling into the compact core, where a black hole of 1 billion solar masses resides. In fact, M104 was one of the earliest galaxies to show evidence for the possible presence of a supermassive black hole in its nucleus. This supermassive black hole may also be the result of a merger of two black holes.

You can find the Sombrero Galaxy, as M104 is aptly called, hovering about 5 1/2° northeast of Eta (η) Corvi, just across the border in Virgo. M104's bright 8th-magnitude glow can be seen through 7 × 35 binoculars. Although it is some 20° south of the Virgo cloud of galaxies, the Sombrero is nonetheless probably an outlying member of that group. Astronomers now believe that the galaxy is 34 million light-years distant, about 70,000 light-years in true physical extent, and receding from us at more than 600 miles per second.

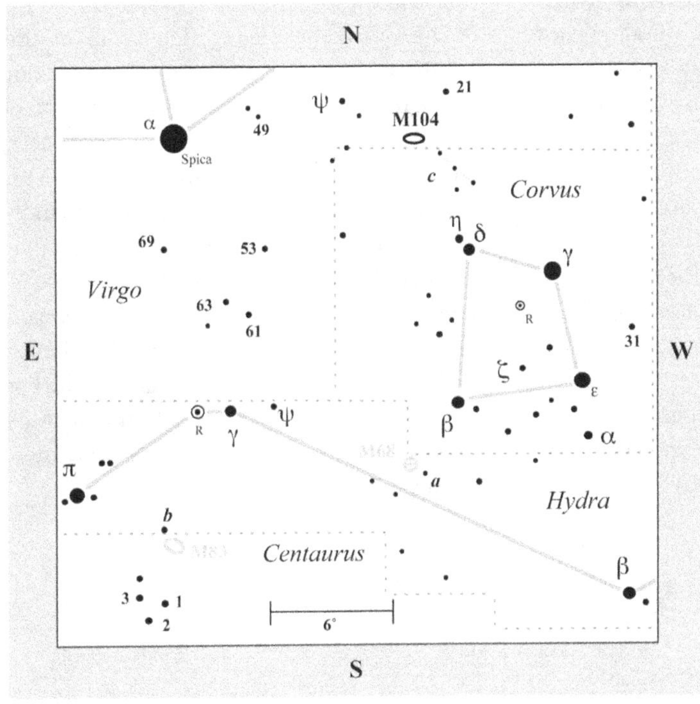

M104

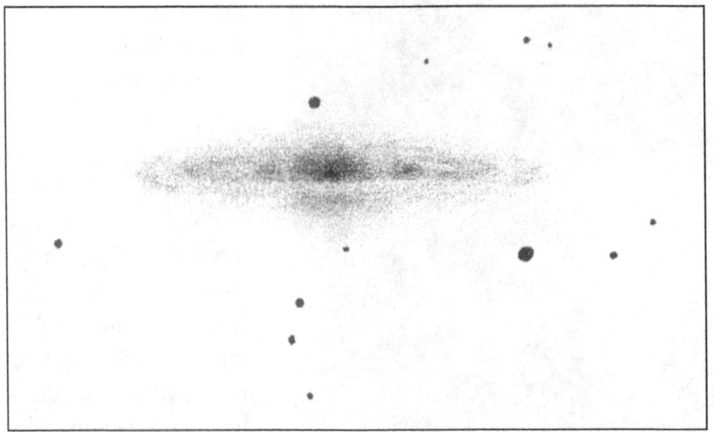

either side. The core is the brightest feature of the three, followed by the western knot, and then the eastern one. With averted vision, the eastern portion of the Sombrero's brim breaks up and flares into a wide brushstroke of light, which shines more brilliantly than the western portion of the brim. Every now and then, the lashlike dust lane wafts into view. High power should reveal the full secrets of these glimmering visions. I find that the galaxy looks best at 130×, when it changes from a misty brew of suggestions into a galaxy with tightly wound spiral arms and clumps of unresolved starlight. Most surprising is that high power shows the nucleus shining with a yellow light. I have found no other reference to this.

The most challenging details lie in the southern portion of the halo, where a faint dome of light connects to the dark lane. Now switch to low power to see if you can discern a soft light enveloping the entire system. With high power, can you resolve the galaxy's individual arms - crescents of light and dark that seem to ripple away from the core? With a little imagination, these crescents can help you see the galaxy in three dimensions, like Saturn and its rings when they are nearly parallel to our line of sight. Few objects in the heavens allow users of small telescopes such an interesting visual perspective.

Some aficionados would prefer that the Messier catalogue end with this stunning object, perhaps feeling that the six subsequent objects (M105 to M110) are visually anticlimactic by comparison. So, how does M104 look in a telescope?

Through the 4-inch, low power shows the galaxy as just a tiny oval glow that begs for more magnification. And, indeed, switching to medium power makes a dramatic difference. The galaxy displays a brilliant core that seems to illuminate the surrounding oval shroud from within, like a distant bonfire seen through a thick fog. A long, bright, needle-like extension runs straight across the major axis of the oval. The sharpness of this line reveals the position of the telltale dark lane, the edge of the Mexican hat's brim. It's quite remarkable to see such detail in a galaxy with only a glance.

With time, the brilliant dome over the northern part of the core appears to be straddled by faint condensations, one on

M105

NGC 3379
Type: Elliptical Galaxy (E1)
Con: Leo

RA: $10^h47.8^m$
Dec: +12°35′
Mag: 9.3
SB: 12.1
Dim: 3.9′ × 3.9′
Dist: 38 million light-years
Disc: Pierre Méchain, 1781

MESSIER: None

NGC: Very bright, considerably large, round, pretty suddenly brighter in the middle, mottled.

MESSIER'S CONTEMPORARY PIERRE Méchain discovered M105 on March 24, 1781. But the object was not included in Messier's list until 1947, when Canadian astronomer Helen Sawyer Hogg suggested adding it (as well as the galaxy M106 and globular star cluster M107) because notations in Messier's copy of the printed catalogue suggested he was aware of them. The galaxy has long been regarded as a dynamically ordinary elliptical (E1) that rotates slowly about its apparent minor axis, though more modern observations with the Multiple Mirror Telescope and Hubble Space Telescope reveal that it may be a nearly face-on S0 galaxy inclined about 90° from edge-on.

The Hubble Space Telescope images, especially, show in astounding detail that the central 5.4 arcseconds of the galaxy have a bright, point-like nucleus surrounded by wispy dark lanes. Spectroscopic HST studies have also shown that the stars near M105's core are in rapid motion around this point-like center, suggesting the presence of a black hole of 50 million solar masses. Such evidence and other similar findings have made astronomers suspect that nearly all galaxies may harbor supermassive black holes that once powered quasars but are now quiet.

This elegant elliptical, a member of the Leo I group of galaxies, is perhaps the purest object in the Messier catalogue. By that I mean that its slightly oval disk shows the least

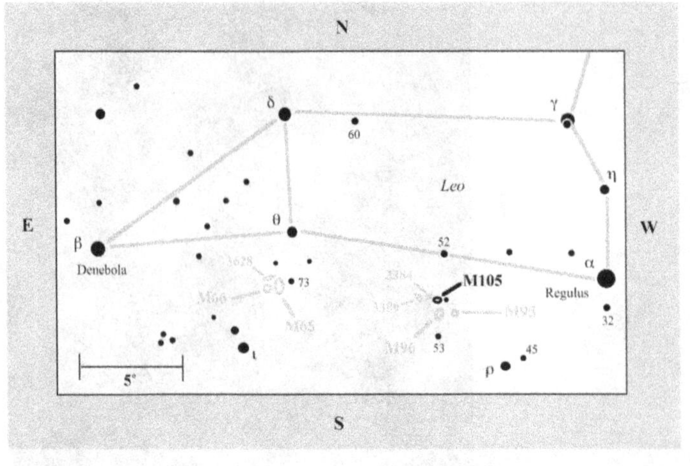

amount of imperfection. Of course, with a diameter of a mere 4′, there's not a lot of room for imperfection.

To our eyes, M105 lingers only 1° north and slightly east of its traveling companions, M95 and M96. In reality, however, M105 is separated from them by a dizzying 400,000 light-years.

The central core of M105 shines brilliantly, "like a 10th-magnitude star," as d'Arrest wrote. Then a bright inner region gives way to an outer halo that fades gradually and uniformly into the cloak of darkness. Only in the very core do I even suspect detail, which amounts to a knot to the southwest of the nucleus and another knot just to the northeast of it; in support of this observation, John Herschel thought he could resolve that glow.

Also traveling with M105 are two other galaxies, NGC 3384 and NGC 3389, both spirals. About 79′ northeast of M105, 10th-magnitude NGC 3384 has a round core and dim extensions that quickly taper off to sharp points. NGC 3389 is far more difficult at 12th magnitude, but use high magnification to search for its elusive, crooked arms. All three share the same low-power field with M95 and M96.

M106

NGC 4258
Type: Mixed Spiral Galaxy (SAB(s)bc)
Con: Canes Venatici

RA: $12^h19.0^m$
Dec: +47°18′
Mag: 8.4
Dim: 20.0′ × 8.4′
SB: 13.8
Dist: 24 million light-years
Disc: Pierre Méchain, 1781

MESSIER: None

NGC: Very bright, very large, very much extended north–south, suddenly brighter in the middle to a bright nucleus.

IN PHOTOGRAPHS, M106 LOOKS LIKE A scarred survivor of galactic violence, and it is. Like M94, this oddly shaped spiral has experienced episodes of violent upheaval in its past. We seem to have caught the nucleus in the throes of an enormous explosion, which has strewn several tens of millions of tons of matter across the plane of the galaxy. Its arms are bruised with rich star-forming regions in a mildly chaotic stew of galactic turbulence.

M106 is a bright barred spiral galaxy that's probably a member of the Ursa Major cloud. We see it inclined only 20° to the line of sight, which partly explains why the dust lanes appear so striking across the central lens. The region inside M106's two bright inner arms is also filled with dust, which forms semispiral chaotic patterns. The galaxy's outer arms look as if they've been stretched like taffy until they nearly snapped free of the main body. Several bursts of star formation appear along them at different stages of their evolution (see the accompanying Hubble Space Telescope image). Radio images reveal still other arms beyond the limp visual appendages.

M106 has long been known to host a quasar-like Seyfert nucleus. A wealth of high-resolution observations have since revealed a rotating ring of water masers, optical jets, and a steeply rising velocity curve, all of which make the galaxy one of the best candidates for harboring a central supermassive black hole. M106's active galactic nucleus must therefore be powered by matter falling into that black hole.

Still, with its belly full of light and two main skeletal arms, M106 is a satisfying view in small-aperture telescopes. In binoculars, it glows faintly 2° south of 3 Canum Venaticorum. Its saucer-shaped nuclear region immediately challenges viewers with its complex, mottled texture. The galaxy has essentially the same orientation in space as the Great Andromeda Galaxy (M31) but only half the mass.

M106

The outer arms of M106 have a distinct S shape and are easy to follow at 23× in the 4-inch. Despite its symmetry, the northern half of the galaxy looks brighter to me than the southern half.

The inner arms also have an S curvature to them, though tighter. So here we have two Ss superposed on one another – one large, one small. Now use high power to study the nuclear region, especially the area around the central condensation. You should see dark matter separating it from the innermost spiral structure. But do you see the tiny S (another one!) of dark material centered on the nucleus? It's almost as if all this galactic violence rippled through the entire system in an orderly fashion! Look also for a series of dark patches aligned with the nucleus along the major axis, just north of the bright core.

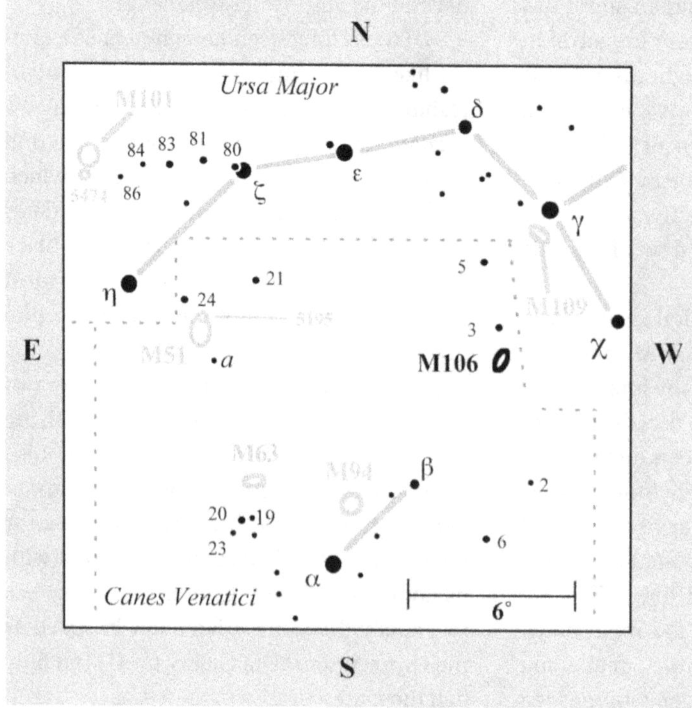

My drawing of M106 is a composite. It is based on views from three nights of observing with low, medium, and high power. The original sketch I made of it on the third and final night showed a star that I did not record during the two previous observing sessions. I have indicated the mystery star with an arrow. Curiously, I have not been able to identify

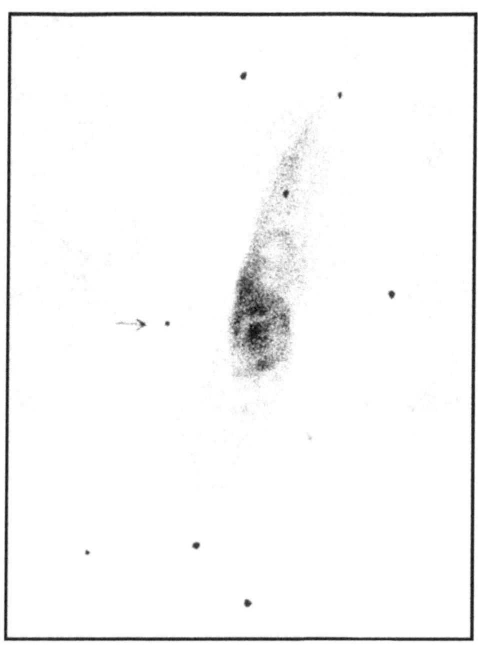

this star on photographs of M106. Could this have been a supernova – yet another incidence of violence in this seemingly troubled galaxy?

M106 has had one supernova, SN 1981K, discovered photographically at magnitude 17, a distant 17″ east and 76″ north of the galaxy's center.

M107

NGC 6171
Type: Globular Cluster
Con: Ophiuchus

RA: $16^h32.5^m$
Dec: -13°03′
Mag: 7.8
Diam: 13′
SB: 13.4
Dist: 20,900 light-years
Disc: Pierre Méchain, 1782

MESSIER: None

NGC: Globular cluster, large, much compressed, round, well resolved.

M107 IS ONE MORE EXAMPLE OF A loose globular cluster whose light appears to be affected by interstellar dust. Located 3° south-southwest of magnitude 2.6 Zeta (ζ) Ophiuchi, this globular, from our perspective, sits virtually on top of the hub of the Milky Way over dusty Scorpius. Long-exposure photographs have revealed several possible obscured regions in the cluster. Regardless of the dust, the globular's pale glow can be detected in binoculars, but not with the naked eye (at least I couldn't see it). M107 shines with a magnitude of 7.8 and measures 13′ in angular diameter.

The cluster is some 10,000 light-years from the Galactic center. It is metal-rich, with each of its stars having, on average, about one-tenth as much metal per unit hydrogen as the Sun, and has an integrated spectral type of G0. It's limping away from us at around 20 miles per second. M107 contains 23 variable stars, all but one of which are RR Lyrae stars.

In 2004, astronomers used M107 as part of an ongoing program to test Newton's law of gravity in the low-acceleration regime using globular clusters. Very Large Telescope spectra for 107 stars within the cluster, combined with data from the literature, revealed a velocity dispersion that remained constant at large radii rather than diminishing the farther the stars are from the nuclear region. Other clusters showed a similar trend, leading researchers to wonder whether this indicates a failure of Newtonian dynamics or some dynamical effect such as tidal heating. But the similarities they found between globular clusters and some elliptical galaxies make them suspect the former explanation.

At low power, M107 instantly appears elongated in the east–west direction, with at least four 11th-magnitude stars around it "in the form of a crucifix," as Smyth recognized. Smyth also inadvertently discovered the dustiness of the surrounding region, calling it a comparative desert. At first glance, the cluster looks rather loose at 23× and shows no strong central condensation. A prolonged gaze, however, will reveal several fairly bright

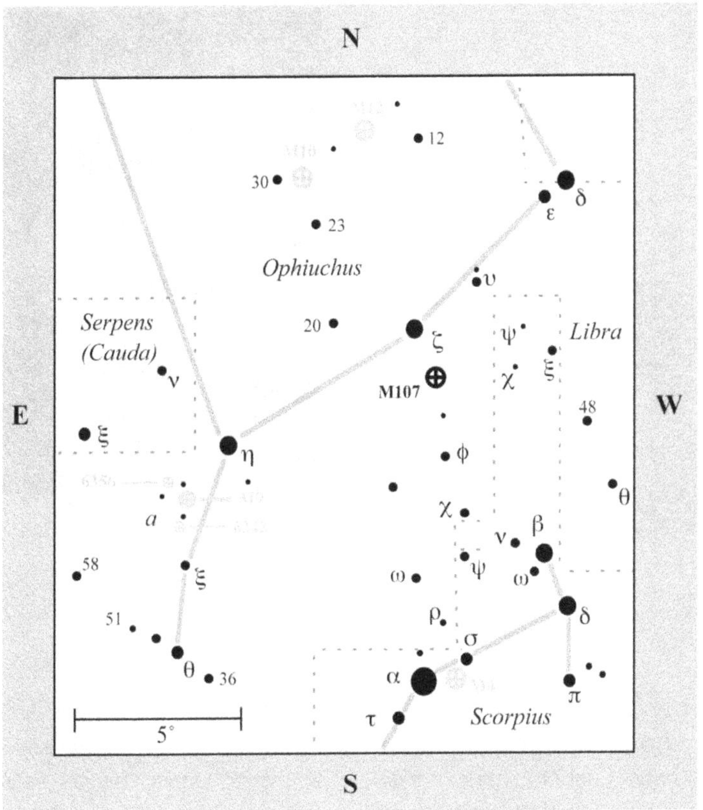

stars populating the ill-defined outer halo, while the rest of the cluster looks granular.

The big surprise comes with moderate and high powers, when the outlying stars stand out boldly against unresolved hazy wisps, which look like ruffled hair. The nucleus is composed of a boxlike arrangement of stellar patches sliced by a dark lane running from north to south. This dark lane, together with the stars bordering it, reminds me, somehow, of a cobblestone path in a morning mist.

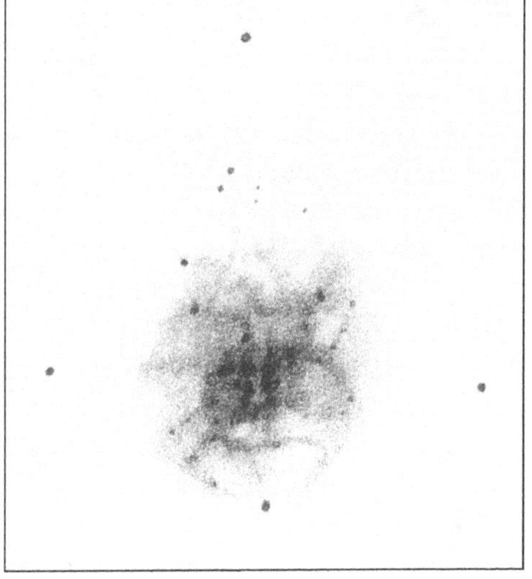

M108

NGC 3556
Type: Spiral Galaxy (SB(s)cd edge-on)
Con: Ursa Major

RA: $11^h11.5^m$
Dec: +55°40′
Mag: 10.0
SB: 13.0
Dim: 8.1′ × 2.1′
Dist: 47 million light-years
Disc: Pierre Méchain, 1781 or 1782

MESSIER: None

NGC: Quite bright, very large, very much extended at position angle 79°, becoming brighter in the middle, mottled.

M108 IS A GORGEOUS EDGE-ON SPIRAL galaxy. It measures 110 light-years across but is only about one-twentieth the mass of M31. Although we see the galaxy nearly edge-on, the central bulge is all but absent. Whatever shines in the nuclear region is masked by turbulent eddies of dark matter lining the galaxy's highly foreshortened arms. The nucleus might have also depleted itself long ago, or it has only periodic bursts of energetic activity that keep it going. Brilliant regions of star formation among the obscuring matter look like signal flares burning in a storm.

M108's nucleus also appears to have star-forming regions, which reveal themselves in the near infrared. These H II regions form a partial ring. Owing to the complexity of the region, astronomers believe that the optical nucleus is most likely displaced from the true nucleus.

Radio observations have detected several interesting gas features, namely one extremely large, expanding half-shell (~23,000 light-years long) on the eastern major axis about 40,000 light-years from the nucleus and extending far beyond the optical disk to the east. Another smaller partial shell is visible about the same distance from the nucleus to the west. The shells most likely originated internally, generated perhaps by winds from a site of massive star formation.

Chandra X-ray Telescope data identify an ultraluminous x-ray source that might result from the accretion of matter from an intermediate-mass black hole. The telescope detected large amounts of diffuse x-ray emission extending about 32,000 light-years radially in the disk and about 13,000 light-years away from the Galactic plane. These emissions may represent various blown-out superbubbles or chimneys of hot gas heated in massive star-forming regions.

On the nights I viewed M108 from the summit of Kilauea volcano, the galaxy was

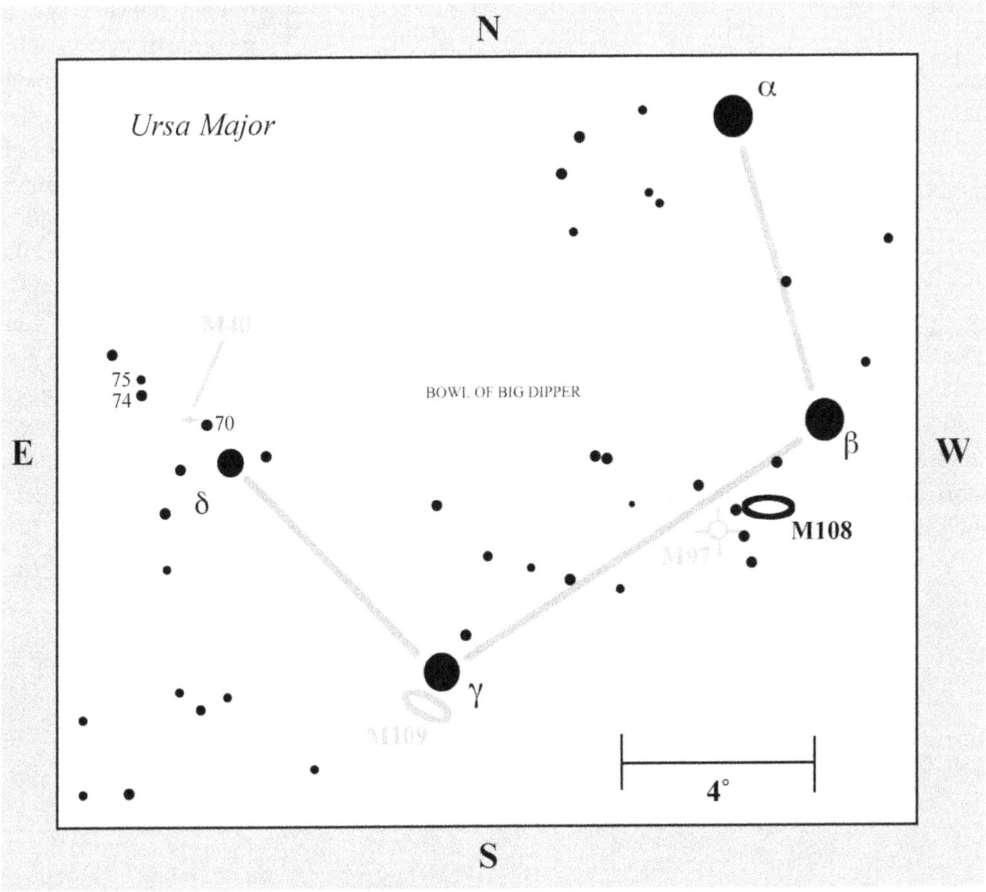

high overhead, so I had to use a star diagonal to relieve the tension in my neck. But that meant I also had to look "down" into the eyepiece. Because I like to observe with both eyes open, I was forced to use a flap on my winter hat to block light from entering the eye I wasn't using. Where was the light coming from? It was the polished glow of the Milky Way reflecting off the ground!

Visible in binoculars as a faint gray streak only two Moon diameters (1°) northwest of the Owl Nebula (M97), or 1° southeast of Merak (Beta [β] Ursae Majoris), M108 is an interesting sight.

With its highly wrinkled texture, this 10th-magnitude system looks as if it is the shed skin of M82 – a galaxy with a similar demeanor – left here to dry. Indeed, when Admiral Smyth turned his telescope on it, all he saw was a "large milky-white nebula with a small star." At 23× in the 4-inch, I can recognize M108's elongated (8′) and highly mottled disk. But moderate and high powers are needed to show its haphazard array of dark lanes, superposed stars, and irregularly bright nebulous patches, which look like they have been pasted together by a three-year-old.

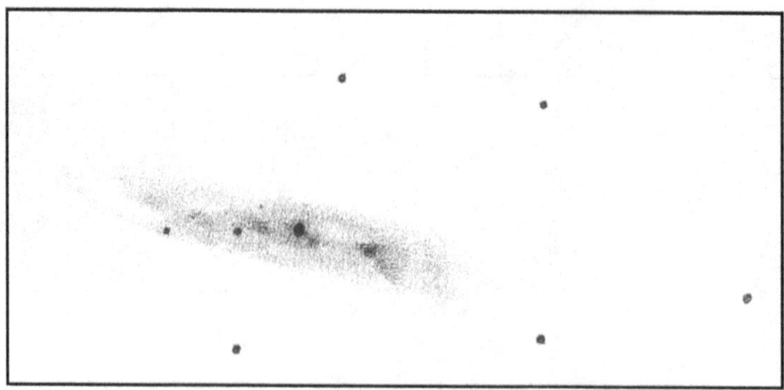

At 130×, one prominent, kinked, dark channel runs along the galaxy's northern rim. More difficult is a long dust lane that follows the southern edge of the galaxy. A beautiful, though delicate, fan of material sprays westward from a faint knot embedded in the western half of M108, and a misty breath of galactic haze shines feebly beneath a 13th-magnitude star in the eastern half.

A fainter glow connects that dim patch to the pseudo-nucleus – a fairly bright foreground star near the galaxy's invisible heart. The real challenge is to see the 14th-magnitude bead immediately southwest of that central pip.

M109

NGC 3992
Type: Barred Spiral Galaxy (SB(rs)bc)
Con: Ursa Major

RA: $11^h57.6^m$
Dec: +53°23′
Mag: 9.8
Dim: 7.6′ × 4.3′
SB: 13.5
Dist: 81 million light-years
Disc: Pierre Méchain, 1781 or 1782

MESSIER: None

NGC: Quite bright, very large, pretty much extended, suddenly brighter in the middle to a bright mottled nucleus.

ALAS, THE MESSIER CATALOGUE ENDS not with a bang but a whimper. The last two objects, M109 and M110, are hardly lens-shattering sights. M109 is a typical barred spiral galaxy; in fact, about one-third of all known spirals are of this nature, characterized by a prominent central bar of stars off of which flow more delicate spiral arms.

M109 displays a very bright and diffuse nucleus and bar with a dark lane. Three main partially resolved filamentary arms wind around the galaxy with some branching. So ordinary is the view that the *Carnegie Atlas of Galaxies* considers its spiral pattern among the most regular known in the sky. Only weak emission from the nucleus has been detected. What's more, even its distribution of neutral hydrogen is … regular! But here's a nice twist: M109 spans a whopping 185,000 light-years of space. It is the most massive spiral of the M81 group. How regular is that?

Glowing at 10th magnitude, M109 (NGC 3992) can be seen in 7 × 35 binoculars 1° southeast of the magnitude 2.4 star Phecda (Gamma [γ] Ursae Majoris) in the bottom of the Big Dipper's bowl.

Whenever I sweep toward M109 and its oval shape enters the field, I get a tingle as if I have encountered a comet, as Méchain must have when he first spotted the object. But, because this glow is so close to 2nd-magnitude Phecda, I wonder for a fleeting moment whether I am just seeing a reflection of that star. A good tap on the telescope tube assures me that I am not.

At low power, the galaxy displays a well-defined core and a halo muddled with irregularities. Medium power proves it to be a more satisfying barred spiral than either M91 in Coma Berenices or M95 in Leo. The central bar of M109 appears prominent, and the nebulous arcs and hints of spiral arms that emanate from it do not require a strong effort.

M109

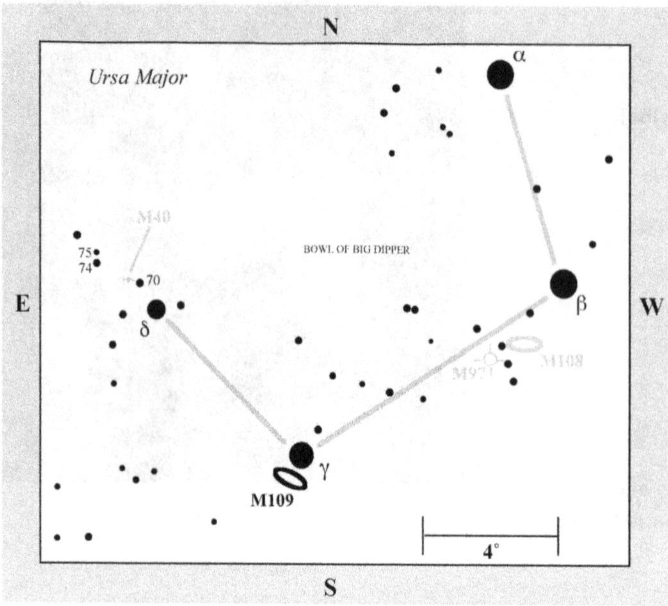

It is noteworthy, however, that neither Kenneth Glyn Jones nor Roger Clark could detect the bar in an 8-inch telescope; this might say something about the galaxy's appearance in less than pristine skies.

One night while I was observing M109, a brief earthquake caused the scope to jitter, which to my delight brought out some of the galaxy's fainter details – especially the crablike pincers extending to the east. The most difficult task is tracing the pale hazes that make up the outer arms. Also, don't confuse the magnitude 12.5 star 50″ north-northwest of the galaxy's core with a supernova; it is just a foreground star, a resident of our own Milky Way Galaxy.

Swing your scope south-southwest from M109 about 1′ and have a look at NGC 3953, a 10th-magnitude barred spiral similar in appearance to M109.

M109: AN ALTERNATE VIEW

NGC 3953 is a pretty barred spiral nearly 1.5° south of 2nd-magnitude Gamma (γ) Ursae Majoris and a little more than 1° south-southwest of M109. Its position near both Gamma Ursae Majoris and M109 has made it an object of controversy.

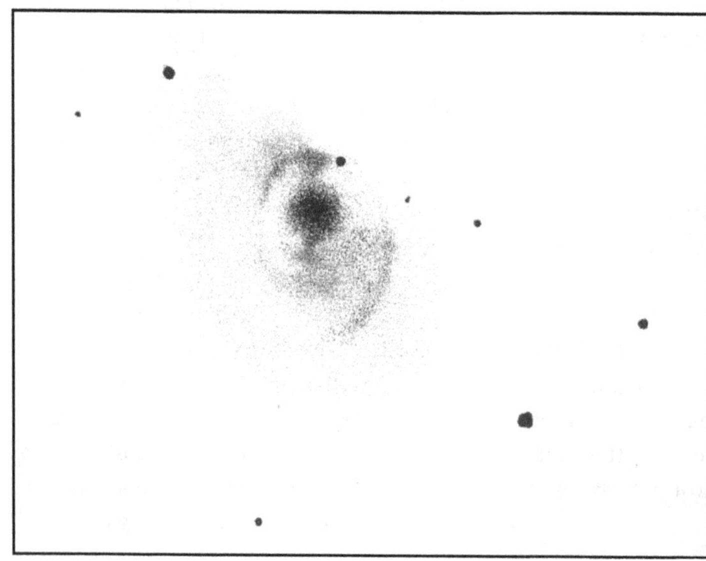

As is well known, the final catalogue of nebulae and star clusters compiled by the French astronomer Charles Messier (1730–1817) contained only 103 objects. The remaining seven objects (one a duplication) were later additions. In 1921, French astronomy popularizer Camille Flammarion added NGC 4594 as M104, on evidence he found in a manuscript addition to Messier's own copy of the *Connaissance des Temps* for 1784. It wasn't until 1953, however, that the famous Harvard astronomical historian Owen Gingerich added NGC 3992 as M109, based on a note he found that Messier made to M97 (the Owl Nebula); without knowledge of this observation, William Herschel rediscovered this object on April 12, 1789, cataloging it as H IV-61.

But is NGC 3992 M109? Henk Bril, an active member of the Royal Dutch Association for Meteorology and Astronomy and 2005 recipient of its prestigious Dr. J. van der Bilt Award, does not think so. Proof, he says, can be found in J. Fortin's 1795 *Atlas Céleste,* which was edited by Joseph Jérome le Français de Lalande and Pierre Méchain, who discovered the nebula attributed to M109. Bril notes that on Plate 6 of the atlas and also on Plate 7, "a nebula is drawn exactly on the position of NGC 3953."

Bril notes that although NGC 3953 is only 0.3 magnitude fainter than M109, it is also about 1.5″ smaller, so it has a slightly higher surface brightness, making it an easier target through small telescopes.

"There is another scenario possible," Bril explains, "Méchain made his discovery on March 12, 1781. Messier observed it on March 24, 1781. But did he actually? Maybe Messier did observe (and as a matter of fact discover) NGC 3992, but mixed it up with Méchain's observation twelve nights earlier."

Bril's more detailed account appears in the March 2007 issue of the Dutch astronomy magazine *Zenit* and the July/August 2007 issue of the French *Astronomie* magazine.

M110

NGC 205
Type: Dwarf Spheroidal Galaxy (companion to M31)
Con: Andromeda

RA: $00^h40.4^m$
Dec: +41°41′
Mag: 8.1
SB: 13.9
Dim: 19.5′ × 12.5′
Dist: 2.5 million light-years
Disc: Messier, 1773

MESSIER: None

NGC: Very bright, very large, much extended toward position angle 165°, very gradually very much brighter in the middle.

We now turn our attention to the final addition to the Messier catalogue, an object recommended for inclusion by Kenneth Glyn Jones as recently as 1967. M110 is the larger of the two prominent satellite galaxies that flank the majestic Andromeda Galaxy (M31), the smaller and closer one being M32. M110 hovers northwest of M31, appearing as a fuzzy oval glow with a total visual magnitude of 8.0 but a much fainter actual surface brightness of 13.9 because of its large size.

As Edwin Hubble explained as early as 1961, "NGC 0205 has two features uncommon to pure elliptical systems. (1) There are two dust patches near the center of the galaxy. [And] (2) A few blue supergiant stars appear to be associated with dust patches in the center of NGC 0205. The intensity gradient of NGC 0205 suggests the presence of an outer envelope similar to that of S01 galaxies. Perhaps this is a transition case."

Actually, M110's elliptical classification status has since been revoked. In 1951, Walter Baade used the 100-inch Hooker reflector at Mount Wilson to become the first to image newly formed, hot, young, high-luminosity stars in an apparently old elliptical galaxy. (Star formation is still taking place in M110 today.) Baade also first resolved the galaxy into stars with the same telescope using red-sensitive plates. More recent imaging has shown M110 to be a normal dwarf spheroidal galaxy with a mass of some 10 billion solar masses and no central black hole.

High magnification reveals a highly mottled nucleus, with the portion closest to M31 detached by an obvious dust lane! A star-like object lies just south of M110's compact nucleus, and it, too, seems detached by dark

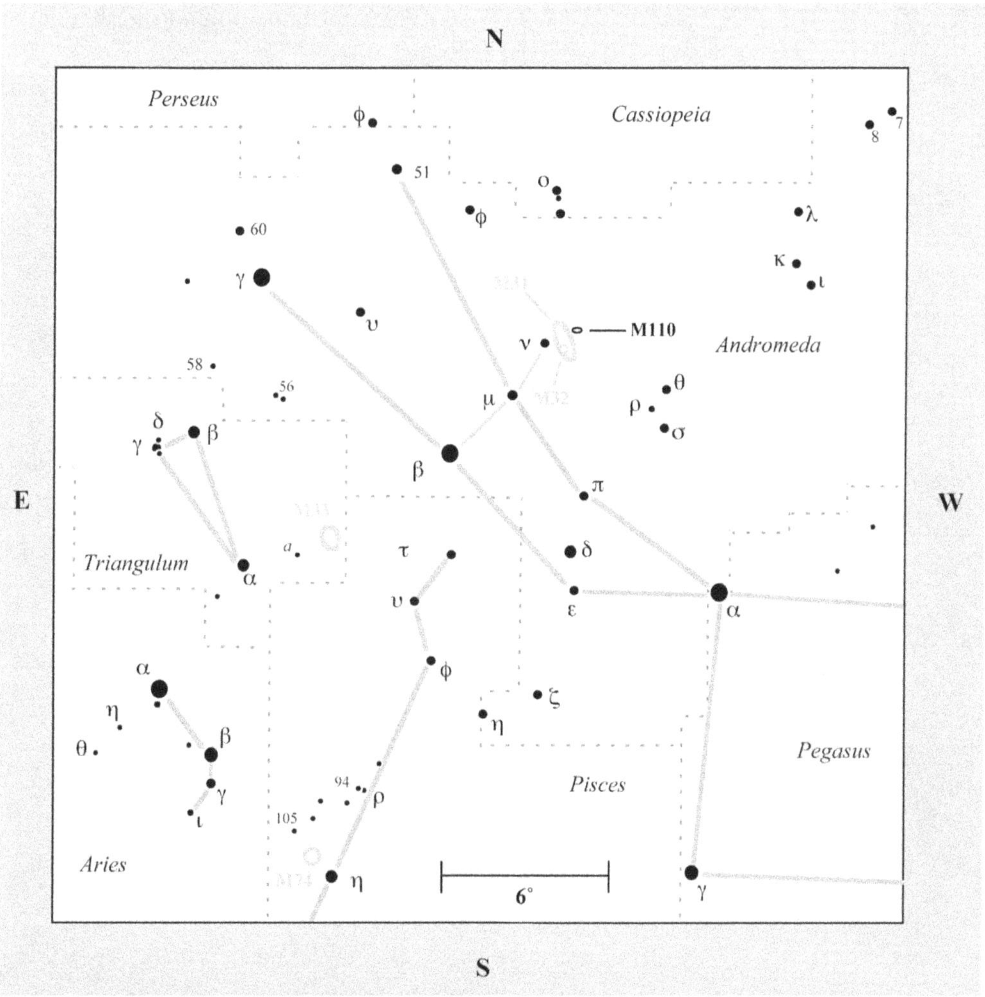

matter, but this might simply be a contrast illusion. Furthermore, the east and west sides of the inner halo exhibit concentrations that are clearly separated from the nucleus. Thus, M110's nucleus appears ringed by arcs of diffuse starlight, which makes the following description of M110 by Heber Curtis of Lick Observatory very intriguing: "The bright central portion ... [shows] traces of rather irregular spiral structure. Nucleus almost stellar. Two small dark patches near brighter central portion."

Although this peculiar galaxy does not have spiral structure, the dark patches do exist and show on photographs made with large-aperture telescopes. But it is possible for you to see them from a dark sky with a 4-inch telescope and high magnification! And M110 tolerates high powers well. See if you get the impression that, like M32, this galaxy has a

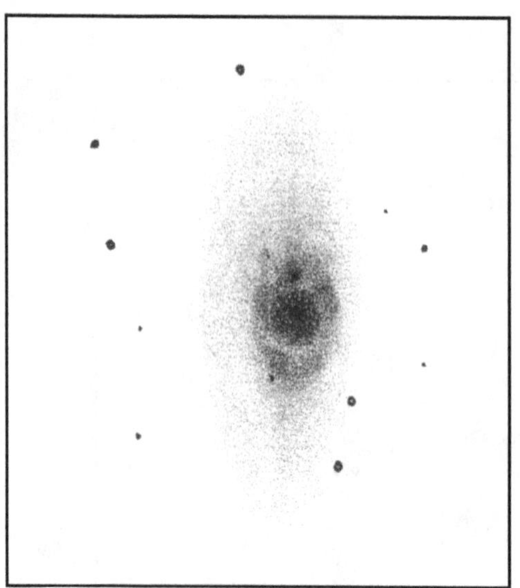

faint extension or bar running north to south through it.

Now, when you look at M110 through your telescope, consider that globular cluster M54 may be the nucleus of such a system being swallowed by our Galaxy.

CHAPTER 5

Some thoughts on Charles Messier

5

Charles Messier is best remembered for the objects he largely tried to avoid and help others avoid. Messier was first and foremost a comet hunter, and he dedicated clear, moonless nights to this pursuit. He was quite successful in his search, having found some 15 comets, though only 12 bear his name (with some exceptions, comet names were not officially used until the twentieth century). In his time, comet hunting was intensely competitive, and Messier was inspired not only by the search but also by the fame new comet discoveries would bring him. The fuzzy, *non*-cometary objects he encountered intrigued him, in part, because they fooled him, if only briefly. Messier created his renowned catalogue largely to benefit *comet hunters*, so they would not confuse these nebulous glows with comets. But it's also possible Messier knew he would reap fame by discovering new nebulae.

Messier's first list of 45 objects, published in the *Memoirs* of the French Academy of Sciences for the year 1771, was well received by his friends and colleagues. In the introduction to the work, astronomer Jerome de Lalande praised Messier, saying the task could only have been undertaken by an indefatigable and experienced observer. William Herschel, in particular, regarded Messier's catalogue as an excellent collection of clusters and nebulae, and it apparently prompted Herschel to undertake his own search for similar deep-sky objects. (Herschel employed much larger telescopes in his survey, which led to the additional discovery of more than 2,000 objects.) Nevertheless, legitimate questions can be raised about why certain objects made it into Messier's catalogue and other seemingly obvious ones did not.

Messier was surprised to resolve stars in several objects described by their discoverers as nebulous patches. But if his aim was to help comet hunters avoid objects that might prove confusing, why would he include *obvious* open clusters, such as M23, M34, M45, and M47? Perhaps, because they appear fuzzy to the naked eye, he did so for the benefit of the naked-eye gazer. (It should be noted that in the mid-1700s spotting a new comet with the unaided eye was not uncommon.)

It would be fruitless to question the observing ability of the man described by King Louis XV as the "ferret of comets." Messier was highly respected for his keen eye and attention to detail. His expertise was not limited to the visual search for comets, as he was an equally skilled observer of, among other things, the planets, stellar occultations, and sunspots. Still, one has to wonder how Messier could have *missed* so many relatively bright nebulous objects during his comet hunts. Anyone who takes the time to sweep the sky with a telescope around the famed M objects cannot help but chance upon some nebulae and clusters that were not included in his catalogue, apparently escaping his gaze. But did they?

We know that of the 110 objects in the catalogue, Messier actually discovered 41 of them; the other objects had been found by others, but Messier recorded their positions and appearance anyway since they could be confused with comets. Once he started the catalogue, Messier wrote, he endeavored to find other nebulosities; he made this something of a project for its own sake, at least in the beginning. But after the first catalogue of 45 objects was published, most of the objects recorded thereafter by Messier were found serendipitously, when particular comets he was following would happen to pass near a nebulous patch. On average, he discovered only about two new catalogue objects per year.

One particularly puzzling omission from the catalogue is the Double Cluster in Perseus.

Why would Messier not include this conspicuous Milky Way object, which appears distinctly comet-like to the naked eye? Yes, the cluster does resolve with the slightest optical aid into a myriad of individual stars, but the same could be said of the Pleiades, the brilliant and easily resolvable open cluster in Taurus, which he catalogued as his forty-fifth entry. His inclusion of the Pleiades makes it hard to argue that he considered some objects too obviously *not* comet-like in appearance to be confused by them, or to bother recording them.

One can only speculate about the absence of the Double Cluster and other bright objects of which we presume Messier should have taken notice, even with his rudimentary scopes (see Appendix A for a discussion on the Double Cluster). And one can perhaps detect a bit of defensiveness in Messier about his catalogue in some remarks he made in the French almanac *Connaissance des Temps* for 1801, just a few years before his death:

> After me, the celebrated Herschel published a catalogue of 2,000 which he has observed. This unveiling of the sky, made with instruments of great aperture, does not help in a perusal of the sky for faint comets. Thus my object[ive] is different from his, as I only need nebulae visible in a telescope of two feet [length]. Since the publication of my catalogue I have observed still others; I will publish them in the future ... for the purpose of making them more easy to recognize, and for those searching for comets to remain in less uncertainty.

It would be informative to know Messier's feelings on this 30 years earlier. I see his remark about his objective being different from Herschel's, and his need to see only nebulae visible in a telescope of two-foot length, as testimony to what I believe was his unwavering purpose, the pursuit of comets.

It is also surprising that Messier could not telescopically resolve the great globular clusters M5 in Serpens Caput and M13 in Hercules. Of course, he was limited not only by the poor location of his observatory but also by the poor quality of the optics in his instruments. He used about a dozen different telescopes ranging from reflectors with inefficient speculum mirrors (the largest having an 8-inch aperture) to simple refractors with apertures of up to 3 1/2 inches. He generally employed magnifications of between 60× and 138×. Compared with today's instruments, all these telescopes were of substandard optical quality.

"It is a pity," wrote Smyth in *Cycle of Celestial Objects*, "that this active and assiduous astronomer could not have been furnished with one of the giant telescopes of the present day. ... One is only surprised that with his methods and means, so much was accomplished." Likewise, Lalande could only lament that the precise and zealous Messier could not live under purer and less cloudy skies, which obviously hampered his ability to find new comets and nebulae.

Some of today's skilled comet hunters claim to have an innate sense of what a new comet will look like through the telescope. Thus, it is the comet hunter's prerogative to skip over or ignore whatever object he or she deems not cometary. It is reasonable to presume that Messier had his own criteria for judging whether an object was sufficiently comet-like in appearance to warrant noting its position, though he did not relate those criteria.

But it should be said that he never endeavored to conduct a comprehensive visual search for noncometary nebulae. An old Chinese proverb tells us that obstacles are what we see when we take our mind off the goal. And Messier's goal was to discover comets, not nebulae or star clusters. In contrast, Herschel's goal *was* to systematically survey the entire sky for deep-sky bounty beyond the limits of Messier's instruments.

Most of the relatively bright non-Messier objects we enjoy today (at least from the Northern Hemisphere) were discovered by Herschel. There can be no doubt that had Messier conducted a similar deliberate search for noncometary objects, his catalogue would have turned out vastly larger and less haphazard than it is. But perhaps that was not Messier's concern, comets were. It is only by the fickleness of fate that none of his comets had the staying power of his catalogue of deep-sky wonders.

CHAPTER 6

Twenty spectacular non-Messier objects

6

Without question, Messier's catalogue contains many of the sky's brightest and most glorious deep-sky objects. But there are a host of other exquisite objects that he did not include, either because they were too far south to be seen from his observing location or because he missed, ignored, or somehow overlooked them. The following are brief descriptions of 20 of my favorite non-Messier objects, in order of right ascension. I only hope these objects will inspire you to continue scanning the skies for your own celestial treasures.

1

NGC 253
Type: Spiral Galaxy (SAB(s)c)
Con: Sculptor

RA: $00^h47.6^m$
Dec: –25°17′
Mag: 7.6
Dim: 30.0′ × 6.9′
Dist: ~10 million light-years

For small-telescope users, the highly elongated spiral galaxy NGC 253 in Sculptor is rivaled in detail only by M31, the Great Andromeda Galaxy. Although NGC 253 shines only as bright as a magnitude 7.6 star (whose light is spread over an area of 30′), its high surface brightness makes it a grand sight through any aperture and with most magnifications. The galaxy displays a highly uniform though mottled texture. Splashes of dark dust are interspersed with the bright patches. Especially prominent is one dark lane to the west of the nucleus. NGC 253 is the brightest member of the Sculptor group of galaxies – the galaxy cluster closest to the Local Group, which includes the Milky Way – and is roughly the same size as M31. The object lacks the obvious star-like core typically seen in many other spiral systems. Indeed, astronomers have found that this galaxy's nucleus seems to be simmering, with gases flowing rapidly away from the nuclear region.

2

NGC 869 and NGC 884
Double Cluster
Type: Open Cluster
Con: Perseus

NGC 869
RA: $02^h19.0^m$
Dec: +57°08′
Mag: 5.3
Diam: 18′
Dist: ~7,300 light-years

Why Messier failed to include the Perseus Double Cluster in his catalogue is a mystery (see Chapter 5). This stunning object is among the finest celestial showpieces in the northern sky. Riding high in the winter sky midway between Delta (δ) Cassiopeiae and Gamma (γ) Persei, this pair of stellar islands looks to the naked eye like a 4th-magnitude nebula spanning 1 1/2 Moon diameters. Binoculars resolve the two clusters, NGC 869 and NGC 884, which are 1/2° apart, but a rich-field telescope provides the best view, revealing some 200 stars piled like rubies and diamonds on black

NGC 884
RA: $02^h22.5^m$
Dec: +57°08′
Mag: 6.1
Diam: 18′
Dist: ~7,250 light-years

velvet. Strings of stars stretch between the couplet, like arms entwining them in an eternal embrace. With an estimated age of only a few million years, the Double Cluster is one of the youngest galactic clusters known.

3

NGC 891
Type: Spiral Galaxy (SA(s)b)
Con: Andromeda

RA: $02^h22.6^m$
Dec: +42°21′
Mag: 9.9
Dim: 13.0′ × 2.8′
SB: 13.7
Dist: ~33 million light-years

Seen exactly edge-on, NGC 891 is the unequivocal rival of NGC 4565 in Coma Berenices. However, the lower surface brightness of this magnitude 9.9 galaxy, which lies 3 1/2° east of the fine double star Gamma^{1,2} ($\gamma^{1,2}$) Andromedae, makes it a less attractive sight for small-telescope users. But I suppose my loyalty to NGC 891 stems from childhood, because an image of the galaxy used to appear in the closing moments of the television classic *Outer Limits*, which had a somewhat mystical effect on me. Like NGC 4565, NGC 891 is disrupted by a thick lane of dark matter. In 1940, Carl Seyfert (of *Seyfert galaxy* fame) discovered that the galaxy's dark lane only *appears* dark because its light is up to a magnitude fainter than the surrounding brightness of the galaxy.

4

NGC 2024
Lips Nebula **or** *Flame Nebula*
Type: Emission Nebula
Con: Orion

RA: $05^h41.9^m$
Dec: -01°51′
Mag: –
Dim: 30′ × 30′
Dist: ~1,500 light-years

This wonderful but overlooked diffuse nebula in Orion's belt is often portrayed as a challenging telescopic object, but it really is not. True, its pale light is spread over an area as large as the full moon, and it lies a mere 15′ northeast of the blazing magnitude 1.8 Zeta (ζ) Orionis. But from a good dark-sky site it shows clearly in 7 × 35 binoculars. Through the 4-inch, this softly glowing cloud is parted by a thick, serrated dark channel. Thus, NGC 2024 looks very much like a slightly open pair of lips. Despite its distinctive telescopic appearance, I prefer the binocular view because the nebula looks devilishly subtle, like the mysterious smile of the woman portrayed in Leonardo da Vinci's *Mona Lisa*.

5

NGC 2392
Eskimo Nebula
Type: Planetary Nebula
Con: Gemini

RA: $07^h29.2^m$
Dec: +20°55′
Mag: 9.2
Diam: 47″
Dist: ~2,900 light-years

A fine 9th-magnitude planetary nebula for small telescopes, NGC 2392 looms in the night sky like a pale apparition on the eastern outskirts of the Gemini Milky Way. It is easily located about 2 1/2° southeast of Delta (δ) Geminorum and 1.6′ south of an 8th-magnitude field star. With low power, it is an abstruse sight, appearing to be little more than a star surrounded by dim haze. But increased magnification and a little patience bring out some titillating details. Most noticeable is the planetary's magnitude 10.5 central star, which seems to burn through a dense, wide shell of greenish-blue gas. With patience, that diffuse shell might appear mottled with dark patches. Photographs show these best; they also reveal another ring of gas some 40″ away. When seen together, the dark patches in the inner shell and the wispy ring of the outer shell suggest the face of an Eskimo, whose dark eyes are peering out of the hood of a fur-lined parka. As with most planetaries, with averted vision the gas surrounding the bright central star appears to swell. The shell, by the way, is expanding at a rate of about 68 miles per second, so its apparent size is growing about 1″ every 30 years. It probably first left the Eskimo's central star some 1,700 years ago, making this planetary one of the youngest known.

6

NGC 2451
Type: Open Cluster
Con: Puppis

RA: $07^h45.4^m$
Dec: -37°57′
Mag: 2.8
Diam: 50′
Dist: 558 light-years

Of the seemingly endless open clusters distributed throughout the Puppis region of the Milky Way band, NGC 2451 has found a special place in my heart. It is a brilliant 3rd-magnitude object 1 1/2 Moon diameters across, with at least 30 colorful stars within the range of binoculars. These stars seem to burn in adoration of the brightest cluster member - the blazing red magnitude 3.6 c Puppis. With a little imagination, the brightest stars can be seen forming a pattern reminiscent of a scorpion with outstretched claws and raised tail. The cluster almost hugs the horizon from midnorthern latitudes, but it is a stunning sight under southern skies.

7

NGC 2477
Type: Open Cluster
Con: Puppis

RA: $07^h52.3^m$
Dec: -38°32′
Mag: 5.8
Diam: 20′
Dist: 3,700 light-years

One of my all-time favorite binocular objects, and yet another surprising omission from Messier's catalogue, is this 6th-magnitude open cluster. Located about 2 1/2° northwest of magnitude 3.3 Zeta (ζ) Puppis, NGC 2477 is the brightest open cluster in that cluster-laden constellation and can be seen with the unaided eye on clear, dark nights. Through binoculars, its compact form appears strikingly like that of a tailless comet, one about the size of the full moon! Telescopically, the cluster is a tight, almost globular-like swarm of some 300 glinting gems. In his book *Star Clusters*, the late Harlow Shapley said NGC 2477 was either a massive open cluster that is 500 million to 1 billion years old or the loosest of globular clusters.

8

NGC 3242
Ghost of Jupiter
Type: Planetary Nebula
Con: Hydra

RA: $10^h24.8^m$
Dec: -18°39′
Mag: 7.3
Diam: 40″
Dist: 2,600 light-years

Here is another unfortunate "miss" by Messier and his contemporaries because undoubtedly NGC 3242 in Hydra is one of the finest examples of a planetary nebula in the heavens. It shines a full magnitude brighter than M57 (the famous Ring Nebula in Lyra), has a magnitude 11.4 central star within range of the smallest of telescopes, and sports a pale blue disk that is about the same apparent size as Jupiter (thus the Ghost of Jupiter nickname). You can locate NGC 3242 with binoculars nearly 2° south of Mu (μ) Hydrae. Moderate-sized telescopes will reveal its ring structure embedded in a larger oval disk of faint light. From dark skies, all its principal features can be seen with a 4-inch telescope. Unlike some planetaries, this one has a high surface brightness, so the nebula can withstand exploration with high magnification. The true diameter of the outer shell may be about 0.6 light-year.

9

NGC 3372
Eta Carinae Nebula
Type: Emission Nebula
Con: Carina

RA: $10^h43.8^m$
Dec: -59°52′
Diam: 2° (var.)
Mag: 4.5 (var.) (O'Meara)
Dist: ~7,500 light-years

Unrivaled in nebulous splendor, the Eta Carinae Nebula is a southern treasure that surpasses even the great Orion Nebula in size and beauty. It is a celestial continent of bright and dark nebulosity spanning more than 2° of sky. To the naked eye, it looks like one of many hazy-looking objects populating the fabulously rich section of the southern Milky Way between Vela and the Southern Cross. But its inconspicuous naked-eye appearance belies the grandeur that awaits telescopic observers: brilliant clouds of twisted gas mix playfully with dark clouds of obscuring matter, one part of which mimics the appearance of a black keyhole. The vaporous swirls in this complex nebula surround one of the sky's most mysterious stars, Eta Carinae, a novalike object 150 times larger and 4 million times brighter than the Sun. Although Eta Carinae now shines at around 6th magnitude, the star is known to experience violent eruptions that cause it to brighten to magnitude -0.8, outshining every star in the sky except for Sirius. That last happened in 1843.

11

NGC 4565
Type: Spiral Galaxy (SA(s)b?)
Con: Coma Berenices

RA: $12^h36.3^m$
Dec: +25°59′
Mag: 9.6
Dim: 14.0′ × 1.8′
SB: 12.9
Dist: 42 million light-years

Nearly 2° east of 17 Comae Berenices lies NGC 4565, the largest and most famous edge-on spiral galaxy in the night sky. In the 4-inch, it is not a stunning sight but an elegant one. Shining at 10th magnitude, the galaxy appears as a slim streak of light with a hazy central bulge that is punctuated by a starlike core. The challenge in small telescopes is to see and trace the dark dust lane that runs along the entire length of this spindle. In photographs taken with large-aperture instruments, this 170,000-light-year-long spiral, which is tilted a mere 4° from edge-on, displays dark arcing festoons of dust silhouetted against the bright central bulge. This material was ejected hundreds of light-years out of the plane of the galaxy, but gravity is drawing it back in.

12

Coalsack Nebula
Type: Dark Nebula
Con: Crux

RA: 12^h52^m
Dec: −63°18′
Mag: –
Dim: 420′ × 240′
Dist: ~550 light-years

Massive and black, the Coalsack Nebula blots out 26 square degrees of Milky Way just east of Alpha (α) Crucis in the Southern Cross; it also lies just south of the rich Jewel Box Cluster. The juxtaposition of these two disparate objects seems almost ironic. To the naked eye, the Coalsack looks like the silhouette of a black hole that is greedily consuming the space around it. By comparison, the Jewel Box Cluster looks like a cache of stars that have been plucked from the Milky Way. The black cloud lies some 500 or 600 light-years away and is one of the dark nebulae closest to our Solar System. Telescopically, the cloud looks shredded, as if millions of dark vapors are wending their way though a forest of dim stars. Seen together with the Jewel Box and the Southern Cross, the Coalsack creates one of the most awe-inspiring sights in the heavens.

13

NGC 4755
Kappa Crucis (Jewel Box)
Type: Open Cluster

Con: Crux
RA: $12^h53.6^m$
Dec: −60°21.4′
Mag: 4.2
Diam: 10′
Dist: ~4,900 light-years

With more than 50 bright and colorful stars packed into an area of sky only 10′ across, NGC 4755 ranks high as one of the most spectacular celestial treasures in the southern sky. The cluster is visible to the naked eye as a slightly swollen 4th-magnitude star (Kappa [κ] Crucis) in the most sought-after southern constellation, Crux, the Southern Cross. Binoculars resolve at least nine of its members, which range in brightness from magnitude 5.7 to 10.5. Small telescopes offer viewers a rich assortment of stellar jewels that shimmer with opalescent light, like a cluster of pearls of various sizes. The bright central region of the Jewel Box Cluster measures about 14 light-years in diameter, while a region spanning twice that distance is populated with fainter stellar members. Its nickname comes from an observation by John Herschel, who called it a "superb piece of fancy jewellery."

14

NGC 5128
Centaurus A **(Radio Source)**
Con: Centaurus
Type: Peculiar Galaxy (S0 pec)

RA: $13^h25.5^m$
Dec: -43°01′
Mag: 6.7
Dim: 16.7′ × 13.0′
Dist: ~12 million light-years

An enormous galaxy with three times the number of stars as reside in our Milky Way Galaxy, this celestial giant 4 1/2° north of Omega Centauri can be located with binoculars. You'll find it shining as a conspicuous 7th-magnitude glow with an apparent diameter equal to that of the full moon. Through a telescope, NGC 5128 looks like two halves of a broken egg – its whitish shell cracked open by a black absorption nebula. When I first swept up this magnificent object in the 4-inch, I thought I had encountered a comet with a strong parabolic hood: the dark lane looked like the shadow of that "comet's" nucleus running down the length of its broad tail. Had Messier found this object, I am certain he would have included it in his catalogue. The galaxy itself is a strong radio emitter, producing a signal 1,000 times more intense than that of our own Milky Way. Indeed NGC 5128 is among the most peculiar galaxies known, with an explosive nucleus that has jettisoned millions of solar masses of material into space. Astronomers believe Centaurus A is what remains of a cosmic collision between two galaxies, an elliptical and a spiral.

15

NGC 5139
Omega Centauri
Type: Globular Cluster
Con: Centaurus

RA: $13^h26.8^m$
Dec: -47°29′
Mag: 3.9
Dia: 55′
Dist: ~16,000 light-years

A million sparkling stars packed into an oval disk slightly larger than the full moon's, that's globular cluster Omega Centauri, arguably the finest specimen of its kind. Shining at magnitude 3.9, Omega Centauri appears to the naked eye as a "fuzzy star" or the head of a comet. It is best seen from the southern half of the United States and points farther south. Messier could not have discovered it from Paris; otherwise, he undoubtedly would have praised its beauty. The globular looks like a sparkling ball of light in binoculars, while a 4-inch telescope shows myriad blue stars bursting out of an ever-tightening core that seems to scintillate with nervous energy. Look for two dark patches on the core.

16

NGC 6231
Type: Open Cluster
Con: Scorpius

RA: $16^h54.0^m$
Dec: −41°47′
Mag: 2.6
Diam: 14′
Dist: 5,200 light-years

NGC 6231 is the most comet-like naked-eye spectacle in the heavens that is not a comet – especially when you include the elongated cluster known as Harvard 12 or Trumpler 24 embracing it to the north. Together, NGC 6231 and Harvard 12 hang above the southern horizon like a bright comet with a broad dust tail. Although these objects have separate catalogue identities, their stars meld. That's because they belong to the same group of high-luminosity stars, known as the "Scorpius OB1 Association," which marks the location of one of our galaxy's spiral arms – the one closer to the Galactic center rather than the one containing the Sun. Certainly, these objects would have been appropriate for Messier's catalogue had he spotted them.

17

NGC 6397
Type: Globular Cluster
Con: Ara

RA: $17^h40.7^m$
Dec: −53°40′
Mag: 5.3
Diam: 31′
Dist: 7,200 light-years

For those who can dip deep into southern skies, NGC 6397 is a most pleasing target for small telescopes. Under dark skies, it is visible to the naked eye as a faint "star" on the eastern fringe of the Milky Way, about 10 1/2° south of Theta (θ) Scorpii, a magnitude 1.9 F-type star in the Scorpion's tail. Whenever, for pleasure, I'm just scanning that region, which is rich in clusters, the alluring glow of this 6th-magnitude wonder inevitably captures my attention. Through a 4-inch refractor, the globular is a clean wash of loosely packed starlight. Many of its members are easily resolved – about two dozen of them shine between magnitude 10 and 12. The cluster also contains many bright red-giant stars that are 500 times more luminous than the Sun. At a distance of 7,200 light-years, NGC 6397 may be the globular cluster closest to our Solar System – some 2.5 times closer than the great Omega Centauri Cluster.

18

Collinder 399
Brocchi's Cluster **or** *Coathanger Cluster*
Type: Asterism
Con: Vulpecula

RA: $19^h26.2^m$
Dec: +20°05′
Mag: 3.6
Diam: 90′
Dist: 423 light-years

Although Collinder 399 is visible to the naked eye as a 4th-magnitude fuzzy patch twice the diameter of the full moon, it looks best in binoculars. Finding it entails hopping just 4° northwest from Alpha (α) Sagittae in the summer Milky Way. The shape of this possible open cluster's dozen or so brightest members looks irresistibly like a coathanger; it is one of the few stellar groupings in the sky that immediately looks like its nickname. All told, this 200-million-year-old aggregation contains about 40 stars without a hint of concentration. Its brightest member shines at 5th magnitude and is visible to the naked eye.

19

NGC 7000
North America Nebula
Type: Emission Nebula
Con: Cygnus

RA: 20^h40^m
Dec: +41°00′
Mag: -
Dim: 480″ × 300″
Dist: ~520 light-years

Just 3° east of blue, magnitude 0.1 Deneb (Alpha [α] Cygni) lies NGC 7000 - the unsung hero of the Cygnus Milky Way. In photographs, it appears as a sprawling cloud of glowing gas bordered by inky opacity, covering an area four times larger than the full moon, and resembling in shape the outline of the North American continent. One misconception, however, is that the nebula itself is visible only in photographs; this is not the case. In fact, it was discovered visually. Its presence can be inferred with the naked eye as an area of enhanced brightness. Binoculars show it unmistakably as a distinct fan-shaped glow with a faint tail to the south, bordered by a dark gulf to the west. A wide-field telescope under a dark sky brings out the full glory of the nebula, including the dark nebulosity forming the "Gulf of Mexico," and the "East" and "West" Coasts. A nebula filter will boost the contrast of this low-surface-brightness emission nebula, making it easier to discern.

19

NGC 6960 and NGC 6992
Veil Nebula **or** *Cirrus Nebula*
Type: Supernova Remnant
Con: Cygnus

NGC 6960
RA: $20^h45.7^m$
Dec: +30°43′
Mag: ~7
Dim: 70′ × 6′
Dist: ~1,500 light-years

NGC 6992
RA: $20^h56.4^m$
Dec: +31°43′
Mag: –
Dim: 60′ × 8′
Dist: ~1,500 light-years

Commonly referred to as the Veil Nebula, the dual arcs of NGC 6960 and NGC 6992 are among the most sought-after sights at summer star parties. In long-exposure photographs, several extended, arcing streamers appear to float against the Milky Way surrounding the star 52 Cygni. NGC 6960, the brightest arc, can be seen easily in binoculars from a dark sky, and both of these NGC nebulae are within range of 7 × 35 binoculars. They appear ghostly in form, like pale images of fractured chicken bones. The Veil streamers are the detritus of a star that exploded some 40,000 years ago.

20

The naked-eye Milky Way

How ironic that I have spent so much time looking for faint details in distant galaxies with my telescope when the most majestic of all galaxies – the Milky Way – reveals itself to me each night with a beauty, shape, and richness of texture that no telescope or spacecraft can capture as well as our naked eyes can.

APPENDIX A
Objects Messier could not find

At the end of his catalogue in the *Connaissance des Temps* for 1784, Messier included a list of objects reported by other astronomers that he had been unsuccessful in locating himself. The list, translated from French by Storm Dunlop, follows. As was his custom, Messier refers to himself in the third person.

Nebulae discovered by various astronomers, which M. Messier has searched for in vain.

Hevelius, in his *Prodronie Astronomie,* gives the position of a nebula located at the very top of the head of Hercules at right ascension 252°24'3", and northern declination 13°18'37".

On 20 June 1764, under good skies, M. Messier searched for this nebula, but was unable to find it.

In the same work, Hevelius gives the positions of four nebulae, one in the forehead of Capricornus, the second preceding the eye, the third following the second, and the fourth above the latter and reaching the eye of Capricornus. M. de Maupertuis gave the position of these four nebulae in his work *Figure of the Stars,* second edition, page 109. M. Derham also mentions them in his paper published in *Philosophical Transactions,* no. 428, page 70. These nebulae are also found on several planispheres and celestial globes.

M. Messier searched for these four nebulae, namely on 27 July and 3 August, and 17 and 18 October 1764, without being able to find them, and he doubts that they exist.

In the same work, Hevelius gives the position of two other nebulae, one this side of the star that is above the tail of Cygnus, and the other beyond the same star.

On 24 and 28 October 1764, M. Messier carefully searched for these two nebulae, without being able to find them. M. Messier did indeed observe, at the tip of the tail of Cygnus, near the star π, a cluster of faint stars, but its position was different from the one reported by Hevelius in his work.

Hevelius also reports, in the same work, the position of a nebula in the ear of Pegasus.

M. Messier looked for it under good conditions during the night of 24 to 25 October 1764, without being able to discover it, unless it is the nebula that M. Messier observed between the head of Pegasus and that of Equuleus. See number 15 in the current catalogue.

M. l'abbé de la Caille, in a paper on the nebulae in the southern sky, published in the Academy volume of 1755, page 194, gives the position of a nebula that resembles, he says, the nucleus of a small comet. Its right ascension on 1 January 1752 is 18°13'41" and its declination is -33°37'5"0.

On 27 July 1764, under an absolutely clear sky, M. Messier searched for this nebula without success. It may be that the instrument that M. Messier was using was not adequate to show it. Subsequently seen by M. Messier. See number 69.

M. de Cassini describes in his *Elements of Astronomy,* page 79, how his father discovered a nebula in the space between Canis Major and Canis Minor, which was one of the finest that could be seen through a telescope.

M. Messier has looked for this nebula on several occasions, under clear skies, without being able to find it, and he assumes that it may have been a comet that had either just appeared, or was in the process of fading. Nothing is so like a nebula as a comet that is just beginning to be visible to instruments.

In Flamsteed's great catalogue of stars, the position of a nebula is given as being in the right leg of Andromeda, at right ascension 23°44' and polar distance [declination] 150°49'15".

M. Messier searched for this on 21 October 1780, using his achromatic telescope, without being able to find it.

APPENDIX B

Why didn't Messier include the Double Cluster in his catalogue?

One mystery that still haunts the halls of astronomy is why the eighteenth-century French comet hunter Charles Messier did not include the Double Cluster in his famous catalogue. "This splendid object ... was certainly known in his day," Robert Burnham Jr. comments in his *Celestial Handbook*, "and he included other bright clusters such as the Praesepe and the Pleiades." So why not the Double Cluster? It's possible that Messier had a very logical reason to omit it. But first a powerful myth must be dispelled.

The Messier catalogue is not a list of the best or the brightest deep-sky objects in the heavens – a common misconception that is the astronomical equivalent of saying a bat is blind. Just as bats are not blind, Messier's catalogue is not a manifest of celestial superlatives. But who hasn't loosely described the Messier objects as the finest in the heavens? In his jacket copy for *The Messier Album*, for instance, Professor Owen Gingerich encourages readers to "find a telescope and a clear night, and begin the chase of the most spectacular sidereal objects of the northern skies."

Without question, the Messier catalogue contains some of the most spectacular deep-sky objects visible in small telescopes. But their inclusion is largely circumstantial. Messier never intended to create a list of the most spectacular objects in the heavens. His was a catalogue of "comet masqueraders," as the late comet discoverer Leslie Peltier called them. Messier started the list in 1758 after he encountered a stationary fuzzy patch near Zeta (ζ) Tauri (M1) while he was looking for the Comet of 1758. As David Levy notes in "Messier and His Catalogue" (Vol. 1, Chapter 1), "Realizing he had been fooled by the sky's version of a practical joke, Messier began to build a catalogue of what he called these 'embarrassing objects.'"

Messier's first list, published in 1774, contained not 110 but 45 "embarrassing" objects. Actually, his original list, compiled nine years earlier, had only 41 objects. "Before submitting the list for publication," Levy explains, "he decided to round it out with a few more objects." These additional objects, in order of their inclusion, are the two components of the Orion Nebula (M42 and M43), the Beehive Cluster (M44) in Cancer, and the Pleiades Cluster (M45) in Taurus. But why not the Double Cluster?

A THEORY

To possibly understand Messier's intent, we have to forget what we now know about the nature of the objects in his list, forget how spectacular these objects look in photographs and CCD images. Instead, we have to journey back in time two and a half centuries and think like Messier – a comet hunter.

Like Messier, I have spent several years hunting for comets. Unlike Messier, I have never discovered one that was not already known. But I have had the privilege to meet many of the sky's "embarrassing objects." Identifying them was relatively easy for me because most are plotted on the star charts I use in the field. Messier and his contemporaries did not have this luxury. And nothing is more unproductive to a comet hunter than sitting behind an eyepiece, waiting to see if an uncharted object will move; the agony increases as twilight advances or the object nears the western horizon. Understandably, then, Messier wanted to "cage" these celestial pests, once and for all, by cataloguing their positions.

When Messier systematically began searching for comets in 1758, competition was scarce. "For about 15 years nearly all comet discoveries were made by Messier," Gingerich writes in *The Messier Album*, "so that he almost considered them his own property." These were not the dim and distant objects astronomers are accustomed to discovering today (most by professional astronomers, who sight them on images well before amateurs have a chance to spy them). In Messier's day, all new comets were swept up visually, usually when the comets were bright and relatively near the Sun. Of the 12 comets discovered between 1758 and 1774 (from the time Messier first started looking until his first list was published), half were found by Messier, half were discovered with the unaided eye, and half were sighted within two weeks of perihelion (the comet's closest approach to the Sun). No comets were found more than about 2 months from perihelion, and the faintest one shined near the limit of naked-eye visibility at magnitude 6.5. These statistics provide us with some valuable insight as to where Messier and other comet hunters concentrated their attention during their searches.

It has long been known that to maximize their chances of a find, comet hunters search a "haystack" of sky centered on the ecliptic, near the position of sunset or sunrise. In the *Guide to Observing Comets* (edited by Daniel W. E. Green and published as a special issue of the *International Comet Quarterly*), veteran comet observer John Bortle explains how these searches are conducted:

> A typical pattern for evening hunting is to begin in late twilight by choosing a point near the horizon and about 45° north or south of the sunset point and sweeping parallel to the horizon until reaching an azimuthal point around 45° on either side of the sun. ... After completing the first sweep, the instrument is raised about one-third to one-half of a field diameter and the observer proceeds to sweep backward to the original starting point. This procedure is continued until one reaches at least 45° altitude above the horizon.

The search pattern is reversed in the morning sky.

Of the 41 deep-sky objects in Messier's original list, only 18 were discovered by Messier. Just as we might expect from a comet hunter, all of his finds were within 45° of the point where the ecliptic met the horizon at the time of discovery; 13 of them were within about 20° of that point. Since it was Messier's intent to catalogue as many of these "embarrassing" objects as possible, he made a thorough search of the literature of previously reported nebulae and, after observing and obtaining accurate positions for them, added these objects to his list. (Another myth is that Messier discovered all the objects in his catalogue.) Messier did not care whether these objects were his discoveries, because they were not comets. He was quite meticulous in acknowledging each discoverer in his catalogue. In fact, he was probably quite happy to do so – their discoveries only benefited him.

One question that immediately surfaces, however, is why Messier included the bright and obvious objects M31 in Andromeda, M6 and M7 in Scorpius, and M8 in Sagittarius. All were well known at the time, some since antiquity.

THE COMET HAYSTACK

One explanation could be that all of these objects lie in the comet "haystack" at various times of the year, and all look nebulous to the naked eye. No skilled observer would confuse, say, M31, for a comet when it was in a dark sky, but what about in a twilit sky?

In Messier's day, twilight comets were not infrequent. The first and only comet discovered in 1758 (when Messier started looking) was sighted from Bourbon Island in the Indian Ocean just 15 days before perihelion; the comet shined at magnitude 2.5 and had a 1.5° tail. Messier's fourth comet find (on March 8, 1766) was in the twilight; the magnitude 6.0 object, which was approaching conjunction with the Sun, never escaped the twilight and was last seen by Messier as a difficult object low over the horizon. On October 13, 1773, one year before his first list was published, Messier discovered another twilight comet; this one shined at magnitude 4.5, about the same brightness as both M6 and M8. (In 1780, Messier would find yet another twilight comet – a telescopic one of magnitude 7.0.) So it was not only reasonable but prudent that Messier include in his catalogue the positions of all comet-like objects, regardless of magnitude, as long as they were in the comet haystack.

Perhaps you're thinking that it's easy to imagine how the compact glows of M6, M7, or M8 could be confused as comets when hiding in the twilight, but M31? Interestingly, in the last 20 years, I know of two cases where respected comet hunters have reported the discovery of a binocular comet in the twilight sky only to learn that they had been fooled by M31. Of course, the small, comet-like form of M32 is almost sitting on top of M31, so how strange it would have been for Messier to include M32 in a catalogue and neglect the obvious elongated mass beside it.

Of the 23 objects not discovered by Messier in his first list, all but one (M40) lie within the magical comet haystack at various times of the year.

But even M40 appears to be an explainable exception. After searching through the lists of previously reported nebulae to make his list as complete as possible, Messier set out to locate each of these objects, some of which he could not find.

"He therefore published as part of his memoir," Gingerich writes, "a number of objects previously reported, but which he could not find. He realized that many of them were nebulous to the naked eye but were mere asterisms when examined with a telescope." This left Messier with a list of 39 comet-like objects. Given Messier's penchant for rounding out his lists, he included in his "final" listing of 1764 the obvious choice of the "nebula" that had fooled not him but John Hevelius in 1660 (M40). The "nebula," as Messier pointed out, is merely "two stars very close to one another and very faint." As many comet hunters today know, close pairs of stars seen at low magnifications can look remarkably like dim comets. Messier's choice to include M40 was not only logical but educational, a warning, if you will, to comet hunters that there are more things in the sky that can masquerade as comets other than nebulae and clusters.

But in January 1765, just as Messier had finished rounding out his list, he discovered by accident (not as part of a systematic search for comets) a cluster of stars (M41) just south of brilliant Sirius. And that is why, four years later, he decided once again to round off his list. The inclusion of M42 seems logical because of M43 (the small nebula north of M42 that has a round, comet-like head, starlike nucleus, and "tail" of nebulosity). Both the Praesepe and the Pleiades star clusters lie nearly on the ecliptic. Although they would not necessarily be problematical to telescopic comet hunters (because they are bright and easily resolved into stars), they still might confuse someone looking in the comet haystack with the unaided eye in the twilight.

THE ONE EXCEPTION

The Double Cluster, then, is the only object visible from Messier's latitude known since classical times that is not included in the original list of 45 objects. The remaining 65 objects in Messier's final catalogue are not part of the mystery. "[The] first group of [45] nebulae was the only part of the catalogue prepared for its own sake. ... Most of the subsequent objects were found and reported in comet searches, so that the remainder of the catalogue is less systematic than the first part," Gingerich explains.

So why not include the Double Cluster? From Messier's latitude of nearly 50° north, the Double Cluster never sets. Even at its lowest point in the northern sky, it would still be almost 20° above the horizon. Its far northerly declination of nearly 60° places it well outside the comet haystack during most of the year. Only in late spring and early fall does it lie on the far northern fringe of the comet haystack. Furthermore, in the twilight, the Double Cluster is not only obvious as two close objects (a dead giveaway of its identity), but its appearance coincides with those of the brightest stars in Cassiopeia. Even on the unlikely chance that Messier were to spot its glow in the early evening twilight, by nightfall he would certainly know its identity.

When the Double Cluster lies close to the comet haystack in the morning sky, it still is not a problem. In the morning sky, comet searches begin before the onset of twilight. The observer starts looking 45° above the horizon and works his way toward the Sun. Since the Double Cluster is circumpolar, its presence and identity would be obvious in a dark sky. Furthermore, as the Earth turns, the cluster would climb ever higher in the sky and well out of the comet haystack.

Meanwhile the comet hunter's sweeps would take him lower and lower as he searched for brighter and brighter comets near the Sun. So only someone who did not know the sky or was unskilled in the trade would be fooled by this bright circumpolar glow; listing the Double Cluster as an object that could be mistaken for a comet by a skilled observer would be folly. Perhaps pride sets the limit, and for that reason, the Double Cluster stands alone as the one that got away.

APPENDIX C

A quick guide to navigating the Coma–Virgo Cluster

Probing the depths of the Coma–Virgo Cluster may seem like a daunting task. Just look at all of those tiny galaxy symbols jammed together near the center of the wide-field map at the back of this book. But don't let the clutter discourage you. The cluster is actually quite easy to navigate. The simplest approach is to start with M58 (see chart on p. 226), one of the brightest members of the Virgo Cluster.

First locate the 3rd-magnitude star Epsilon (ε) Virginis. Five degrees to its west is 5th-magnitude Rho (ρ) Virginis; Rho is easy to confirm in binoculars because it is the brightest star in the middle of an arc of three stars oriented north to south. Two degrees west of Rho you will find the star 20 Virginis. M58 forms the northern apex of an equilateral triangle with Rho and 20 Virginis. M58 is easy to confirm because it is only 7′ to the west of an 8th-magnitude star.

M58 is a member of what I call the "Great Wall of Galaxies" – a strong line of six Messier galaxies, oriented slightly northwest to southeast, that spans 5° of sky. Pairs of galaxies punctuate either end of the wall, making identification of these objects easy. Using M58 as your reference point, move about 1 1/2° to the east and slightly south, where you will find your first pairing, M59 and M60, separated by only 30′. M60, which is the brightest galaxy in this string, has a very faint companion, NGC 4647, to the north. Now return to M58 and continue that line an equal distance to the northwest. There you will come upon the bright, round galaxy M87. One degree farther is the other galaxy pairing, M84 and M86.

The Great Wall also forms the baseline of a coathanger asterism of galaxies that includes M88, M89, M90, and M91. M90 also forms the northern apex of an equilateral triangle with our reference galaxy, M58, and M87. Furthermore, M89 is near the center of that triangle. However, because M90 also is the more obvious of the two, try for it first. Not only is it bright, but it is an oblique spiral galaxy and clearly looks different from the other, more elliptical hazes.

To find M88, simply move the telescope 1 1/2° to the northeast of M90. M88 is easy to identify because it is another fine spiral and there is an obvious double star at its southeastern tip.

M91 lies less than 1° due east of M88. But be careful here not to mistake M91 for NGC 4571 just to its southeast. Because there are no other galaxies in this immediate region, you can identify M91 by moving the telescope to the southeast to pick up NGC 4571 (or vice versa).

The final three galaxies – M98, M99, and M100 – should present no problems because they reside near three binocular stars, the brightest of which is 6 Comae Berenices. All you have to do is locate those stars with your binoculars, point your telescope to them, and you're home free. See how simple it can be?

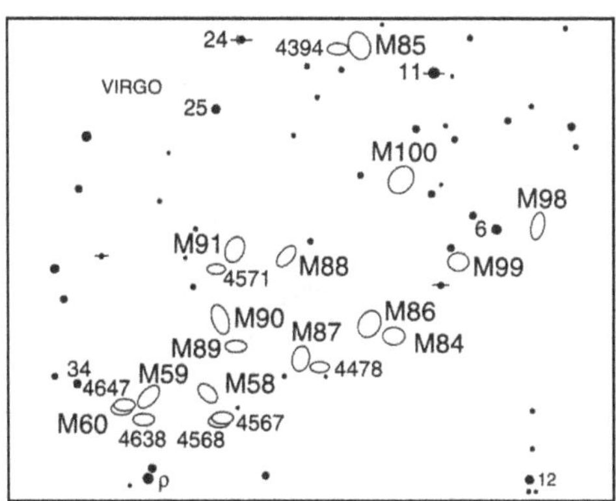

APPENDIX D
Messier marathons

Each spring, amateur astronomers around the world run a marathon – not the grueling 26-mile-long race but a visual race through the night sky to glimpse all 109 Messier objects in a single night. Messier marathons, as they are called, are held during late March and early April, the only time of the year when all the Messier objects are visible between dusk and dawn, and the only time of year when the Sun lies in a region of sky devoid of these celestial treasures. On the first clear, moonless night, marathoners start searching low in the western sky at dusk and then hop from one Messier object to the next until they've exhausted all the objects (or themselves) by dawn. Aside from clear skies, success requires a decent knowledge of the stars and constellations, efficient use of a telescope, and the ability to read star charts and confirm the appearance and position of each Messier object. These skills prove critical especially during twilight hours, when inevitably some targets must be found.

To the best of my knowledge, the Messier marathon originated in Spain, when a group of amateur astronomers set out to attempt the task in the 1960s. Several amateur astronomers across the globe independently conceived the notion a decade later. In the March 1979 issue of *Sky & Telescope* magazine, the late Deep-Sky Wonders columnist Walter Scott Houston described how, in the United States, Tom Hoffelder of Florida began marathoning with amateurs in the mid-1970s, while the Amateur Astronomers of Pittsburgh independently started the activity in 1977. On the West Coast, Donald Machholz, a leading comet discoverer, dreamed up the marathon idea in September 1978. Needless to say, serendipity struck like wildfire, and the activity spread worldwide. Today, astronomy clubs routinely sponsor Messier marathons, as well as variations on the theme, such as binocular marathons, CCD-imaging competitions, and CCD versus visual showdowns. One club even puts on a Messier event in which amateur astronomers compete in a cycling race, stopping occasionally to hunt down Messier objects with portable telescopes!

Messier marathons are popular for several reasons. Besides being fun and challenging, they provide astronomy clubs with an annual activity for their members, and they provide a "proving ground" for testing your observing techniques and search methods. And, let's face it, successful completion of a Messier marathon, just like the running race, earns you "bragging rights" – and possibly a T-shirt or lapel pin. But also, participants often come away from the mad dash around the heavens having learned something useful, whether about the sky, their telescope, or themselves.

This brings me to why I have never participated in a Messier marathon. It's not that there is anything wrong with the activity – as I have mentioned, it sounds like a lot of fun, especially if done with friends. But, as you have seen in this book, I much prefer to dote on objects, often spending hours at a time studying a single one – which would put me at a definite disadvantage in a race! To me it is a philosophical thing, and maybe I'm just a hopeless romantic. But dashing around the sky simply to tally how many of these extraordinary cosmic marvels you can spot in a night is tantamount to sprinting through the halls and galleries of the Louvre just to say you've seen all its famous masterpieces.

But if you are so inclined and are interested in learning more about Messier marathons, I encourage you to contact your local planetarium or astronomy club, many of which sponsor such events. You might also get a copy of the book by Don Machholz titled *Messier Marathon Observer's Guide: Handbook and Atlas*, which offers suggestions on how to prepare for and conduct a marathon and contains other useful information.

In addition to hosting Messier marathons, some astronomy clubs use March and the Messier objects to launch larger programs designed to educate the public about astronomy. By celebrating a Messier Month, clubs bring the magic of Messier and his catalogue to schoolchildren and the greater public through slide shows and special observing events. The idea also showcases the Messier objects with the simple message that the splendors of the universe are readily accessible to anyone who enjoys gazing into the night sky – whether it's with the naked eye, binoculars, or a telescope. Messier-specific activities sometimes also run in conjunction with Astronomy Day celebrations, which are held around the same time of year. And there are ample products to distribute or display to inspire the public about these deep-sky splendors. As part of any Messier celebration, special training sessions may be held, in which skilled observers not participating in a marathon help beginners to locate Messier objects.

Image credits

CHAPTER 1

Young Messier portrait Wikipedia
Comet C/2006 P1 (McNaught) Stephen James O'Meara
Messier at telescope drawing *Bulletin de la Société Astronomique de France,* 1929
Facsimile page *Bulletin de la Société Astronomique de France,* 1929
Old Messier portrait *Bulletin de la Société Astronomique de France,* 1929
Andromeda Nebula drawing *Mémoires de l'Académie des sciences de l'Institut de France,* 1807

CHAPTER 2

Galaxy schematic Stephen James O'Meara
Sketches of increased time behind the eyepiece Stephen James O'Meara

CHAPTER 3

Genesis refractor Courtesy of Al Nagler, Tele Vue Optics
Self-portrait Stephen James O'Meara
Globular cluster schematic *Observing Handbook and Catalogue of Deep-Sky Objects,* 1989
Mario Motta

CHAPTER 4

Images of M1–M110 Mario Motta
Additional credits are as follows:
M1 in x-ray NASA/CXC/SAO/F. Seward
M1 (HST) NASA/ESA/ASU/J. Hester & A. Loll
M4 (Ground) Kitt Peak National Observatory/NOAO; courtesy M. Bolte (UCSC)
M4 (HST) NASA/Harvey Richer (UBC)
Star-forming region in M8 J. I. Arias & Rodolfo H. Barbá (U. de La Serena)/ICATE-CONICET/Gemini Observatory/AURA
M9 NASA/ESA
M10 (HST) NASA, ESA & the Hubble Heritage Team (STScI/AURA)
M13 (HST) NASA, ESA & the Hubble Heritage Team (STScI/AURA)
M16 detail (HST) NASA/ESA/ESO/James Long
M17 detail (HST) NASA/ESA/J. Hester (ASU)
M17 detail (Spitzer) NASA/JPL-Caltech/Penn State
M27 detail (HST) NASA, ESA & the Hubble Heritage Team (STScI/AURA)
M30 (HST) NASA/ESA
Nucleus of M31 (HST) NASA/ESA/T. Lauer (NOAO)
M41 detail Mario Motta
Orion nebula (HST) NASA/ESA/M. Robberto (STScI/ESA) and the HST Orion Treasury Project Team
Trapezium schematic *Burnham's Celestial Handbook*
M51 NASA/ESA/S. Beckwith (STScI)/The Hubble Heritage Team (STScI/AURA)

M62 detail (HST) STScI
M63 (Spitzer) NASA/JPL-Caltech
M64 (HST) NASA, ESA & the Hubble Heritage Team (STScI/AURA)
M66 detail (HST) NASA, ESA & the Hubble Heritage Team (STScI/AURA) in collaboration with Davide De Martin & Robert Gendler
Leo Triplet Mario Motta
M68 detail (HST) NASA/ESA/Hubble
M70 detail (HST) NASA/ESA/Hubble
M71 detail (HST) NASA/ESA/Hubble
M74 detail (HST) NASA/ESA & the Hubble Heritage (STScI/AURA)
M77 detail (HST) NASA/ESA/A. van der Hoeven
M78 detail and McNeil's Nebula (HST) NASA/ESA/ESO/John Corban
M80 (HST) NASA/ESA & the Hubble Heritage (STScI/AURA)
M81 (HST) NASA/ESA & the Hubble Heritage Team STScI/AURA) with acknowledgments to A. Zezas & J. Huchra (Harvard-Smithsonian CfA)
M82 detail (HST) NASA/ESA & the Hubble Heritage Team (STScI/AURA) with acknowledgments to J. Gallagher (U. Wisconsin), M. Mountain (STScI), & P. Puxley (NSF)
M83 detail (HST) NASA/ESA & the Hubble Heritage (STScI/AURA)
M87 NASA, ESA, & the Hubble Heritage Team (STScI/AURA)
M100 detail (HST) NASA/ESA/STScI
M101 (HST) NASA/ESA with acknowledgements to K. D. Kuntz (GSFC), F. Bresolin (U. Hawaii), J. Trauger (JPL), J. Mould (NOAO), & Y.-H. Chu (U. Illinois, Urbana)
M104 (HST) Vicent Peris (OAUV / PTeam)/MAST/STScI/AURA/NASA
M105 Mario Motta
M106 detail (HST) NASA/ESA/R. Gendler (for the Hubble Heritage Team) with acknowledgment to J. GaBany

Alternate name and object index

www.ingramcontent.com/pod-product-compliance
Lightning Source LLC
Chambersburg PA
CBHW080615191224
19258CB00005BB/138